# Elementary Geometry

# *Elementary Geometry*
## Third Edition

**R. David Gustafson**
**Peter D. Frisk**
Rock Valley College
Rockford, Illinois

John Wiley & Sons, Inc.  New York  Chichester  Brisbane  Toronto  Singapore

ACQUISITIONS EDITOR, Robert Macek
DESIGNER, Laura Nicholls
PRODUCTION SUPERVISOR, Linda Muriello
MANUFACTURING MANAGER, Denis Clarke
COPY EDITOR, Deborah Herbert
PHOTO RESEARCHER, Stella Kupferberg
ILLUSTRATION, Dean Gonzales
COVER PHOTO, Marjory Dressler

*Library of Congress Cataloging in Publication Data:*

Gustafson, R. David (Roy David), 1936–
    Elementary geometry / R. David Gustafson, Peter D. Frisk.—3rd ed.
      p.  cm.
    Rev. ed. of: Elementary plane geometry. 2nd. ed. c1985.
    Includes index.
    ISBN 0-471-51002-5 (acid-free)
    1. Geometry.  I. Frisk, Peter D., 1942–
II. Gustafson, R. David (Roy David), 1936–  Elementary geometry.  III. Title.
QA455.G888  1991
516.2′2—dc20
                                                      90-44125
                                                       CIP

Printed in the United States of America

10  9  8

# Preface

As with the second edition, this book is intended primarily for college students or for individuals who wish a quick review of high school geometry. For these students, the instructor must present the subject matter in an extremely short time—usually in a one-semester, three-hour course. To do so, it is necessary to get to the heart of the subject quickly and to condense the material so that it can be completed in about 45 lessons.

For these reasons this book is concise and basically traditional. Some mathematical elegance and rigor have been sacrificed to minimize the possibility of the discussion becoming mired down in preliminaries and detail. For example, many statements have been postulated that could have been proved.

Techniques of geometric construction are still introduced early in the course. Our experience shows that students enjoy the construction problems. From a mathematical viewpoint, the constructions answer many questions of existence.

The concise nature of the material permits the instructor to reteach concepts as necessary and still have ample time for testing. The first 10 chapters cover most of the topics included in a standard high-school geometry course. Some instructors may prefer to teach Chapter 11, "Mathematical Logic," at the beginning of the course.

Although the philosophy of this book remains the same, this third edition has been extensively revised to comply with the suggestions of many users and reviewers. Because many users use this book to train prospective teachers and because more standardized tests are including questions from geometry, we have decided to use a notation that formalizes the distinction between equality and congruence and among a line, ray, and line segment. The improved notation also emphasizes the distinction between a line segment and its measure and an angle and its measure.

Other major changes are as follows:

1. The chapter on mathematical logic has been completely rewritten. The chapter now includes a discussion of truth tables and the logical basis for the discovery of non-Euclidean geometries. A discussion of hyperbolic and elliptic geometry is also included.
2. The chapter on analytic geometry has been expanded. More theorems are discussed and proved with coordinate geometry.

3. The material on parallel lines and parallelograms has been divided into two chapters. Dividing this long chapter in two speeds up the pace of the course and keeps students from being overwhelmed by too much material.

4. The chapter on numerical trigonometry has been condensed and is now included as an application of similar triangles.

5. More problems have been included, many of which are algebraic in nature.

6. The section on surface areas and volume has been expanded.

7. A set of review exercises has been added at the end of each chapter.

We thank our colleagues Gary Schultz, James Yarwood, and Darrell Ropp for their helpful suggestions. We also thank Carol B. Lacampagne and Michael J. Failla for their thoughtful reviews of the second edition and their suggestions for the third edition. We thank Betty Fernandez, Stephanie Piccirilli, and Joan Zarate Martinez for their help in the preparation of the manuscript. Finally, we thank Robert Macek, our editor at Wiley, Stacey Memminger, Linda Muriello, and the members of the Wiley production staff for their encouragement and for their help in producing this book.

R. David Gustafson
Peter D. Frisk

# Contents

**1**

Euclid (*about 300* B.C.). *Euclid first collected the geometric facts known to the ancient Greeks and put them into an organized form in his work,* Elements. The Elements *consists of 13 books and includes all of the important mathematics of early Greece. When asked by a student if geometry was practical, Euclid told his slave, "Give him threepence, since he must make gain out of what he learns"* (Brown Brothers).

# Fundamental Concepts

The word *geometry* comes from the Greek words *geo*, meaning earth, and *metron*, meaning measure. Geometry has its origins in early Egypt where earth measure was made necessary by the annual flooding of the Nile River and the subsequent need for yearly surveying. The Greek mathematician Euclid, who lived about 300 B.C., is considered to be the first person to collect the isolated facts known about geometry and put them into an organized form. His work produced 13 volumes, called the *Elements.* Five of these volumes deal with two-dimensional geometry, 3 deal with three-dimensional geometry, and the rest deal with geometric solutions to what we would consider to be algebraic problems.

In this course we will not only learn many geometric facts, but we will also learn how these facts are tied together in a logical and coherent way. In geometry, as in all of mathematics, the thread that ties one idea to another is called **deductive reasoning**. This type of reasoning is quite different from **inductive reasoning**, the type of reasoning used in the science classroom.

# 1.1 INDUCTIVE AND DEDUCTIVE REASONING

In the scientific laboratory, a scientist conducts an experiment and observes a certain outcome. After repeating the experiment several times, the scientist will generalize the findings into what seems to be a true *If . . . , then . . .* sentence. Some *If . . . , then . . .* sentences are

- *If* I heat water to 212°, *then* it will boil.
- *If* I let go of the weight, *then* it will fall.
- *If* I touch the fire, *then* I will be burned.

This type of reasoning which draws general conclusions from specific observations is called **inductive reasoning**.

Although inductive reasoning is the foundation of all experimental science, it does have one drawback. Inductive reasoning attempts to say something about *all* cases after examining only a *few* cases. For example, many people have tried to find a cure for the common cold. They have tried remedies ranging from wonder drugs to hot toddys, and each attempt has failed. From this evidence, however, it would be incorrect to conclude that no cure will ever be found. There is always the possibility that the next attempt will succeed.

As we have said, inductive reasoning moves from the *specific case* to the *general case*. Some more examples of inductive reasoning include

- Every horse I've ever seen has four legs. Therefore, *all* horses have four legs.
- Every dog I've ever seen wags its tail. Therefore, *all* dogs wag their tails.
- Every bird I've ever seen can fly. Therefore, *all* birds can fly.

Anyone who has ever seen an ostrich or a penguin will see the weakness of inductive reasoning.

In geometry, we will be concerned with a different kind of reasoning, called **deductive reasoning**. As opposed to inductive reasoning, deductive reasoning moves from the general case to the specific case. For example, a general rule of arithmetic states that if $x$ and $y$ represent any two numbers, then

$$x + y = y + x$$

From this general property we use deductive reasoning to conclude that these specific cases are true

$$721 + 62 = 62 + 721$$

and

$$5869 + 4279 = 4279 + 5869$$

without actually doing the calculations. If we know the spelling rule that says

> *The letter Q is always followed by the letter u.*

there is no doubt as to the second letter of the words *quarter* and *quiet*, or about the fourth letter of the word *inquire*. Whenever we apply a general principle to a particular instance, we are using deductive reasoning.

Deductive reasoning is characterized by its four components: *undefined terms*, *defined terms*, *postulates*, and *theorems*.

## UNDEFINED AND DEFINED TERMS

For a sentence to have meaning, each word in the sentence must have meaning. However, this cannot be accomplished by defining every word. For example, suppose we wish to define the word *obstruction*. A dictionary might say that *obstruction* means a *barrier*. If we do not already know the meaning of the word *barrier* and look it up, the dictionary might say that a *barrier* is a *hindrance*. According to one dictionary, a *hindrance* is an *impediment*, and an *impediment* is an *obstruction*. We are now back where we started, wondering about the definition of the word *obstruction*.

Because it is impossible to find an unending string of synonyms for any given word, we are forced to admit that certain words must be left as undefined. Such words are called **undefined terms**.

After accepting some words as undefined terms, we can define other words in a more formal way. For example, after accepting the word *obstruction* as an undefined term, we can formally define the word *obstacle* as an obstruction. Words to which we give formal definitions are called **defined terms**.

## POSTULATES AND THEOREMS

The next step is to accept some general statements about these defined and undefined terms. General statements that we accept as starting points are called **axioms** or **postulates**. In algebra, the word *axiom* is used more often. In geometry, the word *postulate* is used more often.

Postulates are statements that are accepted as true without proof. Sometimes postulates are thought of as self-evident truths. For example, the Declaration of Independence puts forth the postulate that "*All men are created equal.*" This is an idea that all Americans accept as true on faith. It certainly cannot be proved.

Armed with these undefined terms, defined terms, and postulates, we are now ready to deduce some results called **theorems**. To prove theorems, we will reason, using the rules of deductive logic, from a given *hypothesis* to a desired *conclusion*. The tools available to prove theorems include the undefined terms, the defined terms, the accepted postulates, and all previously proved theorems.

A major goal of this course, in addition to learning the facts of geometry, is to learn how to reason deductively and to gain an appreciation of the deductive-reasoning process.

## EXERCISE 1.1

1. To determine if the sum of two odd integers is always an even integer, Martha does the following additions:

$$3 + 5 = 8$$
$$7 + 9 = 16$$
$$11 + 23 = 34$$

After observing that each sum is an even integer, she concludes that the sum of two odd integers is *always* an even integer. Is Martha's conclusion correct? Did she use inductive or deductive reasoning?

2. To determine if all odd integers are prime numbers, Corey considered three odd integers and found that

7 is a prime number

11 is a prime number

13 is a prime number

On the basis of these observations, Corey concluded that *all* odd integers are prime numbers. Is Corey's conclusion correct? Did he use inductive or deductive reasoning?

3. Daniel remembered the **transitive property** from algebra:

**If $a = b$ and $b = c$, then $a = c$.**

Daniel concluded that Bob and Jim must be the same age because they are both the same age as Judy. Is Daniel's conclusion correct? Did he reason inductively or deductively?

4. Juan remembered the **distributive property** from algebra:

**$a(b + c) = ab + ac$**

Juan concluded that $5(1010) = 5050$ because

$$5(1010) = 5(1000 + 10) = 5000 + 50 = 5050$$

Is Juan's conclusion correct? Did Juan reason inductively or deductively?

5. List the four elements of a deductive-reasoning system.

6. Attempt to define the word *bucket*. What synonyms for *bucket* did you use? How many synonyms for *bucket* can you think of?

7. Attempt to define the word *good*. What synonyms for *good* did you use? How many synonyms for *good* can you think of?

8. Explain why undefined terms are necessary in a deductive-reasoning system.

**9.** In a democracy the purpose of government is to serve the people. Under communism the purpose of the people is to serve the government. Why do you think it is so hard for these different ideologies to reach agreements?

**10.** Discuss the problem of creating a general rule in the field of English grammar. For example, write a rule concerning the use of *ie* or *ei* in proper spelling. Then see if your rule applies to the words *biennial, friend, ceiling, weigh, neighbor, atheism, protein, seize,* and *weird.* Why is a general rule so difficult to formulate?

## 1.2 SOME UNDEFINED TERMS AND BASIC DEFINITIONS

We have seen why it is impossible to provide a formal definition for each term to be used in a deductive-reasoning system. Thus, there are some terms in this book that we will not attempt to define. Some of these undefined terms are

> *set, point, line, plane, between, closed, figure, side, length, width, depth, intersection, interior, exterior,* and *surface*

Although no attempt will be made to define these terms in a formal way, we all have some intuitive idea of what they mean.

A *set* refers to a collection of objects such as a set of dishes, a set of chairs, or a set of encyclopedias. In geometry, we will often speak of a set of *points*.

Euclid thought of a *point* as a *figure* that has position but no *length, width,* or *depth.* Points are often represented by small dots drawn on paper, such as the point $A$ represented in Figure 1.1*a*. Points are always named with capital letters.

A *line* is a figure that has length, but no width or depth. Figure 1.1*b* shows a line passing through two points $B$ and $C$. The line $BC$ is often denoted by the symbol $\overleftrightarrow{BC}$. A line can be thought of as a set of points that continues forever in opposite directions.

A *plane* is a flat *surface* that has length and width, but no depth. For example, an oil slick on a calm pond or the surface of the blackboard in a mathematics classroom could represent a plane. In Figure 1.1*c*, $\overleftrightarrow{EF}$ lies in the plane $AB$.

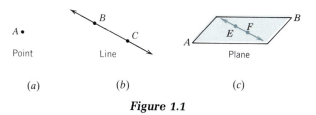

$A \bullet$

Point

Line

Plane

*(a)*              *(b)*              *(c)*

**Figure 1.1**

Figure 1.2 shows a line passing through points $A$, $B$, and $C$. On the line, the point $B$ is *between* points $A$ and $C$. However, point $A$ is not between points $B$ and $C$, and point $C$ is not between points $A$ and $B$. No points that do not lie on $\overleftrightarrow{AC}$ are considered to be between points $A$ and $C$.

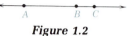

**Figure 1.2**

Figure 1.3 shows three figures. The figure shown in Figure 1.3*a* is a *closed figure* with three *sides*, all of equal *length*. Sides $AC$ and $BC$ have point $C$ in common. We say that sides $AC$ and $BC$ intersect at point $C$, or that point $C$ is the *intersection* of sides $AC$ and $BC$. The figure shown in Figure 1.3*b* has four sides, all of different lengths. This figure is not closed. Figure 1.3*c* shows a closed figure with five sides. Point $F$ is in the *interior* of the five-sided figure, and point $G$ is in its *exterior*.

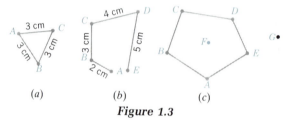

**Figure 1.3**

After we accept some words as fundamental undefined terms, we can use them to define other terms.

---

**Definition 1.1** A **triangle** is a closed three-sided figure.

---

A triangle is shown in Figure 1.3*a*.

As with all definitions, the definition of a triangle is reversible. The definition of a triangle implies each of the following two statements:

- *If a figure is a triangle, then it is a closed three-sided figure.*
- *If a figure is closed and three-sided, then it is a triangle.*

---

**Definition 1.2** The **vertices** of a triangle are the points of intersection of its sides.

---

The vertices of the triangle shown in Figure 1.4 are the points $A$, $B$, and $C$.

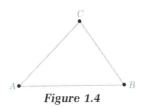

**Figure 1.4**

---

**Definition 1.3**

An **equilateral triangle** is a triangle with all sides of equal length.

An **isosceles triangle** is a triangle with at least two sides of equal length. The third side is called its **base**.

A **scalene triangle** is a triangle with all sides of different length.

---

An equilateral triangle, an isosceles triangle, and a scalene triangle are shown in Figure 1.5.

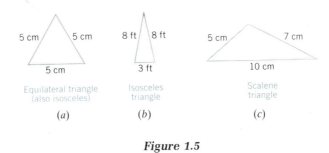

**Figure 1.5**

The distance around a triangle is called its **perimeter**.

---

**Definition 1.4** The **perimeter of a triangle** is the sum of the lengths of its three sides.

---

**EXAMPLE 1**

Find the perimeter of the triangle shown in Figure 1.5c.

Solution

The lengths of the sides of the triangle are 5 cm, 7 cm, and 10 cm. Because the perimeter $P$ of a triangle is the sum of the lengths of its sides, we have

$$P = 5 + 7 + 10$$
$$= 22$$

The perimeter of the triangle is 22 cm.

**EXAMPLE 2**

The base of an isosceles triangle is 5 cm long and its perimeter is 21 cm. Find the length of each of the other two sides.

Solution

Let the length of each of its sides of equal length be represented by $x$, as in Figure 1.6. Because the perimeter of the triangle is 21 cm, we have

$$x + x + 5 = 21$$
$$2x + 5 = 21 \qquad \text{Combine terms.}$$
$$2x = 16 \qquad \text{Add } -5 \text{ to both sides.}$$
$$x = 8 \qquad \text{Divide both sides by 2.}$$

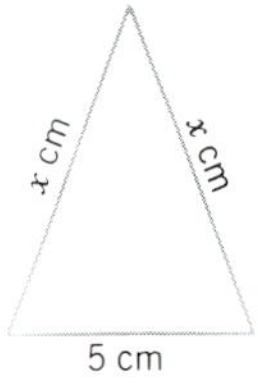

5 cm

**Figure 1.6**

Each of the other two sides have a length of 8 cm.    ■

A line can be thought of as a set of points that continues forever in both directions. Two new figures, called a **ray** and a **line segment**, can be defined as portions of a line.

---

**Definition 1.5** A **ray** is a portion of a line that begins at some point, say $A$, and continues forever in one direction.

Point $A$ is called the **endpoint** of the ray.

---

A ray $AB$ with endpoint $A$ and a ray $BA$ with endpoint $B$ are shown in Figures 1.7a and 1.7b, respectively. These figures illustrate that each ray has exactly one endpoint. The ray $AB$ with endpoint $A$ is denoted by the symbol $\overrightarrow{AB}$, and the ray $BA$ with endpoint $B$ is denoted by the symbol $\overrightarrow{BA}$.

---

**Definition 1.6** Let $A$ and $B$ be two points on a line. The **line segment** $AB$ is the portion of the line that consists of points $A$ and $B$ and all points that are between $A$ and $B$.

The points $A$ and $B$ are called the **endpoints** of the line segment.

---

The line segment $AB$ shown in Figure 1.7c illustrates that a line segment has two endpoints, point $A$ and point $B$.

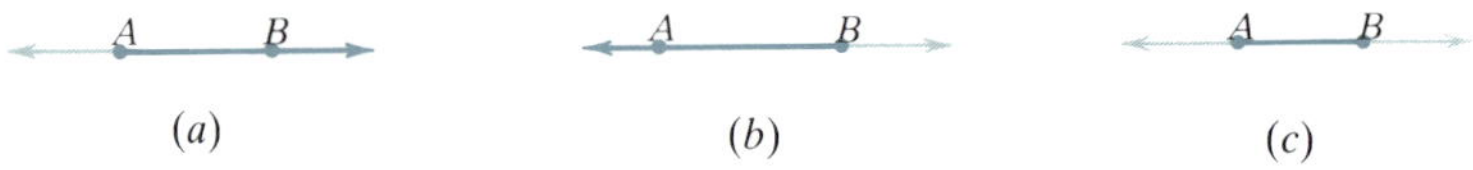

(a)                              (b)                              (c)

**Figure 1.7**

There is a distinction between a line segment and its measure. In Figure 1.8, the line segment $AB$ (denoted as $\overline{AB}$) is the set of all points on the line between the points $A$ and $B$, including $A$ and $B$. The measure of $\overline{AB}$ (denoted as $m(\overline{AB})$) is a number that tells its length in units such as inches or meters.

In Figure 1.8, we find $m(\overline{AB})$ by placing a ruler alongside the segment and subtracting the smaller number from the larger:

$$m(\overline{AB}) = 4 - 1$$

Read $m(\overline{AB})$ as "the measure, or length, of line segment $\overline{AB}$."

$$= 3$$

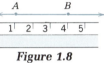

**Figure 1.8**

Thus, $m(\overline{AB}) = 3$ units.

When two line segments such as $\overline{AB}$ and $\overline{CD}$ have equal measures, we say the line segments are **congruent**, and we write

$$\overline{AB} \cong \overline{CD}$$

Read as "line segment $AB$ is congruent to line segment $CD$."

---

**Definition 1.7** Two line segments are **congruent** if, and only if, they have the same measure.

---

The definition of congruent line segments contains the words *if, and only if*. This terminology emphasizes the fact that the definition is reversible. It is equivalent to the following two *If . . . , then . . .* sentences.

- If two line segments are congruent, then they have equal measures.
- If two line segments have equal measures, then they are congruent.

Thus, whenever we say that two line segments are congruent, we also mean they have equal lengths. Whenever we say that two line segments have equal lengths, we also mean they are congruent.

We now consider a postulate involving points and lines. Remember that postulates are general statements that are accepted as true without proof.

---

**Postulate 1.1** Two points determine one and only one line.

---

This first postulate guarantees that at least one line exists that passes through the two points $A$ and $B$. It also guarantees that no more than one line can pass through these two points. Thus, points $A$ and $B$ determine exactly one line. See Figure 1.9.

**Figure 1.9**

## EXERCISE 1.2

In Exercises 1–4, refer to Illustration 1.

**1.** Is point $C$ between points $A$ and $D$?

**2.** Is point $B$ between points $C$ and $D$?

**3.** Is point $D$ between points $A$ and $C$?

**4.** Is point $C$ between points $A$ and $B$?

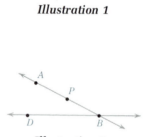

*Illustration 1*

In Exercises 5–6, refer to Illustration 2.

*Illustration 2*

**5.** Is point $P$ between points $B$ and $D$?

**6.** Is point $P$ between points $A$ and $B$?

**7.** The definition of an equilateral triangle is a reversible statement. What does this mean?

**8.** The definition of scalene triangle is a reversible statement. What does this mean?

**9.** Is the following statement a reversible statement?

   *"If I am a woman, then I am a human being."*

**10.** Is the following statement a reversible statement?

   *"If I am alive, then my heart is beating."*

**11.** Can a scalene triangle be isosceles? Explain.

**12.** Can an isosceles triangle be scalene? Explain.

**13.** Find the perimeter of a triangle with sides of length 6 in., 8 in., and 10 in.

**14.** Find the perimeter of an isosceles triangle with a base 8 m long and each of the other two sides 12 m long.

**15.** Find the perimeter of an equilateral triangle with one side 25 cm long.

**16.** A triangle with sides of length 25 ft and 36 ft has a perimeter of 100 ft. Find the length of the third side.

**17.** An isosceles triangle with two sides 52.3 cm long has a perimeter of 142.7 cm. Find the length of its base.

**18.** The perimeter of an equilateral triangle is 79 ft. Find the length of each side.

**19.** The base of an isosceles triangle is half the length of one of its sides with equal length. The perimeter of the triangle is 89.5 in. Find the length of its base.

**20.** The base of an isosceles triangle is 3 less than twice the length of one of its sides with equal length. The perimeter of the triangle is 37 cm. Find the length of its base.

**21.** How many endpoints does a ray have?

**22.** How many endpoints does a line segment have?

**23.** How many endpoints does a line have?

**24.** How many lines can contain the same point?

**25.** How many rays can have the same endpoint?

**26.** How many line segments can share an endpoint?

**27.** How many line segments can be drawn connecting two points?

**28.** How many line segments can be drawn joining exactly two of the points shown in Illustration 3?

**29.** How many line segments can be drawn joining exactly two of the points shown in Illustration 4?

**30.** How many line segments can be drawn joining exactly two of the points shown in Illustration 5?

*Illustration 3*                    *Illustration 4*                    *Illustration 5*

In Exercises 31–36, refer to Illustration 6 and find each measure.

**31.** m($\overline{AC}$)    **32.** m($\overline{BE}$)    **33.** m($\overline{CE}$)    **34.** m($\overline{BD}$)    **35.** m($\overline{CD}$)    **36.** m($\overline{DE}$)

In Exercises 37–40, refer to Illustration 6.

**37.** Is $\overline{AC} \cong \overline{CE}$?              **38.** Is $\overline{BE} \cong \overline{DE}$?

**39.** Is $\overline{CE} \cong \overline{CD}$?              **40.** Is $\overline{BD} \cong \overline{DE}$?

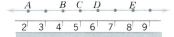

*Illustration 6*

## 1.3 ANGLES

A very important figure in the study of geometry is the **angle**.

> *Definition 1.8* An **angle** is a figure formed by two rays with a common endpoint. The common endpoint is called the **vertex** of the angle. The rays are called **sides** of the angle.

There are three different ways to name an angle. One way is to use letters to label three points on the angle—the vertex and a point on each side. We can call the angle in Figure 1.10*a* ∠*BAC* (read as "angle *BAC*") or ∠*CAB* (read as "angle *CAB*"). When we use three letters to name an angle, it is important that the letter name of the angle's vertex be placed between the other two letters.

A second way to name an angle is to use the name of the vertex only. The angle in Figure 1.10*a* can be called ∠*A*. This simple way of naming an angle can be used when only one angle exists that could have that name. A third way to name an angle is to write a small number or a small letter inside the angle. The angle in Figure 1.10*a* can be called ∠1.

Although the two line segments in Figure 1.10*b* are not rays with a common endpoint, they still determine an angle *D*. This is because segments $\overline{DE}$ and $\overline{DF}$ could be extended beyond *E* and *F* to form two rays.

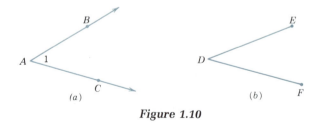

(a)  (b)

**Figure 1.10**

Line segments are commonly measured in inches, feet, yards, meters, centimeters, and so on. Angles are often measured in degrees, denoted with the symbol °. The approximate number of degrees in an angle can be found by using a protractor such as the one shown in Figure 1.11.

If a circle is divided into 360 equal parts and lines are drawn from the center of the circle to two consecutive points of division, the angle formed by those lines measures 1°. In Figure 1.11,

$$m(\angle ABC) = 30°$$   Read as "The measure of angle *ABC* is 30 degrees."

$$m(\angle ABD) = 60°$$

$$m(\angle ABE) = 110°$$
$$m(\angle ABF) = 150°$$

and

$$m(\angle ABG) = 180°$$

If we read the protractor from the left and read the inside measurements, we see that $m(\angle GBF) = 30°$.

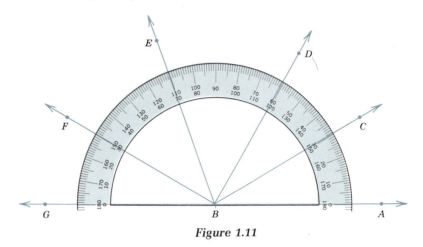

**Figure 1.11**

When two angles such as $\angle ABC$ and $\angle GBF$ have equal measures, we say the angles are **congruent**, and we write

$$\angle ABC \cong \angle GBF \qquad \text{Read as "}\angle ABC \text{ is congruent to } \angle GBF.\text{"}$$

---

**Definition 1.9** Two angles are **congruent** if, and only if, they have the same measure.

---

The preceding definition is equivalent to the following two *If* ... , *then* ... sentences.

- If two angles are congruent, then they have equal measures.
- If two angles have equal measures, then they are congruent.

Thus, whenever we say the two angles are congruent, we also mean they have equal measures. Whenever we say that two angles have equal measures, we also mean they are congruent.

Angles can be classified by their size.

---

**Definition 1.10**

An **acute angle** is an angle whose measure is greater than 0°, but less than 90°.

A **right angle** is an angle whose measure is 90°.

An **obtuse angle** is an angle whose measure is greater than 90° but less than 180°.

A **straight angle** is an angle whose measure is 180°.

---

**EXAMPLE 1**

Consider the angles shown in Figure 1.12.

(a)   ∠AOB and ∠COD are acute angles.

(b)   ∠BOC is a right angle.

(c)   ∠AOC and ∠BOD are obtuse angles.

(d)   ∠AOD is a straight angle.

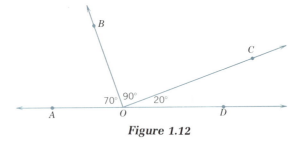

**Figure 1.12**

**EXAMPLE 2**

Refer to Figure 1.13.

(a)   ∠AFB, ∠BFC, ∠BFD, and ∠CFD are acute angles because the measure of each angle is less than 90°.

(b)   ∠EFA and ∠AFC are right angles because the measure of each angle is 90°.

(c)   ∠EFB, ∠EFD, and ∠AFD are obtuse angles because the measure of each angle is greater than 90° but less than 180°.

(d)   ∠EFC is a straight angle because its measure is 180°.

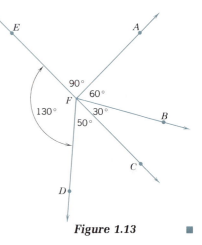

**Figure 1.13**

## COMPLEMENTARY AND SUPPLEMENTARY ANGLES

Certain pairs of angles play important roles in geometry. Two such pairs of angles are **complementary** and **supplementary angles**.

---

***Definition 1.11*** Two angles are called **complementary angles** if the sum of their measures is 90°. If two angles are complementary, either angle is called the **complement** of the other.

---

***Definition 1.12*** Two angles are called **supplementary angles** if the sum of their measures is 180°. If two angles are supplementary, either angle is called the **supplement** of the other.

---

**EXAMPLE 3**

(a) Angles measuring 60° and 30° are complementary angles because the sum of their measures is 90°. See Figure 1.14*a*. An angle of 30° is the complement of an angle of 60°, and an angle of 60° is the complement of an angle of 30°.

(b) Angles measuring 130° and 50° are supplementary because the sum of their measures is 180°. See Figure 1.14*b*. An angle of 130° is the supplement of an angle of 50°, and an angle of 50° is the supplement of an angle of 130°.

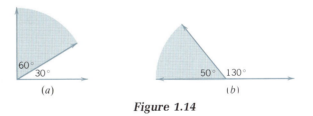

(a)                    (b)

**Figure 1.14**

(c) Because the sum of two angles measuring 70° and 50° is neither 90° nor 180°, the angles are neither complementary nor supplementary. See Figure 1.15*a*.

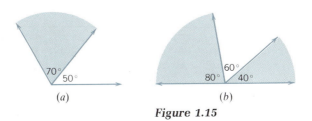

(a)                    (b)

**Figure 1.15**

(d)    Angles measuring 40°, 60°, and 80° are not supplementary angles even though the sum of their measures is 180°. This is because there are *three* angles whose sum is 180°. The definition of supplementary angles requires *two* angles whose sum is 180°. See Figure 1.15*b*.    ■

*EXAMPLE 4*

Find (a) the complement and (b) the supplement of an angle measuring 35°.

Solution

(a)    To find the complement of a 35° angle, subtract 35° from 90° to obtain 55°. Because the sum of the measures of a 35° angle and a 55° angle is 90°, the complement of a 35° angle is a 55° angle.

(b)    To find the supplement of a 35° angle, subtract 35° from 180° to obtain 145°. Because the sum of the measures of a 35° angle and a 145° angle is 180°, the supplement of a 35° angle is a 145° angle.    ■

*EXAMPLE 5*

Two supplementary angles measure $(2x + 60)°$ and $(x - 30)°$. Find $x$ and the number of degrees in each angle.

Solution

Because the angles are supplementary, the sum of their measures is 180°. So set the sum of their measures equal to 180 and solve for $x$:

$$2x + 60 + x - 30 = 180$$
$$3x + 30 = 180 \qquad \text{Combine terms.}$$
$$3x = 150 \qquad \text{Add } -30 \text{ to both sides.}$$
$$x = 50 \qquad \text{Divide both sides by 3.}$$

Once you know the value of $x$, you can compute the measure of each angle:

$$(2x + 60)° = [2(50) + 60]° = (100 + 60)° = 160°$$
$$(x - 30)° = (50 - 30)° = 20°$$

The angles measure 160° and 20°.    ■

## ADJACENT ANGLES

Angles that are placed side by side are called **adjacent angles**.

---

**Definition 1.13** Two angles are called **adjacent angles** if, and only if, they have a common vertex and a common side lying between them.

---

**EXAMPLE 6**    Consider the angles shown in Figure 1.16.

    (a)   In Figure 1.16*a*, ∠1 and ∠2 are adjacent angles because they have a common vertex at point *P* and the common side $\overline{PB}$ lying between them. ∠1 and ∠2 are side by side.

    (b)   In Figure 1.16*b*, ∠*APB* and ∠*APC* are not adjacent angles. While it is true they have a common vertex, point *P*, their common side $\overline{PA}$ is not between the angles. These angles are not side by side. One angle is on top of the other.

    (c)   In Figure 1.16*c*, ∠*APB* and ∠*BEC* are not adjacent angles because they do not have a common vertex.

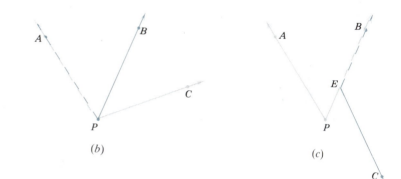

**Figure 1.16**

There are special triangles that derive their names from the types of angles they contain.

---

**Definition 1.14**

A **right triangle** is a triangle with one right angle.
An **acute triangle** is a triangle with three acute angles.
An **obtuse triangle** is a triangle with one obtuse angle.

---

A right triangle is shown in Figure 1.17*a*, an acute triangle is shown in

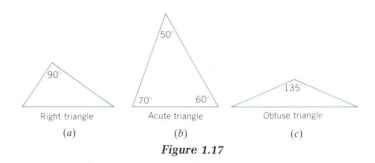

**Figure 1.17**

Figure 1.17*b*, and an obtuse triangle is shown in Figure 1.17*c*. The longest side of a right triangle is called its **hypotenuse**.

## EXERCISE 1.3

In Exercises 1–4, refer to Illustration 1.

**1.** Give another name for ∠*ABD*.

**2.** Give another name for ∠2.

**3.** Give another name for ∠3.

**4.** Give another name for ∠*DCE*.

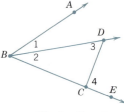

*Illustration 1*

In Exercises 5–10, refer to Illustration 2.

**5.** Find m(∠*AOB*).

**6.** Find m(∠*AOE*).

**7.** Find m(∠*BOE*).

**8.** Find m(∠*BOD*).

**9.** Is ∠*EOD* ≅ ∠*AOB*?

**10.** Is ∠*BOA* ≅ ∠*BOD*?

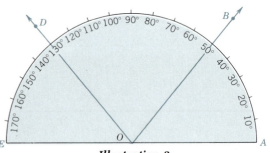

*Illustration 2*

**11.** Is the measure of an angle determined by the lengths of its sides? Explain.

**12.** Is the measure of an angle determined by its vertex? Explain.

In Exercises 13–20, refer to Illustration 3 and classify each angle as an acute angle, a right angle, an obtuse angle, or a straight angle.

**13.** ∠*AGC*       **14.** ∠*EGA*

**15.** ∠*FGD*       **16.** ∠*BGA*

**17.** ∠*BGE*       **18.** ∠*AGD*

**19.** ∠*DGC*       **20.** ∠*DGB*

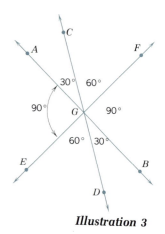

*Illustration 3*

**21.** Find the complement of a 30° angle.

**22.** Find the supplement of a 30° angle.

**23.** Find the supplement of a 105° angle.

**24.** Find the complement of a 75° angle.

**25.** Explain why an angle measuring 105° cannot have a complement.

**26.** Explain why an angle measuring 210° cannot have a supplement.

**27.** Find the supplement of the complement of an 85° angle.

**28.** Find the complement of the supplement of a 115° angle.

**29.** Can two complementary angles be congruent?

**30.** Can two complementary angles be supplementary?

**31.** Two angles are both congruent and complementary. Find their measures.

**32.** Two angles are both congruent and supplementary. Find their measures.

**33.** m($\angle 1$) is five times its supplement. Find m($\angle 1$).

**34.** m($\angle 1$) is two degrees greater than three times its complement. Find m($\angle 1$).

**35.** Are two acute angles measuring $(x + 47)°$ and $(43 - x)°$ complementary?

**36.** Are two angles measuring $(5x + 87)°$ and $(93 - 5x)°$ supplementary?

**37.** If an acute angle measures $(3x + 2)°$, find its complement.

**38.** If an obtuse angle measures $(5x - 2)°$, find its supplement.

**39.** Two angles with measures of $(4x - 20)°$ and $(2x + 14)°$ are complementary. Find $x$.

**40.** Two angles with measures of $(3x + 12)°$ and $(4x + 14)°$ are supplementary. Find $x$.

In Exercises 41–44, refer to Illustration 4 and tell whether each statement is true. If a statement is false, explain why.

**41.** $\angle AGB$ is adjacent to $\angle BGC$.

**42.** $\angle DGC$ is adjacent to $\angle DGB$.

**43.** $\angle FGB$ is adjacent to $\angle FGC$.

**44.** $\angle EGC$ is adjacent to $\angle CGB$.

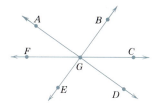

**Illustration 4**

In Exercises 45–46, refer to Illustration 5.

**45.** Explain why $\angle ADC$ and $\angle ADB$ are not adjacent angles.

**46.** Explain why $\angle CBD$ and $\angle ADB$ are not adjacent angles.

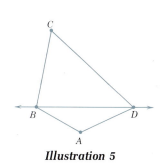

**Illustration 5**

In Exercises 47–50, classify each triangle as a right triangle, an acute triangle, or an obtuse triangle.

**47.**           **48.**           **49.**           **50.**

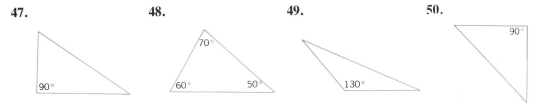

**51.** What angle do the hands of a clock make at 3:00 P.M.?

**52.** What angle do the hands of a clock make at 4:00 P.M.?

**53.** What angle do the hands of a clock make at 6:00 A.M.?

**54.** What angle do the hands of a clock make at 8:00 A.M.?

# 1.4 BASIC GEOMETRIC CONSTRUCTIONS

One of the problems in geometry is that of determining methods for drawing various geometric figures with certain properties. These methods, called **geometric constructions**, allow only certain tools. One tool, called a **straightedge**, is used for drawing lines. A straightedge differs from a ruler in that it cannot be used to measure distances. The only time we can draw a line with a straightedge and pencil is when we know two points through which to draw the line. Another tool, called a **compass**, can be used to transfer the length of a line segment to another place, or to draw circles, or portions of circles, called **arcs**.

We begin by considering the procedures for copying a line segment and for finding the midpoint of a line segment. These constructions involve only a compass and a straightedge.

Suppose we have a line segment $\overline{AB}$ and wish to duplicate it on the line $l$ shown in Figure 1.18.

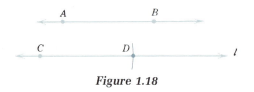

*Figure 1.18*

We first put the point of the compass on point $A$ and the pencil point of the compass on point $B$. After fastening the compass so that it is rigid, we transfer the segment $\overline{AB}$ by placing the point of the compass at point

*C* and drawing an arc with the pencil point. The intersection of this arc and the line determines point *D* so that $\overline{AB} \cong \overline{CD}$.

We now discuss how to divide a line segment into two line segments with equal length.

---

**Definition 1.15** To **bisect a line segment** means to divide it into two congruent segments.

---

To bisect the line segment $\overline{AB}$ shown in Figure 1.19, we set the compass to a length that is greater than one-half the length of segment $\overline{AB}$. Placing the point of the compass at point *A*, we make two large arcs, one above segment $\overline{AB}$ and another below it. Without changing the compass setting, we then place the compass point at *B* and again draw large arcs above and below segment $\overline{AB}$. Finally, we draw a line through the two points where the arcs intersect. This line will divide segment $\overline{AB}$ into two line segments so that

$$\overline{AD} \cong \overline{DB}$$

Later we will prove that this construction works.

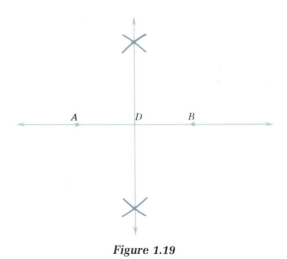

*Figure 1.19*

---

**Definition 1.16** A line that divides a line segment into two congruent line segments is called a **bisector of the line segment**.

The point that divides a line segment into two congruent line segments is called the **midpoint of the line segment**.

The preceding construction demonstrates that every line segment has a midpoint. We will assume that a line segment does not have more than one midpoint.

---

**Postulate 1.2**  A line segment has only one midpoint.

---

Just as there is a procedure for copying a line segment, there is a procedure for copying an angle. To construct an angle that is congruent to angle $F$ (shown in Figure 1.20) with its vertex at point $B$, we set the compass at a convenient setting, place the point of the compass at point $F$, and draw an arc intersecting the sides of $\angle F$ at $E$ and $G$. Without changing the compass setting, we place the compass point at $B$ and draw a similar arc intersecting line $l$ at point $C$. We then set the compass so that its point is at $G$ and its pencil point is at $E$. Without changing this setting, we place the compass point at $C$ and draw a small arc intersecting the larger arc at point $A$. Finally we draw ray $\overrightarrow{BA}$ to obtain $\angle ABC$. It is true that $\angle ABC \cong \angle EFG$. Later we will prove this construction works.

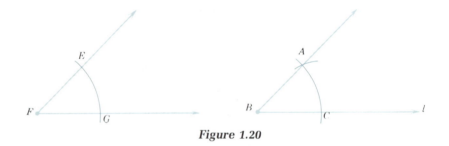

**Figure 1.20**

---

**Definition 1.17**  An **angle bisector** is a ray that divides an angle into two congruent angles.

---

Refer to Figure 1.21. To divide $\angle B$ into two congruent angles, we place the point of the compass on point $B$ and draw an arc intersecting the sides of the angle at $A$ and $C$. We then set the compass for some convenient length, place the compass point on $A$, and draw an arc in the interior of the angle. Now, without changing the compass setting, we place the point of the compass on $C$ and draw a second arc in the interior of the angle, intersecting the first arc at some point $E$. Finally, we draw the ray $\overrightarrow{BE}$ forming $\angle ABE$ and $\angle EBC$. It is true that

$$\angle ABE \cong \angle EBC$$

Later we will prove this construction works.

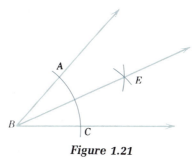

*Figure 1.21*

The previous construction demonstrates the existence of an angle bisector. We will assume that every angle has exactly one angle bisector.

---

**Postulate 1.3** Every angle has exactly one angle bisector. (The bisector of an angle is unique.)

---

If two lines meet and form right angles, they are called **perpendicular lines**. An equivalent statement, and often a more useful one, is given as the definition of perpendicular lines.

---

**Definition 1.18** Two lines are **perpendicular** if, and only if, they meet and form congruent adjacent angles.

---

In Figure 1.22 line $l$ is perpendicular to line $r$ if, and only if, $\angle 1 \cong \angle 2$. If line $l$ is perpendicular to line $r$, we can write $l \perp r$, where the symbol $\perp$ is read as "is perpendicular to."

We now consider how to construct a line that is perpendicular to a given line through a point on the given line. We refer to Figure 1.23 in which

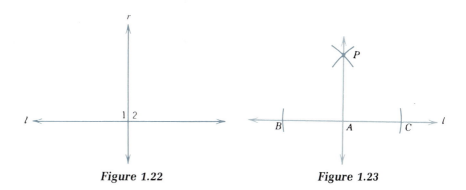

*Figure 1.22*                    *Figure 1.23*

point $A$ is a point on line $l$. To construct a line perpendicular to line $l$ throughout point $A$, we set the compass at a convenient length, place the point of the compass at $A$, and make arcs intersecting line $l$ at points $B$ and $C$. After increasing the compass to some fixed setting, we place the compass point first at $B$ and then at $C$ and draw arcs above the line. These arcs will intersect at some point $P$. Finally, we draw line $\overleftrightarrow{AP}$ which is perpendicular to line $l$. Later we will prove this construction works.

The previous construction demonstrates that a perpendicular to a line $l$ through a given point on line $l$ exists.

One important perpendicular to a line segment is the perpendicular bisector of a line segment.

**Definition 1.19** A **perpendicular bisector** of a line segment is a line that both bisects the segment and is perpendicular to it.

The construction technique we use to bisect a line segment produces the perpendicular bisector of the line segment. We will assume that the perpendicular bisector of any line segment is unique.

**Postulate 1.4** Every line segment has one and only one perpendicular bisector.

One important line related to a triangle is an altitude.

**Definition 1.20** An **altitude of a triangle** is a line segment drawn from one vertex of the triangle perpendicular to the opposite side or, if necessary, to the extension of the opposite side.

The three altitudes of a triangle $ABC$ are shown in Figure 1.24. The symbol ⌐ indicates an angle of 90°. Later we will prove that the three altitudes of a triangle always meet in a point, as the figure suggests.

To construct an altitude of a triangle, we must develop a method for constructing a perpendicular to a line from a point not on the line. To do this, we refer to Figure 1.25 which shows a line $l$ and a point $A$ that is not on line $l$. We first place the point of the compass on $A$ and draw an arc intersecting line $l$ in two points, say points $B$ and $C$. Without changing the setting of the compass, we place the point of the compass on $B$ and draw an arc below line $l$. Without changing the setting of the compass, we then place the point of the compass on $C$ and draw a second arc below line $l$. We will call the intersection of these arcs point $D$. Finally, we draw the line $\overleftrightarrow{AD}$ passing through points $A$ and $D$. This line passes through point $A$ and is perpendicular to line $l$. Later we will prove this construction works.

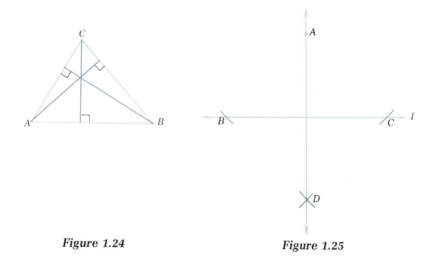

**Figure 1.24**                    **Figure 1.25**

The previous construction demonstrates that a perpendicular to a line from a point not on the line exists. We will assume that only one such perpendicular exists.

---

**Postulate 1.5** The perpendicular to a line through a point not on the line is unique.

---

Another special line in a triangle is the median.

---

**Definition 1.21** A **median of a triangle** is a line segment drawn from one vertex of the triangle to the midpoint of the opposite side.

---

The three medians of a triangle $ABC$ are shown in Figure 1.26. Later we will prove that the three medians of any triangle always meet in a point, as suggested by the figure. In the exercises, you will be asked to construct a median of a triangle.

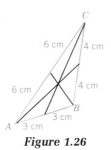

**Figure 1.26**

## EXERCISE 1.4

1. Explain the distinction between a straightedge and a ruler.

2. On a piece of paper, draw a line segment $\overline{AB}$ about 2 in. long and draw a point $C$ not on $\overline{AB}$. Use a compass and a straightedge to construct a line segment $\overline{CD}$ so that $\overline{AB} \cong \overline{CD}$.

3. On a piece of paper, draw a line segment $\overline{AB}$. Use a compass and a straightedge to locate the midpoint of $\overline{AB}$.

4. On a piece of paper, draw a line segment $\overline{AB}$. Use a compass and a straightedge to divide $\overline{AB}$ into four congruent segments.

5. On a piece of paper, draw a line segment $\overline{AB}$. Use a compass and a straightedge to construct an equilateral triangle with each side congruent to $\overline{AB}$.

6. On a piece of paper, draw an angle and use a compass and a straightedge to bisect it.

7. On a piece of paper, draw an angle and use a compass and a straightedge to construct a congruent angle that is adjacent to it.

8. On a piece of paper, draw an angle and use a compass and a straightedge to divide it into four congruent angles.

9. Draw a triangle and bisect each of its three angles. What do you discover about these bisectors?

10. Draw a triangle $ABC$ as in Illustration 1 and construct the altitude from point $C$ to side $\overline{AB}$.

11. Draw a triangle $ABC$ as in Illustration 1 and construct the median from point $B$ to side $\overline{AC}$.

12. Draw a triangle $ABC$ as in Illustration 2 and construct the median from point $C$ to side $\overline{AB}$.

13. Draw a triangle $ABC$ as in Illustration 2 and construct the altitude from point $C$ to the extension of side $\overline{AB}$.

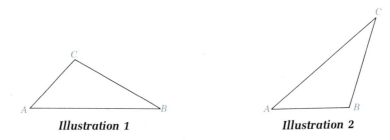

**Illustration 1**     **Illustration 2**

14. Draw a triangle $ABC$ as in Illustration 2 and construct the perpendicular bisector of each side. What do you discover about these bisectors?

**15.** Draw a line *l*. At some point on line *l* construct a perpendicular to line *l*. Call this line *t*. At some point on line *t*, construct a perpendicular to line *t*. Call this line *r*. What do you discover about lines *l* and *r*?

**16.** On a piece of paper draw a line segment $\overline{AB}$. Use a compass and a straightedge to construct a square with $\overline{AB}$ as one side. (*Hint*: A square has four right angles and four congruent sides.)

**17.** On a piece of paper, draw an acute angle and construct that angle's complement.

**18.** Construct an angle that measures 45°.

## 1.5 INTRODUCTION TO PROOF

Previously we listed several undefined terms, defined several other terms, and accepted as true several general statements about these terms without proof. In this section we will begin to prove theorems. To prove these theorems, we will need to use many postulates from algebra.

---

**Postulate 1.6** If *a*, *b*, and *c* are any three quantities, then

$a = a$     **(The reflexive law)**

If $a = b$, then $b = a$.     **(The symmetric law)**

If $a = b$ and $b = c$, then $a = c$.     **(The transitive law)**

---

The reflexive law states that any quantity is equal to itself. For example,

$$5 = 5 \qquad 7x = 7x \qquad \text{and} \qquad m(\angle ABC) = m(\angle ABC)$$

The symmetric law states that if one quantity is equal to a second quantity, then the second quantity is equal to the first. For example,

$$\text{If } x = 3, \text{ then } 3 = x.$$

and

$$\text{If } m(\overline{AB}) = 10 \text{ cm, then } 10 \text{ cm} = m(\overline{AB}).$$

The transitive law states that if one quantity is equal to a second quantity, and if the second is equal to a third, then the first quantity is equal to the third. For example,

$$\text{If } m(\angle ABC) = 5° \text{ and } 5° = m(\angle DEF), \text{ then } m(\angle ABC) = m(\angle DEF).$$

---

**Postulate 1.7** If equal quantities are added to or subtracted from equal quantities, the results are equal quantities.

---

---

**Postulate 1.8** If equal quantities are multiplied by equal quantities or divided by nonzero equal quantities, the results are equal quantities.

---

**Postulate 1.9  The Substitution Law.**    A quantity can be substituted for its equal in any mathematical expression to obtain an equivalent mathematical expression.

---

We have used Postulates 1.7 and 1.8 extensively to solve equations in algebra.

**EXAMPLE 1**

Solve the equation $\dfrac{3}{5} x - 2 = 4$.

**Solution**

$$\frac{3}{5} x - 2 = 4$$

$$\frac{3}{5} x - 2 + 2 = 4 + 2 \qquad \text{Because } 2 = 2, \text{ we can add 2 to both sides of the equation to obtain another equation.}$$

$$\frac{3}{5} x = 6 \qquad \text{Combine terms.}$$

$$5 \left( \frac{3}{5} x \right) = 5(6) \qquad \text{Because } 5 = 5, \text{ we can multiply both sides of the equation by 5 to obtain another equation.}$$

$$3x = 30 \qquad \text{Simplify.}$$

$$\frac{3x}{3} = \frac{30}{3} \qquad \text{Because } 3 = 3, \text{ we can divide both sides of the equation by 3 to obtain another equation.}$$

$$x = 10 \qquad \text{Simplify.}$$

To check the solution, we can use Postulate 1.9 and substitute 10 for $x$ in the original equation and show that both sides equal 4.

$$\frac{3}{5} x - 2 = 4$$

$$\frac{3}{5} (10) - 2 = 4 \qquad \text{Substitute 10 for } x$$

$$3(2) - 2 = 4$$

$$6 - 2 = 4$$

$$4 = 4$$

The solution $x = 10$ does satisfy the original equation.    ■

> **Postulate 1.10** If $A$, $B$, and $C$ are three points on the same line and $B$ is between $A$ and $C$, then
> $$m(\overline{AC}) = m(\overline{AB}) + m(\overline{BC})$$
> $$m(\overline{BC}) = m(\overline{AC}) - m(\overline{AB})$$
> $$m(\overline{AB}) = m(\overline{AC}) - m(\overline{BC})$$

In Figure 1.27, points $A$, $B$, and $C$ are on the same line, $B$ is between $A$ and $C$, $m(\overline{AC}) = 8$, $m(\overline{AB}) = 2$, and $m(\overline{BC}) = 6$. To illustrate Postulate 1.10, we see that

$$m(\overline{AC}) = m(\overline{AB}) + m(\overline{BC})$$
$$8 = 2 + 6$$
$$8 = 8$$
$$m(\overline{BC}) = m(\overline{AC}) - m(\overline{AB})$$
$$6 = 8 - 2$$
$$6 = 6$$

**Figure 1.27**

and

$$m(\overline{AB}) = m(\overline{AC}) - m(\overline{BC})$$
$$2 = 8 - 6$$
$$2 = 2$$

> **Postulate 1.11** If $A$, $B$, and $C$ determine $\angle ABC$ and if point $D$ is in the interior of the angle, then
> $$m(\angle ABC) = m(\angle ABD) + m(\angle DBC)$$
> $$m(\angle ABD) = m(\angle ABC) - m(\angle DBC)$$
> $$m(\angle DBC) = m(\angle ABC) - m(\angle ABD)$$

In Figure 1.28, points $A$, $B$, and $C$ determine $\angle ABC$, point $D$ is in the interior of the angle, $m(\angle ABC) = 50°$, $m(\angle ABD) = 20°$, and $m(\angle DBC) = 30°$. To illustrate Postulate 1.11, we see that

$$m(\angle ABC) = m(\angle ABD) + m(\angle DBC)$$
$$50° = 20° + 30°$$
$$50° = 50°$$
$$m(\angle ABD) = m(\angle ABC) - m(\angle DBC)$$
$$20° = 50° - 30°$$
$$20° = 20°$$

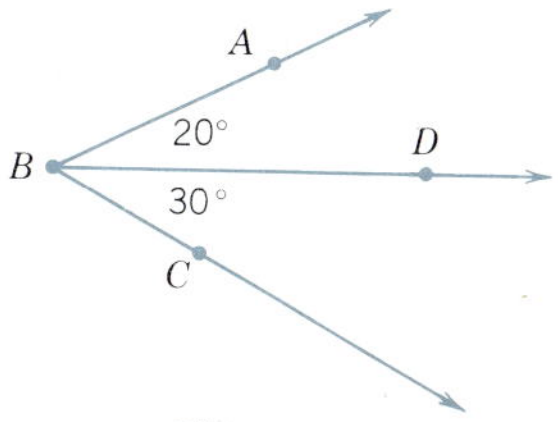

**Figure 1.28**

and

$$m(\angle DBC) = m(\angle ABC) - m(\angle ABD)$$
$$30° = 50° - 20°$$
$$30° = 30°$$

We are now ready to prove some theorems. Each of these theorems will be stated as an *If ... , then ...* sentence. The clause following the word *If* is called the **hypothesis**. It provides the given information. The clause following the word *then* is called the **conclusion**. It indicates what needs to be proved. In writing each proof we will follow a two-column format, in which all statements are listed in one column and all supporting reasons for these statements are listed in a second column.

Study the model proofs given in the text and pattern your proofs after them. For supporting reasons, you may use the given information, any previously accepted definitions and postulates, and any previously proven theorems.

---

***Theorem 1.1*** **The Addition Theorem for Line Segments.**   If point $B$ lies between points $A$ and $C$ on line segment $\overline{AC}$, point $E$ lies between points $D$ and $F$ on line segment $\overline{DF}$, $m(\overline{AB}) = m(\overline{DE})$, and $m(\overline{BC}) = m(\overline{EF})$, then

$$m(\overline{AC}) = m(\overline{DF})$$

---

*Given*: $B$ lies between $A$ and $C$ on segment $\overline{AC}$.

$E$ lies between $D$ and $F$ on segment $\overline{DF}$.

$m(\overline{AB}) = m(\overline{DE})$.

$m(\overline{BC}) = m(\overline{EF})$.

*Prove*: $m(\overline{AC}) = m(\overline{DF})$.

*Proof*:

| STATEMENTS | REASONS |
|---|---|
| **1.** $B$ lies between $A$ and $C$ on segment $\overline{AC}$. | **1.** Given. |
| **2.** $E$ lies between $D$ and $F$ on segment $\overline{DF}$. | **2.** Given. |
| **3.** $m(\overline{AB}) = m(\overline{DE})$. | **3.** Given. |
| **4.** $m(\overline{BC}) = m(\overline{EF})$. | **4.** Given. |

| STATEMENTS | REASONS |
|---|---|
| 5. $m(\overline{AB}) + m(\overline{BC})$ $= m(\overline{DE}) + m(\overline{EF})$. | 5. If equal quantities are added to equal quantities, the results are equal quantities. |
| 6. $m(\overline{AB}) + m(\overline{BC}) = m(\overline{AC})$. | 6. Postulate 1.10. |
| 7. $m(\overline{DE}) + m(\overline{EF}) = m(\overline{DF})$. | 7. Postulate 1.10. |
| 8. $m(\overline{AC}) = m(\overline{DF})$. | 8. Substitution.    □ |

---

**Theorem 1.2 The Subtraction Theorem for Line Segments.** If point $B$ lies between points $A$ and $C$ on line segment $\overline{AC}$, point $E$ lies between points $D$ and $F$ on line segment $\overline{DF}$, $m(\overline{AC}) = m(\overline{DF})$, and $m(\overline{BC}) = m(\overline{EF})$, then

$$m(\overline{AB}) = m(\overline{DE})$$

---

The proof of Theorem 1.2 is left as an exercise. Model your proof after the proof given for Theorem 1.1.

---

**Theorem 1.3 The Angle-Addition Theorem.** If point $D$ is in the interior of $\angle ABC$, point $P$ is in the interior of $\angle RST$, $m(\angle ABD) = m(\angle RSP)$, and $m(\angle DBC) = m(\angle PST)$, then

$$m(\angle ABC) = m(\angle RST)$$

---

*Given:* $D$ is in the interior of $\angle ABC$.

$P$ is in the interior of $\angle RST$.

$m(\angle ABD) = m(\angle RSP)$.

$m(\angle DBC) = m(\angle PST)$.

*Prove:* $m(\angle ABC) = m(\angle RST)$.

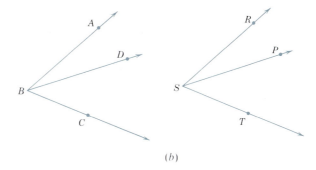

(b)

*Proof*:

| STATEMENTS | REASONS |
| --- | --- |
| **1.** $D$ is in the interior of $\angle ABC$. | **1.** Given. |
| **2.** $P$ is in the interior of $\angle RST$. | **2.** Given. |
| **3.** $m(\angle ABD) = m(\angle RSP)$. | **3.** Given. |
| **4.** $m(\angle DBC) = m(\angle PST)$. | **4.** Given. |
| **5.** $m(\angle ABD) + m(\angle DBC)$ $= m(\angle RSP) + m(\angle PST)$. | **5.** If equal quantities are added to equal quantities, the results are equal quantities. |
| **6.** $m(\angle ABD) + m(\angle DBC)$ $= m(\angle ABC)$. | **6.** Postulate 1.11. |
| **7.** $m(\angle RSP) + m(\angle PST)$ $= m(\angle RST)$. | **7.** Postulate 1.11. |
| **8.** $m(\angle ABC) = m(\angle RST)$. | **8.** Substitution.    □ |

---

**Theorem 1.4 The Angle-Subtraction Theorem.**   If point $D$ is in the interior of $\angle ABC$, point $P$ is in the interior of $\angle RST$, $m(\angle ABC) = m(\angle RST)$, and $m(\angle ABD) = m(\angle RSP)$, then

$$m(\angle DBC) = m(\angle PST)$$

---

The proof of Theorem 1.4 is left as an exercise. Model your proof after the one given for Theorem 1.3.

---

**Theorem 1.5**   If $B$ is the midpoint of the line segment $\overline{AC}$, then $B$ divides $\overline{AC}$ into two segments, each half as long as $\overline{AC}$.

---

*Given*: $B$ is the midpoint of segment $\overline{AC}$.

*Prove*: $m(\overline{AB}) = \dfrac{m(\overline{AC})}{2}$.

$m(\overline{BC}) = \dfrac{m(\overline{AC})}{2}$.

*Proof*:

| STATEMENTS | REASONS |
|---|---|
| **1.** $B$ is the midpoint of $\overline{AC}$. | **1.** Given. |
| **2.** $\overline{AB} \cong \overline{BC}$. | **2.** A midpoint divides a line segment into two congruent line segments. |
| **3.** $m(\overline{AB}) = m(\overline{BC})$. | **3.** Congruent line segments have equal measures. |
| **4.** $m(\overline{AC}) = m(\overline{AB}) + m(\overline{BC})$. | **4.** Postulate 1.10. |
| **5.** $m(\overline{AC}) = m(\overline{AB}) + m(\overline{AB})$. | **5.** Use Postulate 1.9 to substitute $m(\overline{AB})$ for $m(\overline{BC})$. |
| **6.** $m(\overline{AC}) = 2\, m(\overline{AB})$. | **6.** Use Postulate 1.9 to substitute $2\, m(\overline{AB})$ for $m(\overline{AB}) + m(\overline{AB})$. |
| **7.** $\dfrac{m(\overline{AC})}{2} = m(\overline{AB})$. | **7.** If equal quantities are divided by nonzero equal quantities, the results are equal. |
| **8.** $m(\overline{AB}) = \dfrac{m(\overline{AC})}{2}$. | **8.** Use the symmetric law. |

In the exercises you will be asked to prove that $m(\overline{BC}) = \dfrac{m(\overline{AC})}{2}$. □

---

**Theorem 1.6 The Line-Segment-Bisector Theorem.** If $B$ is the midpoint of the line segment $\overline{AC}$, $E$ is the midpoint of line segment $\overline{DF}$, and $m(\overline{AC}) = m(\overline{DF})$, then

$$m(\overline{AB}) = m(\overline{DE})$$

---

*Given*: $B$ is the midpoint of segment $\overline{AC}$.

$E$ is the midpoint of segment $\overline{DF}$.
$m(\overline{AC}) = m(\overline{DF})$.

*Prove*: $m(\overline{AB}) = m(\overline{DE})$.

*Proof*:

| STATEMENTS | REASONS |
|---|---|
| **1.** $B$ is the midpoint of $\overline{AC}$. | **1.** Given. |
| **2.** $m(\overline{AB}) = \dfrac{m(\overline{AC})}{2}$. | **2.** Theorem 1.5. |
| **3.** $E$ is the midpoint of $\overline{DF}$. | **3.** Given. |
| **4.** $m(\overline{DE}) = \dfrac{m(\overline{DF})}{2}$. | **4.** Theorem 1.5. |
| **5.** $m(\overline{AC}) = m(\overline{DF})$. | **5.** Given. |
| **6.** $\dfrac{m(\overline{AC})}{2} = \dfrac{m(\overline{DF})}{2}$. | **6.** If equal quantities are divided by nonzero equal quantities, then the results are equal. |
| **7.** $m(\overline{AB}) = m(\overline{DE})$. | **7.** Substitute $m(\overline{AB})$ for $\dfrac{m(\overline{AC})}{2}$, and substitute $m(\overline{DE})$ for $\dfrac{m(\overline{DF})}{2}$. $\qquad\square$ |

---

**Theorem 1.7**  If ray $\overrightarrow{BD}$ bisects $\angle ABC$, then it divides $\angle ABC$ into two angles, each with half the measure of $\angle ABC$.

---

The proof of Theorem 1.7 is left as an exercise. Model your proof after the proof of Theorem 1.5.

---

**Theorem 1.8  The Angle-Bisector Theorem.**  If ray $\overrightarrow{BD}$ bisects $\angle ABC$, ray $\overrightarrow{SP}$ bisects $\angle RST$, and $m(\angle ABC) = m(\angle RST)$, then

$$m(\angle ABD) = m(\angle RSP)$$

---

The proof of Theorem 1.8 is left as an exercise. Model your proof after the proof of Theorem 1.6.

## EXERCISE 1.5

In Exercises 1–12, use Postulates 1.7 and 1.8 to solve each equation.

**1.** $x + 3 = 5$      **2.** $y - 2 = 7$      **3.** $\dfrac{y}{5} = 3$      **4.** $3y = 21$

**5.** $2x - 4 = 12$     **6.** $3z + 2 = 23$     **7.** $\frac{2}{5}x + 1 = 7$     **8.** $\frac{5}{6}y - 2 = 13$

**9.** $\frac{3}{4}u - 3 = 9$                     **10.** $\frac{4}{5}t + 5 = 17$

**11.** $2r - 5 = 1 - r$                     **12.** $5s - 13 = s - 1$

**13.** Refer to Illustration 1 and illustrate Postulate 1.10 by showing that

$$m(\overline{AC}) = m(\overline{AB}) + m(\overline{BC}).$$
$$m(\overline{AD}) = m(\overline{AE}) - m(\overline{DE}).$$
$$m(\overline{AB}) = m(\overline{AD}) - m(\overline{BD}).$$

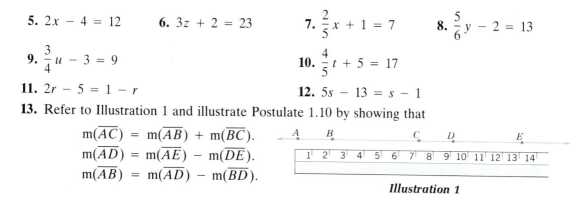

*Illustration 1*

**14.** Refer to Illustration 2 and illustrate Postulate 1.11 by showing that

$$m(\angle ABD) = m(\angle ABC) + m(\angle CBD).$$
$$m(\angle EBF) = m(\angle DBF) - m(\angle DBE).$$
$$m(\angle ABE) = m(\angle ABC) + m(\angle CBE).$$

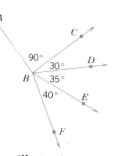

*Illustration 2*

**15.** *Prove Theorem 1.2*: If point $B$ lies between points $A$ and $C$ on line segment $\overline{AC}$, point $E$ lies between points $D$ and $F$ on line segment $\overline{DF}$, $m(\overline{AC}) = m(\overline{DF})$ and $m(\overline{BC}) = m(\overline{EF})$, then $m(\overline{AB}) = m(\overline{DE})$.

**16.** *Prove Theorem 1.4*: If point $D$ is in the interior of $\angle ABC$, point $P$ is in the interior of $\angle RST$, $m(\angle ABC) = m(\angle RST)$ and $m(\angle ABD) = m(RSP)$, then $m(\angle DBC) = m(\angle PST)$.

**17.** In Theorem 1.5, prove that $m(\overline{BC}) = \dfrac{m(\overline{AC})}{2}$.

**18.** *Prove Theorem 1.7*: If ray $\overrightarrow{BD}$ bisects $\angle ABC$, then ray $\overrightarrow{BD}$ divides $\angle ABC$ into two angles, each with half the measure of $\angle ABC$.

**19.** *Prove Theorem 1.8*: If ray $\overrightarrow{BD}$ bisects $\angle ABC$, ray $\overrightarrow{SP}$ bisects $\angle RST$ and $m(\angle ABC) = m(\angle RST)$, then $m(\angle ABD) = m(\angle RSP)$.

In Exercises 20–21, refer to Illustration 3.

**20.** If $m(\overline{AB}) = m(\overline{CD})$, prove that $m(\overline{AC}) = m(\overline{BD})$.

**21.** If $m(\overline{AC}) = m(\overline{BD})$, prove that $m(\overline{AB}) = m(\overline{CD})$.

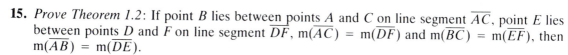

*Illustration 3*

In Exercises 22–23, refer to Illustration 4.

**22.** If m($\angle AOC$) = m($\angle BOD$), prove that m($\angle AOB$) = m($\angle COD$).
**23.** If m($\angle AOB$) = m($\angle COD$), prove that m($\angle AOC$) = m($\angle BOD$).

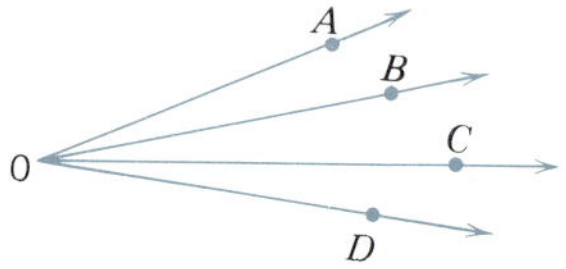

*Illustration 4*

# 1.6 THEOREMS INVOLVING SUPPLEMENTARY, COMPLEMENTARY, AND VERTICAL ANGLES

In this section we will prove many theorems about supplementary and complementary angles. Because these theorems will be used extensively throughout the book, be sure you understand them.

> **Theorem 1.9** If two angles are both supplementary and congruent, then they are right angles.

*Given*: $\angle 1$ is supplementary to $\angle 2$.

$\angle 1 \cong \angle 2$.

*Prove*: $\angle 1$ and $\angle 2$ are right angles.

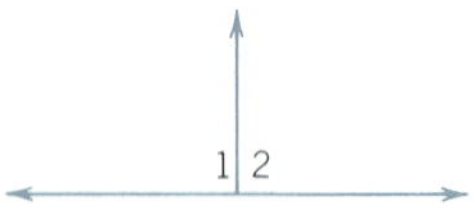

*Proof*:

| STATEMENTS | REASONS |
| --- | --- |
| **1.** $\angle 1$ is supplementary to $\angle 2$. | **1.** Given. |
| **2.** m($\angle 1$) + m($\angle 2$) = 180°. | **2.** If two angles are supplementary, then the sum of their measures is 180°. |
| **3.** $\angle 1 \cong \angle 2$. | **3.** Given. |
| **4.** m($\angle 1$) = m($\angle 2$). | **4.** If two angles are congruent, they have equal measures. |
| **5.** m($\angle 1$) + m($\angle 1$) = 180°. | **5.** Substitute m($\angle 1$) for m($\angle 2$) in Step 2. (Postulate 1.9) |
| **6.** 2 m($\angle 1$) = 180°. | **6.** Combine terms. |

| STATEMENTS | REASONS |
|---|---|
| **7.** $m(\angle 1) = 90°$. | **7.** If equal quantities are divided by nonzero equal quantities, the results are equal. |
| **8.** $m(\angle 2) = 90°$. | **8.** Substitute $m(\angle 2)$ for $m(\angle 1)$ in Step 7. (Postulate 1.9) |
| **9.** $\angle 1$ and $\angle 2$ are right angles. | **9.** If angles have measures of 90°, then they are right angles.   □ |

---

**Theorem 1.10**  Supplements of the same angle are congruent.

---

*Given*:  $\angle A$ is supplementary to $\angle C$.

$\angle B$ is supplementary to $\angle C$.

*Prove*:  $\angle A \cong \angle B$.

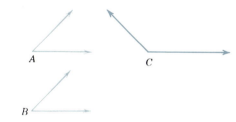

*Proof*:

| STATEMENTS | REASONS |
|---|---|
| **1.** $\angle A$ is supplementary to $\angle C$. | **1.** Given. |
| **2.** $m(\angle A) + m(\angle C) = 180°$. | **2.** If two angles are supplementary, then the sum of their measures is 180°. |
| **3.** $\angle B$ is supplementary to $\angle C$. | **3.** Given. |
| **4.** $m(\angle B) + m(\angle C) = 180°$. | **4.** If two angles are supplementary, then the sum of their measures is 180°. |
| **5.** $m(\angle A) + m(\angle C)$ $= m(\angle B) + m(\angle C)$. | **5.** If two quantities are equal to the same quantity, then they are equal to each other. (Transitive law) |

| STATEMENTS | REASONS |
| --- | --- |
| **6.** m($\angle C$) = m($\angle C$). | **6.** A quantity is equal to itself. (Reflexive law) |
| **7.** m($\angle A$) = m($\angle B$). | **7.** If equal quantities are subtracted from equal quantities, then the results are equal. |
| **8.** $\angle A \cong \angle B$. | **8.** If two angles have the same measure, then they are congruent.    □ |

There is a similar theorem for complements of the same angle.

---

**Theorem 1.11**  Complements of the same angle are congruent.

---

You will be asked to prove Theorem 1.11 in the exercises. Model your proof after the proof given for Theorem 1.10.

---

**Theorem 1.12**  Supplements of congruent angles are congruent.

---

*Given*:  $\angle C$ is supplementary to $\angle A$.
$\angle D$ is supplementary to $\angle B$.
$\angle A \cong \angle B$.

*Prove*:  $\angle C \cong \angle D$.

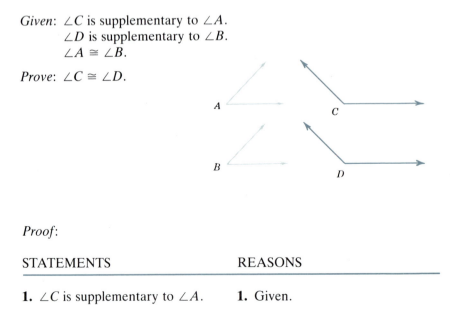

*Proof*:

| STATEMENTS | REASONS |
| --- | --- |
| **1.** $\angle C$ is supplementary to $\angle A$. | **1.** Given. |

| STATEMENTS | REASONS |
|---|---|
| **2.** $m(\angle A) + m(\angle C) = 180°$. | **2.** If two angles are supplementary, then the sum of their measures is 180°. |
| **3.** $\angle D$ is supplementary to $\angle B$. | **3.** Given. |
| **4.** $m(\angle B) + m(\angle D) = 180°$. | **4.** If two angles are supplementary, then the sum of their measures is 180°. |
| **5.** $m(\angle A) + m(\angle C)$ $= m(\angle B) + m(\angle D)$. | **5.** If two quantities are equal to the same quantity, then they are equal to each other. (Transitive law) |
| **6.** $\angle A \cong \angle B$. | **6.** Given. |
| **7.** $m(\angle A) = m(\angle B)$. | **7.** If two angles are congruent, they have equal measures. |
| **8.** $m(\angle C) = m(\angle D)$. | **8.** If equal quantities in Step 7 are subtracted from the equal quantities in Step 5, then the results are equal quantities. |
| **9.** $\angle C \cong \angle D$. | **9.** If two angles have equal measures, they are congruent.    □ |

There is a similar theorem for complements of angles with equal measures.

---

**Theorem 1.13** Complements of congruent angles are congruent.

---

You will be asked to prove Theorem 1.13 in the exercises. Model your proof after the proof of Theorem 1.12.

---

**Theorem 1.14** Two adjacent angles whose noncommon sides form a line are supplementary.

---

In Figure 1.29 $\angle ABD$ and $\angle DBC$ are adjacent angles whose noncommon sides form a line. Because of Theorem 1.14, they are supplementary. You will be asked to prove Theorem 1.14 in the exercises.

If two lines intersect as in Figure 1.30, they form several different pairs of angles. Angles that are nonadjacent angles, such as ∠1 and ∠2, or ∠3 and ∠4, are called **vertical angles**.

| Figure 1.29 | Figure 1.30 |

**Definition 1.22** Nonadjacent angles formed by two intersecting lines are called **vertical angles**.

It is easy to prove the following theorem.

**Theorem 1.15** Vertical angles are congruent.

*Given*: Line *l* and line *m* intersect and form the pair of vertical angles ∠1 and ∠2.

*Prove*: ∠1 ≅ ∠2.

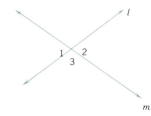

*Proof*:

| STATEMENTS | REASONS |
| --- | --- |
| **1.** Line *l* and line *m* intersect and form the pair of vertical angles ∠1 and ∠2. | **1.** Given. |

| STATEMENTS | REASONS |
|---|---|
| **2.** $\angle 1$ and $\angle 3$ are adjacent angles. | **2.** If two angles have the same vertex and a common side between them, they are adjacent angles. |
| **3.** $\angle 1$ is supplementary to $\angle 3$. | **3.** Two adjacent angles whose noncommon sides form a line are supplementary. |
| **4.** $\angle 2$ and $\angle 3$ are adjacent angles. | **4.** If two angles have the same vertex and a common side between them, they are adjacent angles. |
| **5.** $\angle 2$ is supplementary to $\angle 3$. | **5.** Two adjacent angles whose noncommon sides form a line are supplementary. |
| **6.** $\angle 1 \cong \angle 2$. | **6.** Supplements of the same angle are congruent. ☐ |

**EXAMPLE 1**

In Figure 1.31, $\overleftrightarrow{AD}$, $\overleftrightarrow{BE}$, and $\overleftrightarrow{CF}$ are lines. Find (a) m($\angle EGD$) and (b) m($\angle AGE$).

Solution

(a) Because $\angle BGA$ and $\angle EGD$ form a pair of vertical angles, they are congruent and have equal measures. Thus,

$$m(\angle EGD) = 100°$$

(b) Because $\angle DGC$ and $\angle AGF$ are vertical angles,

$$m(\angle AGF) = 50°$$

Thus,

$$m(\angle AGE) = m(\angle AGF) + m(\angle FGE)$$
$$= 50° + 30°$$
$$= 80°$$

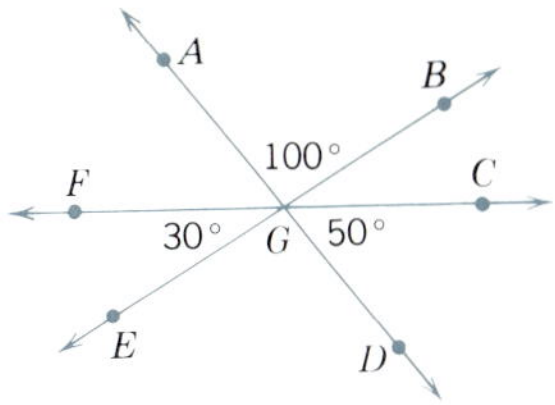

**Figure 1.31**

## EXERCISE 1.6

In Exercises 1–2, refer to the intersecting lines in Illustration 1.

**1.** Name each pair of vertical angles.

**2.** Name each pair of adjacent angles.

In Exercises 3–4, refer to the intersecting lines in Illustration 2.

**3.** Find m(∠1).                    **4.** Find m(∠2).

In Exercises 5–6, refer to the intersecting lines in Illustration 3.

**5.** Find m(∠1).                    **6.** m(∠1) = m(∠2). Find m(∠3).

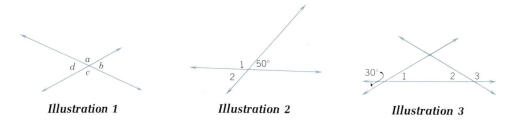

*Illustration 1*          *Illustration 2*          *Illustration 3*

In Exercises 7–10, refer to the intersecting lines in Illustration 4, in which m(∠1) = 50°.

**7.** Find m(∠4).                    **8.** Find m(∠3).

**9.** Find m(∠1) + m(∠2) + m(∠3).    **10.** Find m(∠2) + m(∠4).

In Exercises 11–14, refer to Illustration 5, in which segments $\overline{AB}$ and $\overline{CD}$ intersect, m(∠1) + m(∠3) + m(∠4) = 180°, and m(∠3) = m(∠4), and m(∠4) = m(∠5).

**11.** Find m(∠1).    **12.** Find m(∠2).    **13.** Find m(∠3).    **14.** Find m(∠6).

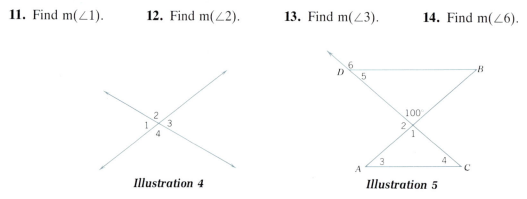

*Illustration 4*                    *Illustration 5*

**15.** If two angles have the same measure, must they be vertical angles? Explain.

**16.** Must angles be supplementary to be vertical angles? Explain.

**17.** Must angles be nonadjacent angles to be vertical angles? Explain.

**18.** Can vertical angles be supplementary angles? Explain.

**19.** Can vertical angles be complementary angles? Explain.

**20.** Can vertical angles be congruent complementary angles? Explain.

**21.** *Prove Theorem 1.11*: Complements of the same angle are congruent.

**22.** *Prove Theorem 1.13*: Complements of congruent angles are congruent.

**23.** *Prove Theorem 1.14*: Two adjacent angles whose noncommon sides form a line are supplementary.

**24.** In Illustration 6, ∠1 and ∠2 are complementary. Prove that ∠3 and ∠4 are complementary.

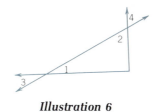

*Illustration 6*

## CHAPTER ONE REVIEW EXERCISES

**1.** To determine if the product of two odd integers is always an odd integer, Sally finds the following products:

$$5 \cdot 7 = 35$$
$$7 \cdot 9 = 63$$
$$9 \cdot 9 = 81$$

After observing that each product is an odd integer, she concludes that the product of two odd integers is always an odd integer. Is Sally's conclusion correct? Did she reason inductively or deductively?

**2.** John remembers the commutative law of multiplication from algebra: $ab = ba$. Thus, he knows that $93{,}760 \cdot 45{,}275 = 45{,}275 \cdot 93{,}760$ without actually doing the multiplications. Is John's conclusion correct? Did he reason inductively or deductively?

In Review Exercises 3–4, refer to Illustration 1.

**3.** Is point $C$ between points $A$ and $D$?

**4.** Is point $B$ between points $C$ and $D$?

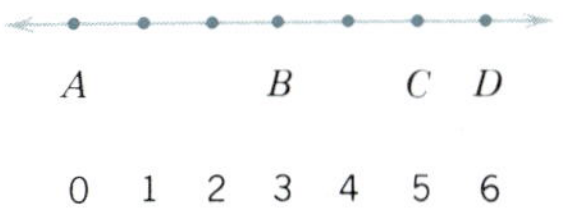

*Illustration 1*

In Review Exercises 5–8, refer to Illustration 1 and find each measure.

**5.** m($\overline{AC}$)      **6.** m($\overline{BC}$)      **7.** m($\overline{AB}$)      **8.** m($\overline{AD}$)

**9.** Find the perimeter of a triangle with sides measuring 8, 12, and 15 cm.

**10.** The perimeter of an isosceles triangle with two sides of length 12.5 m is 40 m. Find the length of its base.

In Exercises 11–22, refer to Illustration 2 and tell whether each statement is true. If a statement is false, explain why.

**11.** The ray $\overrightarrow{GF}$ has point $G$ as its endpoint.

**12.** The line segment $\overline{AG}$ has no endpoints.

**13.** The line $\overleftrightarrow{CD}$ has three endpoints.

**14.** Point $D$ is the vertex of $\angle DGB$.

**15.** The vertex of $\angle EGC$ is point $G$.

**16.** $\overline{EF}$ is perpendicular to $\overline{AB}$.

**17.** m($\angle AGC$) = m($\angle BGD$)

**18.** m($\angle AGF$) = m($\angle BGE$)

**19.** $\angle FGB \cong \angle EGA$

**20.** $\angle AGC \cong \angle FGC$

**21.** $\overrightarrow{GE}$ bisects $\angle AGB$.

**22.** $\overrightarrow{GC}$ bisects $\angle AGF$.

*Illustration 2*

In Review Exercises 23–28, refer to Illustration 3, in which $AB$ is a line.

**23.** Name each right angle.

**24.** Name each acute angle.

**25.** Name each obtuse angle.

**26.** Name each straight angle.

**27.** Name each pair of adjacent angles.

**28.** Explain why there are no vertical angles in Illustration 3.

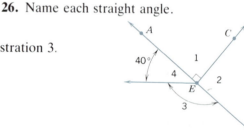

*Illustration 3*

**29.** Find the complement of an angle that measures 35°.

**30.** Find the supplement of an angle that measures 35°.

**31.** Are angles measuring $(x - 40)°$ and $(50 - x)°$ complementary?

**32.** Two angles with measures of $3x°$ and $(2x + 80)°$ are supplementary. Find $x$.

**33.** Draw a line segment and use a compass and a straightedge to divide it into four equal segments.

**34.** Draw an angle and bisect it.

**35.** Draw a line segment and a point not on it. Construct a perpendicular to the line segment passing through the point.

**36.** Draw a triangle and construct one of its medians.

**37.** Refer to Illustration 4. If $\overline{AB} \cong \overline{AC}$ and $\overline{AD} \cong \overline{AE}$, prove $\overline{BD} \cong \overline{CE}$.

**38.** Refer to Illustration 5. If $B$ is the midpoint of $\overline{AC}$, $D$ is the midpoint of $\overline{CE}$, and $\overline{AC} \cong \overline{CE}$, prove $\overline{AB} \cong \overline{DE}$.

**Illustration 4**          **Illustration 5**

In Exercises 39–44, refer to Illustration 6, in which *l*, *m*, and *n* are lines.

**39.** Find m(∠1).                    **40.** Find m(∠2).
**41.** Find m(∠3).                    **42.** Find m(∠4).
**43.** Are ∠2 and ∠4 complementary?
**44.** Are ∠3 and ∠4 supplementary?

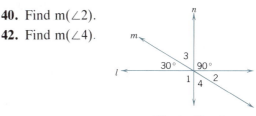

**Illustration 6**

## CHAPTER ONE TEST

In Exercises 1–25, classify each of the following statements as true or false. In this book a statement that is sometimes true and sometimes false is classified as a false statement.

**1.** Postulates are statements that have been proved.
**2.** The points of intersection of the sides of a triangle are called *vertices* of the triangle.
**3.** In a construction problem, the only tools allowed are a compass, pencil, and straightedge.
**4.** A bisector of an angle divides the angle into two congruent angles.
**5.** An acute angle is greater than a right angle.
**6.** In Illustration 1, ∠ADB and ∠ADC are a pair of adjacent angles.
**7.** In Illustration 1, ∠ADB and ∠BDC are a pair of complementary angles.
**8.** A line segment has only one bisector.

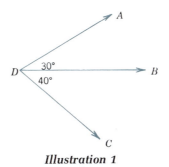

**Illustration 1**

 **9.** If two lines intersect, they form two pairs of vertical angles.

**10.** An angle of 150° has no complement.

**11.** If two angles are complementary, they are also supplementary.

**12.** If two angles are complementary, their supplements are congruent.

**13.** An acute triangle has three acute angles.

**14.** If one line segment bisects a second line segment, the segments are perpendicular.

**15.** Adjacent angles can be supplementary.

**16.** In Illustration 2, $\angle ABC$ and $\angle ABD$ are complementary.

**17.** In Illustration 2, $\angle ABC$ and $\angle CBD$ are adjacent angles.

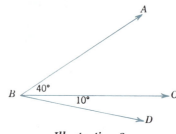

*Illustration 2*

**18.** An equilateral triangle has only one altitude.

**19.** An isosceles triangle can also be a scalene triangle.

**20.** A perpendicular to a line segment also bisects the line segment.

**21.** An isosceles triangle can be a right triangle.

**22.** Complements of complementary angles are congruent.

**23.** An equilateral triangle is also isosceles.

**24.** An angle has only one bisector.

**25.** The bisector of an obtuse angle divides the angle into two acute angles.

# 2

*Thales (about 640–550 B.C.). Thales was a very successful and widely traveled Greek merchant. In his travels, Thales learned the "practical geometry" known to the early Egyptians and helped bring the knowledge back to Greece. Thales has been called "the father of Greek mathematics, astronomy, and philosophy." One of his pupils was Pythagoras (Radio Times Hulton Picture Library).*

# Congruent Triangles and Basic Theorems

T he word *congruent* means *in agreement with*. It comes from the Latin words *con* (with) and *gruere* (to agree). Congruent figures are figures that agree with each other. In fact, they are identical twins. They are the same size (have the same area) and the same shape. In this chapter we shall prove many theorems that involve congruent triangles.

## 2.1 CONGRUENT TRIANGLES

Every triangle is made up of exactly six parts. They are the three angles and the three sides that form the triangle. In Figure 2.1, the six parts of $\triangle ABC$ (read as "triangle $ABC$") are $\angle A$, $\angle B$, $\angle C$, $\overline{BC}$, $\overline{CA}$, and $\overline{AB}$.

In Figure 2.2, $\triangle DEF$ and $\triangle ABC$ are congruent because they are the same size and the same shape. In fact, if $\triangle DEF$ were placed on top of

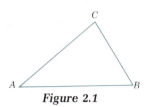

*Figure 2.1*

△*ABC*, the two triangles would coincide because their corresponding parts would coincide.

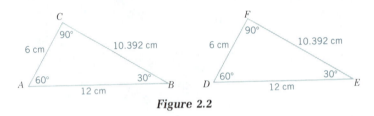

*Figure 2.2*

The congruent triangles *DEF* and *ABC* have six pairs of corresponding parts:

$$\angle A \text{ corresponds to } \angle D$$
$$\angle B \text{ corresponds to } \angle E$$
$$\angle C \text{ corresponds to } \angle F$$
$$\overline{BC} \text{ corresponds to } \overline{EF}$$
$$\overline{AC} \text{ corresponds to } \overline{DF}$$

and

$$\overline{AB} \text{ corresponds to } \overline{DE}$$

To say that △*ABC* is congruent to △*DEF*, we write

$$\triangle ABC \cong \triangle DEF$$

In △*ABC* of Figure 2.2, we say that side $\overline{CB}$ is opposite ∠*A*, and that ∠*A* is opposite side $\overline{CB}$. Likewise, side $\overline{AC}$ and ∠*B* are opposite each other, and side $\overline{AB}$ and ∠*C* are opposite each other. In △*DEF*, side $\overline{EF}$ and ∠*D* are opposite each other, side $\overline{DF}$ and ∠*E* are opposite each other, and side $\overline{DE}$ and ∠*F* are opposite each other. If two angles in a pair of triangles are corresponding angles, then the sides opposite those angles form a pair of corresponding sides. In a similar way, corresponding angles are always opposite a pair of corresponding sides.

---

**Definition 2.1** Two triangles are called *congruent triangles* if, and only if, all of their corresponding parts are congruent.

To prove that two triangles are congruent triangles, we could use Definition 2.1 after showing that all pairs of corresponding parts are congruent. However, this would be more work than is necessary. If we choose the parts in a special way, we can establish that two triangles are congruent by using only three pairs of corresponding parts. The following three postulates indicate which three pairs of congruent corresponding parts are necessary to prove triangles congruent.

---

**Postulate 2.1** If three sides of one triangle are congruent, respectively, to three sides of a second triangle, then the triangles are congruent. ($SSS \cong SSS$).

---

To illustrate Postulate 2.1, we refer to the triangles in Figure 2.3. Because the corresponding sides of the two triangles have equal measures, the three sides of $\triangle ABC$ are congruent, respectively, to the three sides of $\triangle DEF$. Thus, by Postulate 2.1, the triangles are congruent.

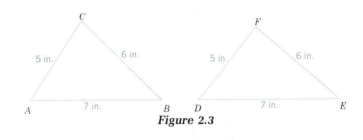

**Figure 2.3**

---

**Postulate 2.2** If two sides and the included angle of one triangle are congruent, respectively, to two sides and the included angle of a second triangle, then the triangles are congruent. ($SAS \cong SAS$).

---

To illustrate Postulate 2.2, we refer to the triangles in Figure 2.4. Because two pairs of corresponding sides and the angle between them have

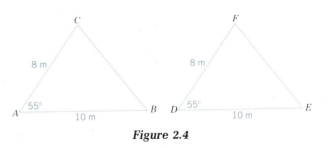

**Figure 2.4**

equal measures, two sides and the angle between them in △*ABC* are congruent to two sides and the angle between them in △*DEF*. Thus, by Postulate 2.2, the triangles are congruent.

---

**Postulate 2.3** If two angles and the included side of one triangle are congruent, respectively, to two angles and the included side of a second triangle, then the triangles are congruent. (*ASA* ≅ *ASA*).

---

To illustrate Postulate 2.3, we refer to the triangles in Figure 2.5. Because two pairs of corresponding angles and the side between them have equal measures, two angles and the side between them in △*ABC* are congruent to two angles and the side between them in △*DEF*. Thus, by Postulate 2.3, the triangles are congruent.

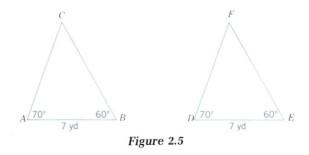

**Figure 2.5**

The following examples will illustrate how Postulates 2.1, 2.2, and 2.3 are used to prove two triangles are congruent. In each example, the plan will be to show that three congruences are true and then apply one of the postulates.

When proving these examples or later theorems, we must be careful not to assume too much about a geometric figure by just looking at it. We must not, for example, assume that two line segments are congruent just because they look that way. Likewise, we must not assume that an angle is a right angle just because it looks like one, or that two lines are perpendicular just because they intersect and look like they form right angles.

**EXAMPLE 1**

*Given*: $\overline{AE}$ and $\overline{BD}$ intersect at *C*.

$\overline{BC} \cong \overline{DC}$.

$\overline{AC} \cong \overline{EC}$.

*Prove*: △*ABC* ≅ △*EDC*.

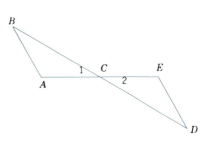

*Proof*:

| STATEMENTS | REASONS |
|---|---|
| **1.** $\overline{AE}$ and $\overline{BD}$ intersect at $C$. | **1.** Given. |
| **2.** $\overline{BC} \cong \overline{DC}$. | **2.** Given. |
| **3.** $\overline{AC} \cong \overline{EC}$. | **3.** Given. |
| **4.** $\angle 1$ and $\angle 2$ are vertical angles. | **4.** A pair of nonadjacent angles formed by two intersecting lines is a pair of vertical angles. |
| **5.** $\angle 1 \cong \angle 2$. | **5.** Vertical angles are congruent. |
| **6.** $\triangle ABC \cong \triangle EDC$. | **6.** If two sides and the included angle of one triangle are congruent, respectively, to two sides and the included angle of a second triangle, then the triangles are congruent. ($SAS \cong SAS$). ∎ |

In the previous chapter we discussed the reflexive, symmetric, and transitive laws of equality. These laws are also true for the relation of congruence. Although we only state a postulate for congruent line segments, the postulate applies to all congruent geometric figures.

---

*Postulate 2.4* If $\overline{AB}$, $\overline{CD}$, and $\overline{EF}$ are three line segments, then

**Reflexive law:** $\overline{AB} \cong \overline{AB}$.
**Symmetric law:** If $\overline{AB} \cong \overline{CD}$, then $\overline{CD} \cong \overline{AB}$.
**Transitive law:** If $\overline{AB} \cong \overline{CD}$ and $\overline{CD} \cong \overline{EF}$, then $\overline{AB} \cong \overline{EF}$.

---

**EXAMPLE 2**

*Given*: $\overline{AD} \cong \overline{CD}$.
$\overline{AB} \cong \overline{CB}$.

*Prove*: $\triangle ABD \cong \triangle CBD$.

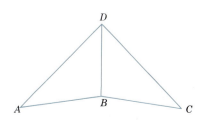

*Proof*:

| STATEMENTS | REASONS |
|---|---|
| **1.** $\overline{AD} \cong \overline{CD}$. | **1.** Given. |
| **2.** $\overline{AB} \cong \overline{CB}$. | **2.** Given. |
| **3.** $\overline{DB} \cong \overline{DB}$. | **3.** A line segment is congruent to itself. (Reflexive law). |
| **4.** $\triangle ABD \cong \triangle CBD$. | **4.** If three sides of one triangle are congruent, respectively, to three sides of a second triangle, then the triangles are congruent. $(SSS \cong SSS)$. ∎ |

**EXAMPLE 3**

*Given*: $\angle 1 \cong \angle 2$.
$\angle 3 \cong \angle 4$.

*Prove*: $\triangle ADC \cong \triangle BDC$.

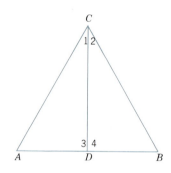

*Proof*:

| STATEMENTS | REASONS |
|---|---|
| **1.** $\angle 1 \cong \angle 2$. | **1.** Given. |
| **2.** $\overline{CD} \cong \overline{CD}$. | **2.** A line segment is congruent to itself. (Reflexive law). |
| **3.** $\angle 3 \cong \angle 4$. | **3.** Given. |
| **4.** $\triangle ADC \cong \triangle BDC$. | **4.** If two angles and the included side of one triangle are congruent, respectively, to two angles and the included side of a second triangle, then the triangles are congruent. $(ASA \cong ASA)$. ∎ |

Sometimes the triangles we wish to prove congruent are overlapping triangles. In this case it is a good idea to use colored pencils to trace the individual triangles. This will help identify the triangles that are to be proved congruent.

**EXAMPLE 4**     *Given*: $\overline{CA} \cong \overline{DB}$.
$\overline{CB} \cong \overline{DA}$.

*Prove*: $\triangle ABC \cong \triangle BAD$.

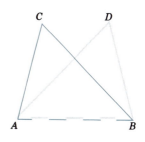

*Proof*:

| STATEMENTS | REASONS |
|---|---|
| **1.** $\overline{CA} \cong \overline{DB}$. | **1.** Given. |
| **2.** $\overline{CB} \cong \overline{DA}$. | **2.** Given. |
| **3.** $\overline{AB} \cong \overline{BA}$. | **3.** A line segment is congruent to itself. (Reflexive law). |
| **4.** $\triangle ABC \cong \triangle BAD$. | **4.** If three sides of one triangle are congruent, respectively, to three sides of a second triangle, then the triangles are congruent. ($SSS \cong SSS$). ■ |

> **Theorem 2.1** If two triangles are congruent to a third triangle, then they are congruent to each other.

In the exercises you will be asked to prove Theorem 2.1.

## EXERCISE 2.1

In Exercises 1–6, refer to Illustration 1. Pairs of congruent sides of congruent triangles are shown with cross marks. Complete each statement involving a correspondence between parts of these two triangles.

**1.** Side $\overline{AC}$ corresponds to side _____.

**2.** Side $\overline{DE}$ corresponds to side _____.

**3.** Side $\overline{BC}$ corresponds to side _____.

**4.** $\angle A$ corresponds to $\angle$ _____.

**5.** $\angle E$ corresponds to $\angle$ _____.

**6.** $\angle F$ corresponds to $\angle$ _____.

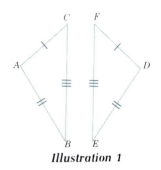

*Illustration 1*

In Exercises 7–12, the given information refers to Illustration 2. In each exercise, indicate whether the triangles are congruent. If so, give the reason why.

**7.** $\angle A \cong \angle D$
$\angle B \cong \angle E$
$\overline{AB} \cong \overline{DE}$

**8.** $\angle A \cong \angle D$
$\angle B \cong \angle E$
$\angle C \cong \angle F$

**9.** $\angle A \cong \angle D$
$\overline{BC} \cong \overline{EF}$
$\overline{AC} \cong \overline{DF}$

**10.** $\overline{AC} \cong \overline{DF}$
$\overline{AB} \cong \overline{DE}$
$\overline{BC} \cong \overline{EF}$

**11.** $\overline{AB} \cong \overline{DE}$
$\angle B \cong \angle E$
$\overline{BC} \cong \overline{EF}$

**12.** $\overline{BC} \cong \overline{EF}$
$\angle C \cong \angle E$
$\angle B \cong \angle F$

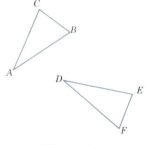

*Illustration 2*

In Exercises 13–16, fill in the missing statements and the missing reasons in each proof.

**13.** *Given:* $\overline{AC} \cong \overline{BC}$.
$\angle 1 \cong \angle 2$.
*Prove:* $\triangle ACD \cong \triangle BCD$.

*Proof:*

| STATEMENTS | REASONS |
|---|---|
| **1.** $\overline{AC} \cong \overline{BC}$. | **1.** _____ ? _____ |
| **2.** _____ ? _____ | **2.** Given. |
| **3.** _____ ? _____ | **3.** A line segment is congruent to itself (Reflexive law). |
| **4.** $\triangle ACD \cong \triangle BCD$. | **4.** _____ ? _____ |

**14.** *Given*: $\overline{AB} \cong \overline{CD}$.
           $\angle 1 \cong \angle 2$.
   *Prove*: $\triangle ABD \cong \triangle CDB$.

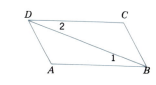

*Proof*:

| STATEMENTS | REASONS |
|---|---|
| **1.** $\overline{AB} \cong \overline{CD}$. | **1.** _____?_____ |
| **2.** _____?_____ | **2.** Given. |
| **3.** $\overline{BD} \cong \overline{DB}$. | **3.** _____?_____ |
| **4.** _____?_____ | **4.** $SAS \cong SAS$. |

**15.** *Given*: $\overline{AE}$ and $\overline{DB}$ intersect at $C$.
           $\angle A \cong \angle E$.
           $\overline{AC} \cong \overline{EC}$.
   *Prove*: $\triangle ABC \cong \triangle EDC$.

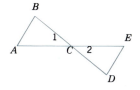

*Proof*:

| STATEMENTS | REASONS |
|---|---|
| **1.** $\overline{AE}$ and $\overline{DB}$ intersect at $C$. | **1.** Given. |
| **2.** _____?_____ | **2.** Given. |
| **3.** $\overline{AC} \cong \overline{EC}$. | **3.** _____?_____ |
| **4.** _____?_____ | **4.** A pair of nonadjacent angles formed by two intersecting lines is a pair of vertical angles. |
| **5.** _____?_____ | **5.** _____?_____ |
| **6.** _____?_____ | **6.** $ASA \cong ASA$. |

**16.** *Given*: $\overline{AB} \cong \overline{DC}$.
           $\overline{AC} \cong \overline{DB}$.
   *Prove*: $\triangle ABC \cong \triangle DCB$.

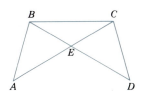

*Proof:*

| STATEMENTS | REASONS |
|---|---|
| **1.** $\overline{AB}$ and $\overline{DC}$. | **1.** _____ ? |
| **2.** _____ ? | **2.** Given. |
| **3.** _____ ? | **3.** _____ ? |
| **4.** $\triangle ABC \cong \triangle DCB$. | **4.** _____ ? |

In Exercises 17–26, write each proof using a two-column format.

**17.** Refer to Illustration 3.

    *Given*: $\overline{AD} \cong \overline{BD}$.

           $\overline{AC} \cong \overline{BC}$.

    *Prove*: $\triangle ADC \cong \triangle BDC$.

**18.** Refer to Illustration 4.

    *Given*: $\overline{AD} \cong \overline{BD}$.

           $\angle 1 \cong \angle 2$.

    *Prove*: $\triangle ADC \cong \triangle BDC$.

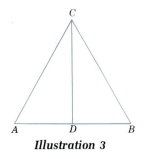

**Illustration 3**

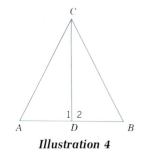

**Illustration 4**

**19.** Refer to Illustration 5.

    *Given*: $\overline{AE}$ and $\overline{BD}$ are line segments intersecting at point $C$.

           $\overline{BC} \cong \overline{DC}$.

           $\angle B \cong \angle D$.

    *Prove*: $\triangle ABC \cong \triangle EDC$.

**20.** Refer to Illustration 6.

    *Given*: $\overline{AE}$ and $\overline{BD}$ are line segments intersecting at point $C$.

           $\overline{AC} \cong \overline{EC}$.

           $\overline{BC} \cong \overline{DC}$.

    *Prove*: $\triangle ABC \cong \triangle EDC$.

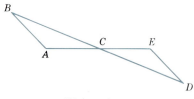

**Illustration 5**

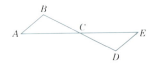

**Illustration 6**

**21.** Refer to Illustration 7.

 *Given:* $\overline{DC} \cong \overline{BA}.$
   $\overline{AD} \cong \overline{CB}.$

 *Prove:* $\triangle ABD \cong \triangle CDB.$

**22.** Refer to Illustration 8.

 *Given:* $\triangle ADE.$
   $\angle A \cong \angle D.$
   $\overline{AE} \cong \overline{DE}.$
   $\overline{AC} \cong \overline{DB}.$

 *Prove:* $\triangle ABE \cong \triangle DCE.$
 (*Hint:* Consider Postulate 1.7)

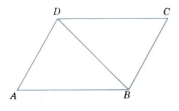

*Illustration 7*

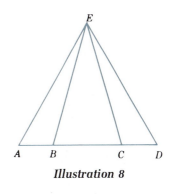

*Illustration 8*

**23.** Refer to Illustration 9.

 *Given:* $\triangle ADE.$
   $\overline{BE} \cong \overline{CE}.$
   $\angle 1 \cong \angle 2.$
   $\overline{AB} \cong \overline{DC}.$

 *Prove:* $\triangle ABE \cong \triangle DCE.$

**24.** Refer to Illustration 10.

 *Given:* $\angle A \cong \angle B.$
   $\overline{AC} \cong \overline{BC}.$

 *Prove:* $\triangle AFC \cong \triangle BEC.$

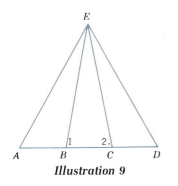

*Illustration 9*

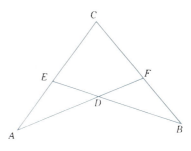

*Illustration 10*

**25.** Refer to Illustration 11.

 *Given:* $\overline{CA} \cong \overline{DB}.$
   $\overline{CB} \cong \overline{DA}.$

 *Prove:* $\triangle ABC \cong \triangle BAD.$

**26.** Refer to Illustration 12.

 *Given:* $\overline{AD} \cong \overline{AC}.$
   $\overline{AB} \cong \overline{AE}.$

 *Prove:* $\triangle ADB \cong \triangle ACE.$

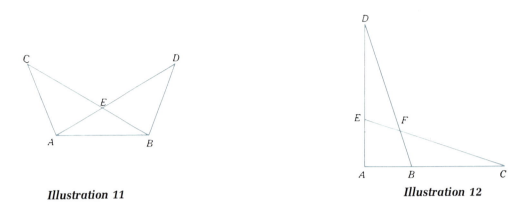

Illustration 11                                    Illustration 12

**27.** *Prove Theorem 2.1*: If two triangles are congruent to a third triangle, then they are congruent to each other. (*Hint*: Use the transitive law of congruence.)

## 2.2 PROVING CORRESPONDING PARTS OF CONGRUENT TRIANGLES CONGRUENT

We can extend the proofs done in Section 2.1 to prove that corresponding parts of two congruent triangles are congruent. To do so, we first prove that two triangles are congruent, and then use the definition of congruent triangles, which states that *all* corresponding parts of congruent triangles are congruent. An abbreviation for this statement is **cpctc** (corresponding parts of congruent triangles are congruent).

**EXAMPLE 1**

*Given*: $\overline{AC} \cong \overline{BC}$.
        $\angle 1 \cong \angle 2$.

*Prove*: $\angle A \cong \angle B$.

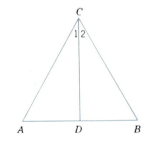

*Proof*:

| STATEMENTS | REASONS |
| --- | --- |
| **1.** $\overline{AC} \cong \overline{BC}$. | **1.** Given. |

| STATEMENTS | REASONS |
|---|---|
| **2.** $\angle 1 \cong \angle 2$. | **2.** Given. |
| **3.** $\overline{CD} \cong \overline{CD}$. | **3.** A line segment is congruent to itself (Reflexive law). |
| **4.** $\triangle ACD \cong \triangle BCD$. | **4.** $SAS \cong SAS$. |
| **5.** $\angle A \cong \angle B$. | **5.** cpctc.   ■ |

*EXAMPLE 2*

*Given*: $\overline{AC} \cong \overline{DB}$.
$\quad\quad\quad \angle 1 \cong \angle 2$.
$\quad\quad\quad \overline{CE} \cong \overline{BE}$.

*Prove*: $\angle A \cong \angle D$.

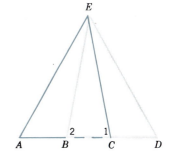

*Proof*:

| STATEMENTS | REASONS |
|---|---|
| **1.** $\overline{AC} \cong \overline{DB}$. | **1.** Given. |
| **2.** $\overline{CE} \cong \overline{BE}$. | **2.** Given. |
| **3.** $\angle 1 \cong \angle 2$. | **3.** Given. |
| **4.** $\triangle ACE \cong \triangle DBE$. | **4.** $SAS \cong SAS$. |
| **5.** $\angle A \cong \angle D$. | **5.** cpctc.   ■ |

We can now prove the construction of the bisector of a line segment given on page 21.

*EXAMPLE 3*

By construction we are given:  $\overline{AC} \cong \overline{BC}$.
$\quad\quad\quad\quad\quad\quad\quad\quad\quad\quad\quad\quad\quad\quad\quad \overline{AD} \cong \overline{BD}$.

*Prove*: $\overline{CD}$ bisects $\overline{AB}$.

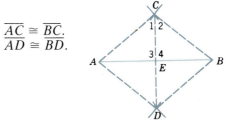

*Proof:*

| STATEMENTS | REASONS |
|---|---|
| **1.** $\overline{AC} \cong \overline{BC}$. | **1.** Given. |
| **2.** $\overline{AD} \cong \overline{BD}$. | **2.** Given. |
| **3.** $\overline{CD} \cong \overline{CD}$. | **3.** Reflexive law. |
| **4.** $\triangle ADC \cong \triangle BDC$. | **4.** $SSS \cong SSS$. |
| **5.** $\angle 1 \cong \angle 2$. | **5.** cpctc. |
| **6.** $\overline{CE} \cong \overline{CE}$. | **6.** Reflexive law. |
| **7.** $\triangle ACE \cong \triangle BCE$. | **7.** $SAS \cong SAS$. |
| **8.** $\overline{AE} \cong \overline{BE}$. | **8.** cpctc. |
| **9.** $\overline{CD}$ bisects $\overline{AB}$. | **9.** If a line divides segment $\overline{AB}$ into two congruent segments, then it bisects segment $\overline{AB}$. ■ |

## EXERCISE 2.2

In Exercises 1–4, fill in the missing statements and missing reasons in each proof.

**1.** *Given:* $\overline{AD} \cong \overline{BD}$.
  $\quad\quad\quad \angle 3 \cong \angle 4$.
  *Prove:* $\angle 1 \cong \angle 2$.

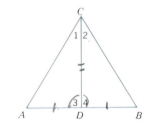

*Proof:*

| STATEMENTS | REASONS |
|---|---|
| **1.** $\overline{AD} \cong \overline{BD}$. | **1.** _____ ? |
| **2.** $\angle 3 \cong \angle 4$   ? | **2.** Given. |
| **3.** $\overline{DC} \cong \overline{DC}$. | **3.** reflexive   ? |
| **4.** $\triangle ADC \cong \triangle BDC$   ? | **4.** $SAS \cong SAS$. |
| **5.** $\angle 1 \cong \angle 2$. | **5.** CPCTC   ? |

**2.** *Given:* $\overline{AB} \cong \overline{CD}$.
  $\quad\quad\quad \overline{BC} \cong \overline{DA}$.
  *Prove:* $\angle 1 \cong \angle 2$.

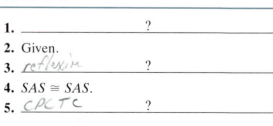

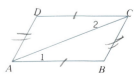

*Proof*:

| STATEMENTS | REASONS |
|---|---|
| 1. $AB \cong CD$ ? | 1. Given. |
| 2. $\overline{BC} \cong \overline{DA}$. | 2. _____? |
| 3. $AC \cong AC$ ? | 3. A line segment is congruent to itself (Reflexive law). |
| 4. $\triangle ABC \cong \triangle CDA$. | 4. $SSS \cong SSS$. |
| 5. $\angle 1 \cong \angle 2$ ? | 5. cpctc. |

**3.** *Given*: $\triangle ADE$.
$\qquad \angle 2 \cong \angle 1$.
$\qquad \overline{BE} \cong \overline{CE}$.
$\qquad \angle BED \cong \angle CEA$.
$\quad$ *Prove*: $\overline{BD} \cong \overline{CA}$.

*Proof*:

| STATEMENTS | REASONS |
|---|---|
| 1. $\triangle ADE$. | 1. _____? |
| 2. $\angle 2 \cong \angle 1$. | 2. _____? |
| 3. _____? | 3. Given. |
| 4. $\angle BED \cong \angle CEA$. | 4. _____? |
| 5. _____? | 5. $ASA \cong ASA$. |
| 6. $\overline{BD} \cong \overline{CA}$. | 6. _____? |

**4.** *Given*: $\overline{BF}$ and $\overline{DE}$ intersect at $C$.
$\qquad \overline{AB} \cong \overline{AD}$.
$\qquad \angle B \cong \angle D$.
$\quad$ *Prove*: $\angle 2 \cong \angle 1$.

*Proof*:

| STATEMENTS | REASONS |
|---|---|
| 1. _____? | 1. Given. |
| 2. $\overline{AB} \cong \overline{AD}$. | 2. _____? |
| 3. $\angle B \cong \angle D$. | 3. _____? |

STATEMENTS

REASONS

_____

**4.** _____?_____

**4.** An angle is congruent to itself (Reflexive law).

**5.** $\triangle ABF \cong \triangle \underline{?}$

**5.** _____?_____

**6.** $\angle 2 \cong \angle 1$.

**6.** _____?_____

In Exercises 5–8, write each proof.

**5.** Refer to Illustration 1.

  *Given:* $\overline{AE}$ and $\overline{BD}$ intersect at $C$.
    $\angle A \cong \angle E$.
    $\overline{AC} \cong \overline{EC}$.
  *Prove:* $\angle B \cong \angle D$.

**6.** Refer to Illustration 2.

  *Given:* $\overline{DB} \cong \overline{EA}$.
    $\overline{AD} \cong \overline{BE}$.
  *Prove:* $\angle DAB \cong \angle EBA$.

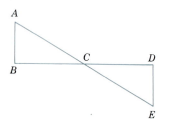

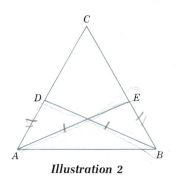

*Illustration 1*

*Illustration 2*

**7.** Refer to Illustration 3.

  *Given:* $\triangle ADE$.
    $\angle 1 \cong \angle 2$.
    $\overline{AB} \cong \overline{DC}$.
    $\overline{BE} \cong \overline{CE}$.
  *Prove:* $\overline{AE} \cong \overline{DE}$.

**8.** Refer to Illustration 4.

  *Given:* $\triangle ABC$.
    $\overline{DC} \cong \overline{EC}$.
    $\angle 1 \cong \angle 2$.
  *Prove:* $\overline{AC} \cong \overline{BC}$.

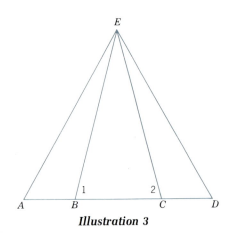

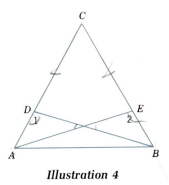

*Illustration 3*

*Illustration 4*

**9.** Prove the construction for copying an angle. Refer to Illustration 5.

*Given*: ∠A.

*By construction*:   $\overline{AC} \cong \overline{DF}$.
$\overline{BC} \cong \overline{EF}$.
$\overline{AB} \cong \overline{DE}$.

*Prove*: ∠A ≅ ∠D.

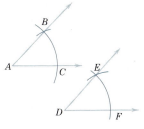

***Illustration 5***

**10.** Prove the construction for bisecting an angle. Refer to Illustration 6.

*Given*: ∠PAQ.

*By construction*:   $\overline{AB} \cong \overline{AC}$.
$\overline{BD} \cong \overline{CD}$.

*Prove*: $\overline{AD}$ bisects ∠A.

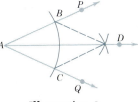

***Illustration 6***

**11.** Prove the construction for a perpendicular to a line through a point on the line. Refer to Illustration 7.

*Given*: Line *l* passing through point *A*.

*By construction*: $\overline{AB} \cong \overline{AC}$.
$\overline{BD} \cong \overline{CD}$.

*Prove*: $\overline{AD} \perp \overline{BC}$.

***Illustration 7***

**12.** Prove the construction for a perpendicular to a line through a point not on the line. Refer to Illustration 8.

*Given*: Line *l* and point *A* not on line *l*.

*By construction*: $\overline{AB} \cong \overline{AC}$.
$\overline{BD} \cong \overline{CD}$.

*Prove*: $\overline{AD} \perp \overline{BC}$.

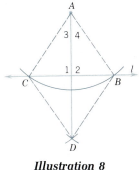

***Illustration 8***

**13.** Refer to Illustration 9.

    *Given*: $\overline{AC}$ bisects $\overline{DB}$ at point $O$.

            $\overline{AC} \perp \overline{BD}$.

    *Prove*: $\overline{DC} \cong \overline{BC}$.

**14.** Refer to Illustration 10.

    *Given*: $\overline{DC} \cong \overline{AB}$.

            $\angle CDA \cong \angle BAD$.

    *Prove*: $\overline{AC} \cong \overline{DB}$.

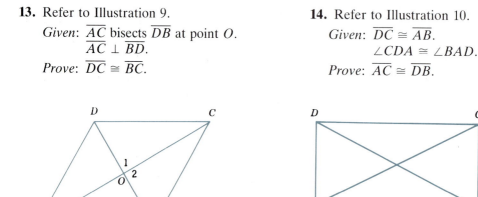

*Illustration 9*                               *Illustration 10*

# 2.3 THEOREMS INVOLVING ISOSCELES TRIANGLES

We are now ready to prove many theorems. Do not be concerned with memorizing the number associated with a theorem, but rather with understanding what a theorem says. To accomplish this, state each theorem in your own words. It may be helpful to start a list of important theorems and definitions in a notebook, adding to it as the work progresses.

There are no specific rules that can be given to prove all theorems. However, the following list of steps is often helpful.

1. Read the theorem carefully, making sure that you know the meaning of each word.

2. Determine the hypothesis and the conclusion. To do so, it may be necessary to write the theorem as an *If... then ...* sentence. The hypothesis (or given information) is contained in the clause following the word *If*. The conclusion (what is to be proved) is contained in the clause following the word *then*.

3. Draw a figure that illustrates each point, line, and angle described in the hypothesis. Letter the vertices of figure and mark the parts that are given to be congruent.

4. Try to determine what information will be necessary to prove the theorem. Then search for the definitions, postulates, and previously proved theorems that will be used to prove the theorem.

5. It is sometimes necessary to draw **auxiliary lines** to aid in establishing a proof. Whenever you draw an auxiliary line, be sure that you can give a reason why it can be drawn.

**6.** Write your proof, giving a reason for each statement.

We will use these steps as we prove the following theorem.

---

**Theorem 2.2** If two sides of a triangle are congruent, then the angles opposite those sides are congruent.

---

We first read the theorem to make sure we understand each word. We then determine the hypothesis and the conclusion. The hypothesis is that two sides of a triangle are congruent. The conclusion we are trying to prove is that the angles opposite those congruent sides are also congruent. We then draw a figure such as the one in Figure 2.6 illustrating a $\triangle ABC$ with sides $\overline{AC}$ and $\overline{BC}$ congruent.

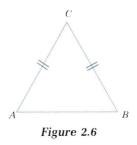

*Figure 2.6*

Finally we search for the appropriate definitions, postulates, and theorems to construct the following written proof, which requires an auxiliary line.

*Given*: $\triangle ABC$ with $\overline{AC} \cong \overline{BC}$.

*Prove*: $\angle A \cong \angle B$.

*Construction*: Construct the bisector of $\angle ACB$.

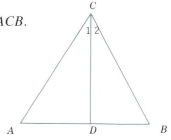

*Proof*:

| STATEMENTS | REASONS |
|---|---|
| **1.** $\triangle ABC$ with $\overline{AC} \cong \overline{BC}$. | **1.** Given. |

| STATEMENTS | REASONS |
|---|---|
| **2.** $\overline{CD}$ bisects $\angle ACB$. | **2.** Every angle has a bisector. |
| **3.** $\angle 1 \cong \angle 2$. | **3.** An angle bisector divides an angle into two congruent angles. |
| **4.** $\overline{CD} \cong \overline{CD}$. | **4.** Reflexive law. |
| **5.** $\triangle ACD \cong \triangle BCD$. | **5.** $SAS \cong SAS$. |
| **6.** $\angle A \cong \angle B$. | **6.** cpctc.                    ☐ |

The triangle of Theorem 2.2 is an isosceles triangle. The angle opposite the base of an isosceles triangle is called the **vertex angle**. The two congruent angles are called **base angles**. Theorem 2.2 is often stated as,

*The base angles of an isosceles triangle are congruent.*

---

**Theorem 2.3**  The bisector of the vertex angle of an isosceles triangle is the perpendicular bisector of the base of the triangle.

---

*Given*:  $\triangle ABC$ is an isosceles triangle with $\overline{AC} \cong \overline{BC}$. $\overline{CD}$ bisects $\angle ACB$.

*Prove*:  $\overline{CD}$ is the perpendicular bisector of $\overline{AB}$.

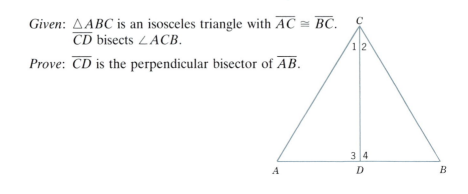

*Proof*:

| STATEMENTS | REASONS |
|---|---|
| **1.** $\triangle ABC$ is an isosceles triangle with $\overline{AC} \cong \overline{BC}$. | **1.** Given. |
| **2.** $\overline{CD}$ bisects $\angle ACB$. | **2.** Given. |
| **3.** $\angle 1 \cong \angle 2$. | **3.** An angle bisector divides an angle into two congruent angles. |
| **4.** $\angle A \cong \angle B$. | **4.** The base angles of an isosceles triangle are congruent. |

| STATEMENTS | REASONS |
|---|---|
| **5.** $\triangle ACD \cong \triangle BCD$. | **5.** $ASA \cong ASA$. |
| **6.** $\angle 3 \cong \angle 4$. | **6.** cpctc. |
| **7.** $\angle 3$ is adjacent to $\angle 4$. | **7.** If two angles have a common vertex and a common side between them, then they are adjacent angles. |
| **8.** $CD \perp AB$. | **8.** Two lines are perpendicular if they meet and form congruent adjacent angles. |
| **9.** $\overline{AD} \cong \overline{BD}$. | **9.** cpctc. |
| **10.** $\overline{CD}$ bisects $\overline{AB}$. | **10.** If a line segment divides a segment into two congruent segments, then it bisects the segment. |
| **11.** $\overline{CD}$ is the perpendicular bisector of $\overline{AB}$. | **11.** Because $\overline{CD}$ is both perpendicular to and bisects $\overline{AB}$, it is the perpendicular bisector of $\overline{AB}$. □ |

---

**Theorem 2.4** The perpendicular bisector of the base of an isosceles triangle passes through the vertex of its vertex angle.

---

*Given*: $\triangle ABC$ is an isosceles triangle with base $\overline{AB}$.
  $\overline{DF}$ is the perpendicular bisector of $\overline{AB}$.

*Prove*: $\overline{DF}$ passes through point $C$.

*Construction*: Construct the angle bisector of the vertex angle, $\angle ACB$.

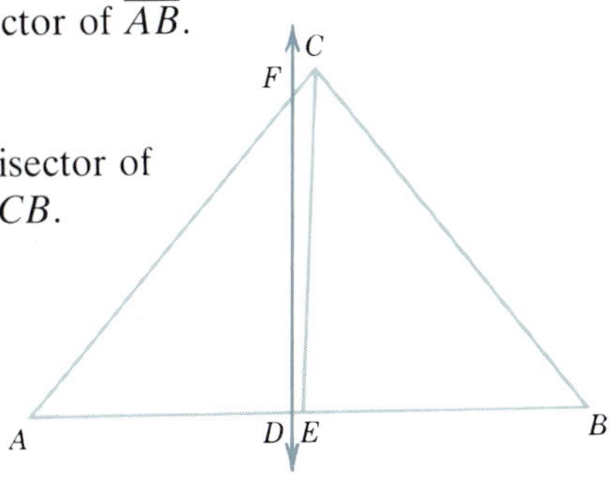

*Proof*:

| STATEMENTS | REASONS |
|---|---|
| **1.** $\triangle ABC$ is an isosceles triangle with base $\overline{AB}$. | **1.** Given. |

| STATEMENTS | REASONS |
|---|---|
| **2.** $\overline{CE}$ is the bisector of $\angle ACB$. | **2.** Every angle has a bisector. |
| **3.** $\overline{CE}$ is the perpendicular bisector of $\overline{AB}$. | **3.** The bisector of the vertex angle of an isosceles triangle is the perpendicular bisector of its base. |
| **4.** $\overline{DF}$ is the perpendicular bisector of $\overline{AB}$. | **4.** Given. |
| **5.** $\overline{CE}$ and $\overline{DF}$ are the same line. | **5.** A line segment has only one perpendicular bisector. |
| **6.** $\overline{DF}$ passes through point $C$. | **6.** $\overline{DF}$ and $\overline{CE}$ are the same line and $\overline{CE}$ passes through point $C$. |
| | □ |

In the previous proof, two lines were shown to coincide. Such a proof is called a **proof by coincidence**.

## EXERCISE 2.3

In Exercises 1–6, find $x$.

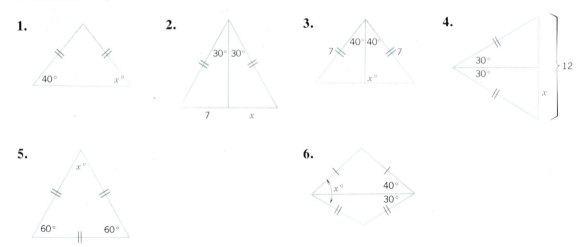

**1.**

**2.**

**3.**

**4.**

**5.**

**6.**

**7.** *Prove*: An equilateral triangle has three congruent angles. (A triangle with three congruent angles is called an **equiangular triangle**.)

**8.** *Prove*: The median drawn to the base of an isosceles triangle is perpendicular to its base.

**9.** *Prove*: The median drawn to the base of an isosceles triangle bisects the vertex angle.

**10.** Refer to Illustration 1.

*Given:* $\angle 1 \cong \angle 2$.
$\overline{AC} \cong \overline{BC}$.

*Prove:* $\angle 3 \cong \angle 4$.

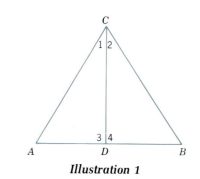

*Illustration 1*

**11.** Refer to Illustration 2.

   *Given:* $\triangle ABC$ is an isosceles triangle with base $\overline{AB}$. $D$ is the midpoint of $\overline{AC}$. $E$ is the midpoint of $\overline{BC}$.

   *Prove:* $\angle 1 \cong \angle 2$.

**12.** Refer to Illustration 3.

   *Given:* $\overline{AC} \cong \overline{BC}$.
             $\overline{AD} \cong \overline{BD}$.

   *Prove:* $\angle CAD \cong \angle CBD$.

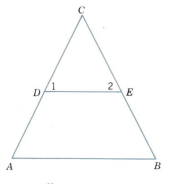

*Illustration 2*

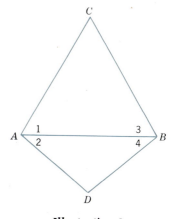

*Illustration 3*

**13.** Refer to Illustration 4.

   *Given:* $\triangle ADE$.
         $\overline{AE} \cong \overline{DE}$.
         $\overline{AB} \cong \overline{DC}$.

   *Prove:* $\angle 1 \cong \angle 2$.

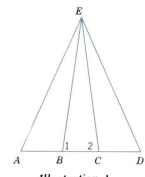

*Illustration 4*

**14.** Refer to Illustration 5.

*Given*: Isosceles △*ABC* with $\overline{AC} \cong \overline{BC}$.
$\overline{CE}$ bisects $\overline{AB}$.

*Prove*: △*ABD* is isosceles.

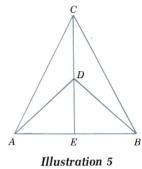

*Illustration 5*

**15.** Refer to Illustration 6.

*Given*: Equilateral △*ABC*.
*D* is the midpoint of $\overline{AB}$.
*E* is the midpoint of $\overline{BC}$.
*F* is the midpoint of $\overline{CA}$.

*Prove*: △*ADF*, △*BED*, and △*CFE* are all congruent triangles, and that △*DEF* is equilateral.

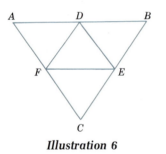

*Illustration 6*

**16.** Refer to Illustration 7.

*Given*: Isosceles triangles *ABC* and *ABD* with common base $\overline{AB}$.

*Prove*: ∠1 ≅ ∠2.

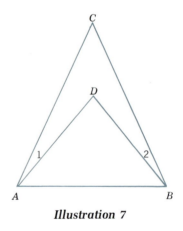

*Illustration 7*

## 2.4 MORE ON ISOSCELES TRIANGLES

Any sentence in the form of an *If... then...* statement is called a **conditional statement**. In Theorem 2.2, we proved the conditional statement

*If a triangle has two congruent sides, then the angles opposite those sides are congruent.*

If the **hypothesis** (the information following the word *If*) and the **conclusion** (the information following the word *then*) are interchanged, a new statement is formed that is called the **converse** of the original statement. The converse of Theorem 2.2 is

> *If a triangle has two congruent angles, then the sides opposite those angles are congruent.*

If a conditional statement is true, its converse is not necessarily true. For example, the statement

> *If two angles are vertical angles, then they are congruent.*

is a true statement. However, its converse

> *If two angles are congruent, then they are vertical angles.*

is not necessarily true. Angles can be congruent without being vertical angles. The conditional statement

> *If Pat Smith is a woman, then Pat Smith is a human being.*

is a true statement. However, its converse may or may not be true. Pat Smith could be a man.

Because the converse of a true conditional statement may or may not be true, we must prove the converse of a theorem just as we had to prove the original theorem.

In many of the exercises and theorems encountered thus far, we have shown certain triangles to be congruent. We have always worked with two triangles, although the two often shared a common side or a common angle, or even overlapped. There is no reason, however, why two distinct triangles need to be present. It is sometimes useful to prove that a triangle is congruent to itself. This is done in the next theorem, which is the converse of Theorem 2.2.

---

**Theorem 2.5** If a triangle has two congruent angles, then the sides opposite those angles are congruent.

---

*Given:* $\triangle ABC$ with $\angle A \cong \angle B$.

*Prove:* $\overline{BC} \cong \overline{AC}$.

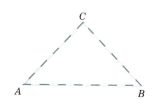

*Plan:* We shall prove $\triangle ABC$ congruent to itself in such a way that side $\overline{BC}$ corresponds to side $\overline{AC}$.

*Proof:*

| STATEMENTS | REASONS |
|---|---|
| **1.** $\angle A \cong \angle B$. | **1.** Given. |
| **2.** $\angle B \cong \angle A$. | **2.** Symmetric law of congruence. |

| STATEMENTS | REASONS |
|---|---|
| **3.** $\overline{AB} \cong \overline{BA}$. | **3.** Reflexive law of congruence. |
| **4.** $\triangle ACB \cong \triangle BCA$. | **4.** $ASA \cong ASA$. |
| **5.** $\overline{AC} \cong \overline{BC}$. | **5.** cpctc.   □ |

**EXAMPLE 1**

*Given*: $\triangle ABE$.
  $\angle A \cong \angle B$.
  $\overline{AD} \cong \overline{BC}$.

*Prove*: $\angle 3 \cong \angle 4$.

*Proof*:

| STATEMENTS | REASONS |
|---|---|
| **1.** $\angle A \cong \angle B$. | **1.** Given. |
| **2.** $\overline{AE} \cong \overline{BE}$. | **2.** In a triangle, sides opposite congruent angles are congruent. |
| **3.** $\overline{AD} \cong \overline{BC}$. | **3.** Given. |
| **4.** $m(\overline{AD}) = m(\overline{BC})$. | **4.** If two segments are congruent, they have equal measures. |
| **5.** $m(\overline{CD}) = m(\overline{CD})$. | **5.** Reflexive law. |
| **6.** $m(\overline{AD}) - m(\overline{CD})$ $= m(\overline{BC}) - m(\overline{CD})$. | **6.** Equal quantities subtracted from equal quantities are equal. |
| **7.** $m(\overline{AC}) = m(\overline{AD}) - m(\overline{CD})$. | **7.** Postulate 1.10. |
| **8.** $m(\overline{BD}) = m(\overline{BC}) - m(\overline{CD})$. | **8.** Postulate 1.10. |
| **9.** $m(\overline{AC}) = m(\overline{BD})$. | **9.** Substitution. |
| **10.** $\overline{AC} \cong \overline{BD}$. | **10.** If two segments have equal measures, they are congruent. |
| **11.** $\triangle ACE \cong \triangle BDE$. | **11.** $SAS \cong SAS$. |
| **12.** $\angle 3 \cong \angle 4$. | **12.** cpctc. ▪ |

## EXERCISE 2.4

In Exercises 1–4, find the value of *x*.

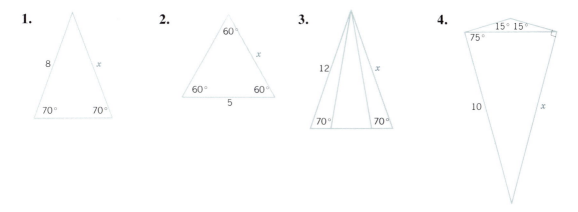

**1.**

**2.**

**3.**

**4.**

In Exercises 5–8, the perimeter of each triangle is 20 m. Find the value of *x* and justify your answer.

**5.**

**6.**

**7.**

**8.**

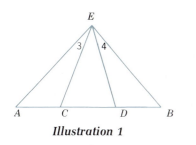

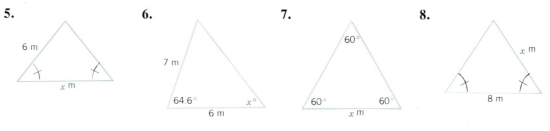

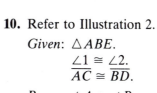

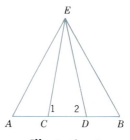

**9.** Refer to Illustration 1.

Given: △*ABE*.
$\overline{AE} \cong \overline{BE}$.
$\overline{AC} \cong \overline{BD}$.

Prove: ∠3 ≅ ∠4.

**10.** Refer to Illustration 2.

Given: △*ABE*.
∠1 ≅ ∠2.
$\overline{AC} \cong \overline{BD}$.

Prove: ∠*A* ≅ ∠*B*.

*Illustration 1*

*Illustration 2*

**11.** Refer to Illustration 3.

   *Given*: $\angle A \cong \angle B$.
   $\overline{AD} \cong \overline{BE}$.

   *Prove*: $\angle CDE \cong \angle CED$.

**12.** Refer to Illustration 4.

   *Given*: $\angle 1 \cong \angle 2$.
   $\angle 3 \cong \angle 4$.

   *Prove*: $\angle CAD \cong \angle CBD$.

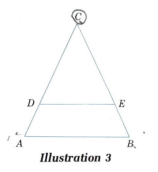

**Illustration 3**

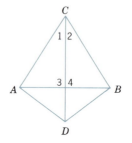

**Illustration 4**

**13.** Refer to Illustration 5.

   *Given*: $\overline{DA}$ bisects $\angle CAB$.
   $\overline{DB}$ bisects $\angle CBA$.
   $\angle 1 \cong \angle 2$.

   *Prove*: $\overline{CA} \cong \overline{CB}$.

**14.** Refer to Illustration 6.

   *Given*: $\angle 1 \cong \angle 2$.
   $\angle 3 \cong \angle 4$.

   *Prove*: $\angle ADC \cong \angle BEC$.

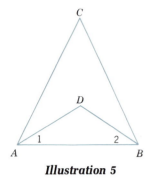

**Illustration 5**

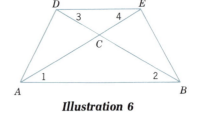

**Illustration 6**

**15.** Refer to Illustration 7.

   *Given*: Isosceles triangle *AFC* with base
   $\overline{AC}$.
   Isosceles triangle *BGD* with base
   $\overline{BD}$.
   Isosceles triangle *BHC* with base
   $\overline{BC}$.

   *Prove*: $\triangle AED$ is isosceles.

**16.** Refer to Illustration 8.

   *Given*: Points *A*, *B*, *C*, and *D* lie on a line.
   $\overline{CE} \cong \overline{AE}$.
   $\angle 2 \cong \angle A$.

   *Prove*: $\triangle BCF$ is isosceles.

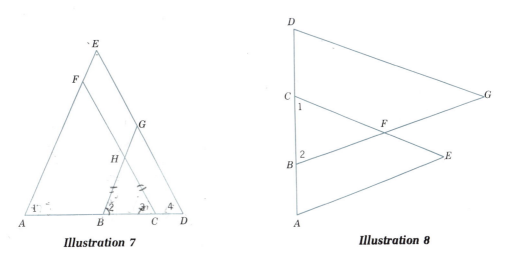

*Illustration 7*                                    *Illustration 8*

# 2.5 ANOTHER WAY TO PROVE RIGHT TRIANGLES CONGRUENT

After we make the following definition we can prove a useful theorem concerning right triangles.

> **Definition 2.2** The side opposite the right angle in a right triangle is called the **hypotenuse** of the right triangle. The other two sides are called **legs** of the right triangle.

> **Theorem 2.6** If the hypotenuse and a leg of one right triangle are congruent, respectively, to the hypotenuse and a leg of a second right triangle, then the right triangles are congruent.

*Given*:  △DEF and △ABC are right triangles
with ∠E and ∠ABC right angles.
$\overline{DF} \cong \overline{AC}$.
$\overline{DE} \cong \overline{AB}$.

*Prove*:  △DEF ≅ △ABC.

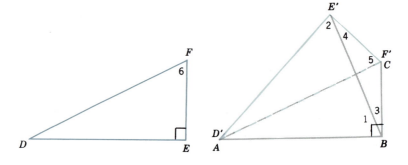

*Construction*:  On side $\overline{AC}$ of $\triangle ABC$ construct $\triangle D'E'F'$ so that $D'F' \cong$ $\overline{DF}$, $\angle 5 \cong \angle 6$, and $\overline{E'F'} \cong \overline{EF}$. Note that $\triangle DEF \cong$ $\triangle D'E'F'$. Draw line segment $\overline{E'B}$.

*Proof*:

| STATEMENTS | REASONS |
|---|---|
| **1.** $\triangle DEF$ and $\triangle ABC$ are right triangles with $\angle E$ and $\angle ABC$ right angles. | **1.** Given. |
| **2.** $\overline{DF} \cong \overline{AC}$; $\overline{DE} \cong \overline{AB}$. | **2.** Given. |
| **3.** $\triangle D'E'F'$ is constructed so that $\overline{D'F'} \cong \overline{DF}$, $\angle 5 \cong \angle 6$, and $\overline{E'F'} = \overline{EF}$. | **3.** Possible construction. |
| **4.** $\triangle D'E'F' \cong \triangle DEF$. | **4.** $SAS \cong SAS$. |
| **5.** $\angle D'E'F' \cong \angle E$. | **5.** cpctc. |
| **6.** $\angle D'E'F'$ is a right angle. | **6.** It is congruent to $\angle E$ and $\angle E$ is a right angle. |
| **7.** $\overline{DE} \cong \overline{D'E'}$. | **7.** cpctc. |
| **8.** $\overline{AB} \cong \overline{D'E'}$. | **8.** Substitution. |
| **9.** $\angle 1 \cong \angle 2$. | **9.** In $\triangle ABE'$, if two sides are congruent, then the angles opposite those sides are congruent. |
| **10.** $\angle ABC \cong \angle D'E'F'$. | **10.** All right angles are congruent. |
| **11.** $\angle 1$ is complementary to $\angle 3$; $\angle 2$ is complementary to $\angle 4$. | **11.** If the sum of the measures of two angles is 90°, the angles are complementary. |
| **12.** $\angle 3 \cong \angle 4$. | **12.** Complements of congruent angles are congruent. |

| STATEMENTS | REASONS |
|---|---|
| **13.** $\overline{E'F'} \cong \overline{BC}$. | **13.** In $\triangle BCE'$, if two angles are congruent, then the sides opposite those angles are congruent. |
| **14.** $\triangle ABC \cong \triangle D'E'F'$. | **14.** $SAS \cong SAS$. |
| **15.** $\triangle ABC \cong \triangle DEF$. | **15.** If two triangles are congruent to the same triangle, then they are congruent to each other. ☐ |

Theorem 2.6 gives us a new method for proving right triangles congruent. This new method is abbreviated as $hl \cong hl$, which is read as

"If the hypotenuse and a leg of one right triangle are congruent, respectively, to the hypotenuse and a leg of another right triangle, then the right triangles are congruent."

Of course, we can still prove right triangles congruent by the $SSS \cong SSS$, $SAS \cong SAS$, and $ASA \cong ASA$ methods.

*EXAMPLE 1*

*Given:* $\angle 1 \cong \angle 2$.
$\overline{AB} \perp \overline{BC}$.
$\overline{AD} \perp \overline{DC}$.

*Prove:* $\angle 3 \cong \angle 4$.

*Proof*:

| STATEMENTS | REASONS |
|---|---|
| **1.** $\angle 1 \cong \angle 2$. | **1.** Given. |
| **2.** $\overline{BC} \cong \overline{DC}$. | **2.** In a triangle, if two angles are congruent, then the sides opposite those angles are congruent. |
| **3.** $\overline{AB} \perp \overline{BC}$; $\overline{AD} \perp \overline{DC}$. | **3.** Given. |
| **4.** $\angle ABC$ and $\angle ADC$ are right angles. | **4.** Perpendicular lines meet and form right angles. |
| **5.** $\triangle ABC$ and $\triangle ADC$ are right triangles. | **5.** If a triangle contains a right angle, then it is a right triangle. |

| STATEMENTS | REASONS |
|---|---|
| **6.** $\overline{AC}$ is the hypotenuse of right triangles *ABC* and *ADC*. | **6.** In a right triangle, the side opposite the right angle is the hypotenuse. |
| **7.** $\overline{AC} \cong \overline{AC}$. | **7.** Reflexive law. |
| **8.** $\triangle ABC \cong \triangle ADC$. | **8.** If the hypotenuse and one leg of one right triangle are congruent, respectively, to the hypotenuse and one leg of a second right triangle, then the right triangles are congruent. $(hl \cong hl)$. |
| **9.** $\angle 3 \cong \angle 4$. | **9.** cpctc.                                   ∎ |

$\angle 3 \cong \angle 4$

## EXERCISE 2.5

**1.** Distinguish between the words *hypotenuse* and *hypothesis*.

**2.** Construct an isosceles right triangle with its congruent legs congruent to segment $\overline{AB}$ shown in Illustration 1.

*Illustration 1*

**3.** Refer to Illustration 2.

*Given*: $\triangle ABC$ is isosceles with base $\overline{AB}$.
$\overline{CD} \perp \overline{AB}$.

*Prove*: $\overline{CD}$ bisects $\overline{AB}$.

**4.** Refer to Illustration 3.

*Given*: $\overline{ED}$ and $\overline{AC}$ intersect at point *B*.
$\overline{AE} \perp \overline{AC}$.
$\overline{CD} \perp \overline{AC}$.
*B* is the midpoint of $\overline{ED}$.
$\overline{AE} \cong \overline{CD}$.

*Prove*: *B* is the midpoint of $\overline{AC}$.

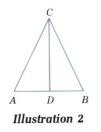

*Illustration 2*

*Illustration 3*

**5.** Refer to Illustration 4.

*Given:* $\overline{AC} \cong \overline{BD}$.
m($\angle CDA$) = m($\angle BAD$) = 90°.

*Prove:* $\overline{AB} \cong \overline{DC}$.

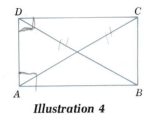

*Illustration 4*

**6.** Refer to Illustration 5.

*Given:* $\overline{DB}$ and $\overline{AC}$ intersect at point $O$.
m($\angle 1$) = m($\angle 2$) = 90°.
$\overline{CD} \cong \overline{CB}$.

*Prove:* $\overline{OD} \cong \overline{OB}$.

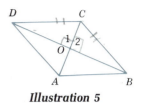

*Illustration 5*

**7.** Refer to Illustration 6.

*Given:* $\triangle ABC$ is isosceles with base $\overline{AB}$.
$\overline{BD} \perp \overline{AC}$.
$\overline{AE} \perp \overline{BC}$.
$\angle 1 \cong \angle 2$.

*Prove:* $\angle 3 \cong \angle 4$.

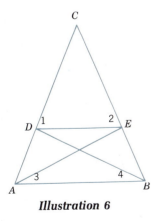

*Illustration 6*

**8.** Refer to Illustration 7.

*Given:* $\triangle ABC$ is isosceles with base $\overline{AB}$.
m($\overline{AD}$) + m($\overline{DF}$) = m($\overline{DC}$).
$\overline{BF} \perp \overline{CD}$.
$\overline{AD} \perp \overline{DC}$.

*Prove:* $\angle 3 \cong \angle 4$.

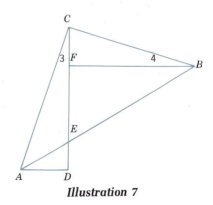

*Illustration 7*

## CHAPTER TWO REVIEW EXERCISES

In Review Exercises 1–6, refer to Illustration 1. Pairs of congruent sides of congruent triangles are shown with cross marks. Complete each statement involving a correspondence between parts of these two triangles.

**1.** Side $\overline{AC}$ corresponds to side _____.

**2.** Side $\overline{DE}$ corresponds to side _____.

**3.** Side $\overline{BC}$ corresponds to side _____.

**4.** $\angle A$ corresponds to $\angle$ _____.

**5.** $\angle E$ corresponds to $\angle$ _____.

**6.** $\angle F$ corresponds to $\angle$ _____.

**7.** Name three ways to prove two triangles congruent.

**8.** Give the meaning of the abbreviation **cpctc**.

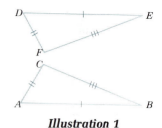

***Illustration 1***

In Review Exercises 9–14, the given information refers to Illustration 2. In each exercise, indicate whether the triangles are congruent. If so, give the reason why.

**9.** $\angle A \cong \angle D$
$\angle B \cong \angle E$
$\overline{AB} \cong \overline{DE}$

**10.** $\angle A \cong \angle D$
$\angle B \cong \angle E$
$\angle C \cong \angle F$

**11.** $\angle A \cong \angle D$
$\overline{BC} \cong \overline{EF}$
$\overline{AC} \cong \overline{DE}$

**12.** $\overline{AC} \cong \overline{DF}$
$\overline{AB} \cong \overline{DE}$
$\overline{BC} \cong \overline{EF}$

**13.** $\overline{AB} \cong \overline{DE}$
$\angle B \cong \angle E$
$\overline{BC} \cong \overline{EF}$

**14.** $\overline{BC} \cong \overline{FE}$
$\angle C \cong \angle E$
$\angle B \cong \angle F$

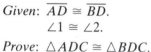

***Illustration 2***

**15.** Refer to Illustration 3.

*Given*: $\overline{AD} \cong \overline{BD}$.
$\angle 1 \cong \angle 2$.

*Prove*: $\triangle ADC \cong \triangle BDC$.

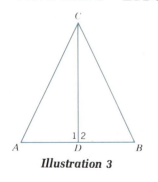

***Illustration 3***

**16.** Refer to Illustration 4.

*Given*: $\overline{DE}$ and $\overline{AB}$ are line segments intersecting at point $C$.
$\overline{AC} \cong \overline{BC}$.
$\angle A \cong \angle B$.

*Prove*: $\angle D \cong \angle E$.

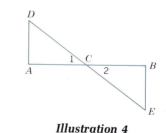

***Illustration 4***

In Review Exercises 17–24, find $x$. Be able to give your reasons.

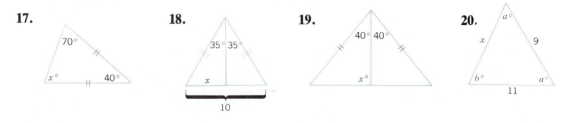

**17.**  70°  $x°$  40°

**18.**  35° 35°  $x$  10

**19.**  40° 40°  $x°$

**20.**  $a°$  $x$  9  $b°$  $a°$  11

**21.**    **22.**    **23.**    **24.**

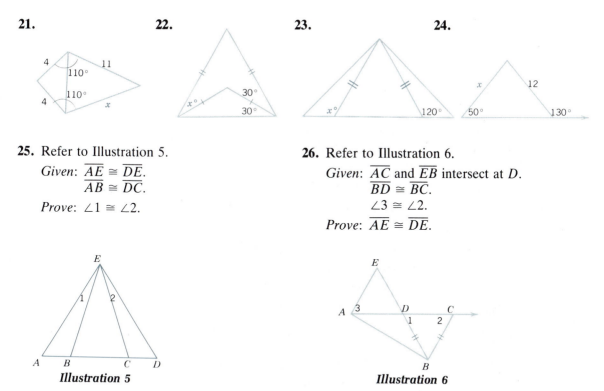

**25.** Refer to Illustration 5.

*Given*: $\overline{AE} \cong \overline{DE}$.
   $\overline{AB} \cong \overline{DC}$.

*Prove*: $\angle 1 \cong \angle 2$.

**26.** Refer to Illustration 6.

*Given*: $\overline{AC}$ and $\overline{EB}$ intersect at $D$.
   $\overline{BD} \cong \overline{BC}$.
   $\angle 3 \cong \angle 2$.
*Prove*: $\overline{AE} \cong \overline{DE}$.

*Illustration 5*

*Illustration 6*

In Review Exercises 27–30, tell whether the triangles in each pair are congruent. Be able to give the reason why.

**27.**    **28.**

**29.**    **30.**

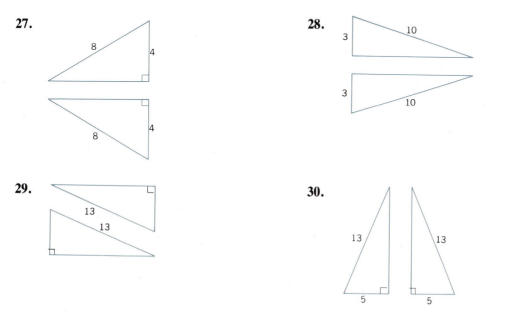

## CHAPTER TWO TEST

In Exercises 1–20, classify each statement as true or false.

1. Two triangles are congruent if three sides of one triangle are congruent, respectively, to three sides of the second triangle.

2. Two triangles are congruent if an angle and a side of one triangle are congruent, respectively, to an angle and a side of the second triangle.

3. If two triangles are congruent, all pairs of corresponding parts are congruent.

4. All equilateral triangles are congruent.

5. The vertex angle of an isosceles triangle can never be congruent to one of the base angles.

6. The bisector of the vertex angle of an isosceles triangle bisects the base of the triangle.

7. A right triangle can never have more than one hypotenuse.

8. If two right triangles have congruent hypotenuses, the triangles are congruent.

9. If an angle appears to be a 45° angle, it may be assumed that the angle is a 45° angle.

10. If a triangle has two congruent angles, it is isosceles.

11. If an angle in one triangle is congruent to an angle in a second triangle, then the sides opposite those angles are congruent.

12. The converse of a conditional statement is sometimes true.

13. An equilateral triangle is always equiangular.

14. A line segment has only one bisector.

15. The *hypothesis* is the clause following the word *If* in an *If... then...* sentence.

16. If two triangles are congruent, then they are isosceles.

17. If two right triangles are congruent, each pair of corresponding acute angles are complementary.

18. Three pieces of information must be known before the congruence of triangles can be shown.

19. The base of an isosceles triangle is always included between a pair of congruent angles of the triangle.

20. If three angles of one triangle are congruent, respectively, to three angles of a second triangle, then the triangles are congruent.

# 3

*Nikolai Ivanovich Lobachevsky (1793–1856). Lobachevsky, a Russian mathematician, was one of the first to recognize the merits of a geometry that denied Euclid's parallel postulate. In his geometry, more than one parallel can be drawn to a given line through a point not on that line. Unfortunately, his work in plane hyperbolic geometry was largely ignored by others until after his death. Non-Euclidean geometries, such as Lobachevsky's, provided Albert Einstein with a basis for his theory of relativity. The work of Lobachevsky and others has made it apparent that Euclidean geometry is only one of many possible geometries (Historical Picture Service, Chicago).*

# Parallel Lines

*I*n the previous chapters we have been concerned directly or indirectly with lines that intersect and form various angles. We now pay attention to lines that do not intersect.

## 3.1 PARALLEL LINES

> **Definition 3.1** Two lines are called **parallel lines** if, and only if, they are in the same plane and do not intersect.

Lines $\overleftrightarrow{AB}$ and $\overleftrightarrow{CD}$ shown in Figure 3.1*a* are parallel lines. If two lines do not intersect and do not lie in the same plane, they are called **skew**

**lines**. For example, the lines $\overleftrightarrow{AB}$ and $\overleftrightarrow{CD}$ drawn on the top and bottom of the box shown in Figure 3.1*b* are skew lines. In this book, however, we will only consider lines that are in the same plane. Thus, if two lines do not intersect, we will consider them to be parallel lines.

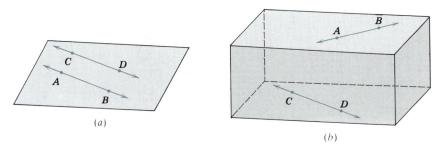

**Figure 3.1**

We will need to know when two lines are parallel, and Definition 3.1 alone is not particularly useful. The next postulate is more helpful.

---

**Postulate 3.1** If two lines are perpendicular to a third line, then they are parallel.

---

When a third line intersects two lines in two different points, the third line is called a **transversal**. In Figure 3.2, line $l_3$ is a transversal intersecting lines $l_1$ and $l_2$.

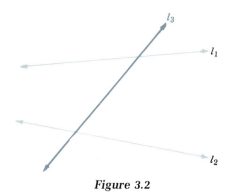

**Figure 3.2**

Several pairs of angles are formed by a transversal and the lines it intersects. The following three definitions classify these pairs of angles.

---

**Definition 3.2** If two lines are cut by a transversal, then nonadjacent angles on opposite sides of the transversal and on the interior of the two lines are called **alternate interior angles**.

---

In Figure 3.3, ∠1 and ∠2 are a pair of alternate interior angles. Angle 3 and ∠4 form another pair of alternate interior angles.

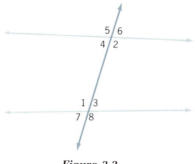

**Figure 3.3**

---

**Definition 3.3** If two lines are cut by a transversal, then angles on the same side of the transversal and in corresponding positions with respect to the lines are called **corresponding angles**.

---

In Figure 3.3, ∠1 and ∠5 are a pair of corresponding angles. So are ∠3 and ∠6, ∠4 and ∠7, and ∠2 and ∠8.

---

**Definition 3.4** If two lines are cut by a transversal, then nonadjacent angles on opposite sides of the transversal and on the exterior of the two lines are called **alternate exterior angles**.

---

In Figure 3.3, ∠5 and ∠8 are a pair of alternate exterior angles. So are ∠7 and ∠6.

A useful abbreviation for the word *parallel* is the symbol ∥. We are now ready to prove some theorems involving parallel lines.

---

**Theorem 3.1** If two lines are cut by a transversal so that a pair of alternate interior angles are congruent, then the lines are parallel.

---

*Given*: Lines $l_1$ and $l_2$ with
line $\overleftrightarrow{AB}$ as a transversal.
$\angle 3 \cong \angle 4$.

*Prove*: $l_1 \parallel l_2$.

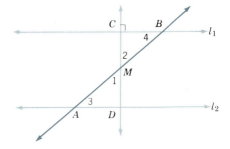

*Construction*: Construct the midpoint of line segment $\overline{AB}$ and call it point $M$. Then construct a perpendicular to line $l_1$ that passes through point $M$.

*Proof*:

| STATEMENTS | REASONS |
|---|---|
| **1.** $l_1$ and $l_2$ are lines with line $\overleftrightarrow{AB}$ as a transversal. | **1.** Given. |
| **2.** $M$ is the midpoint of $\overline{AB}$. | **2.** Every line segment has a midpoint. |
| **3.** $\overline{MA} \cong \overline{MB}$. | **3.** A midpoint divides a line segment into two congruent line segments. |
| **4.** $\overline{CD} \perp l_1$. | **4.** There exists a perpendicular to a line from a point not on the line. |
| **5.** $\angle BCM$ is a right angle. | **5.** Perpendicular lines meet and form right angles. |
| **6.** m($\angle BCM$) = 90°. | **6.** The measure of a right angle is 90°. |
| **7.** $\angle 1$ and $\angle 2$ are vertical angles. | **7.** Two nonadjacent angles formed by two intersecting lines are vertical angles. |
| **8.** $\angle 1 \cong \angle 2$. | **8.** Vertical angles are congruent. |
| **9.** $\angle 3 \cong \angle 4$. | **9.** Given. |
| **10.** $\triangle ADM \cong \triangle BCM$. | **10.** ASA $\cong$ ASA. |
| **11.** $\angle ADM \cong \angle BCM$. | **11.** cpctc. |
| **12.** m($\angle ADM$) = m($\angle BCM$). | **12.** Congruent angles have equal measures. |
| **13.** m($\angle ADM$) = 90°. | **13.** Substitution. |
| **14.** $\angle ADM$ is a right angle. | **14.** An angle with a measure of 90° is a right angle. |

| STATEMENTS | REASONS |
|---|---|
| **15.** $\overline{CD} \perp l_2$. | **15.** If two lines intersect and form right angles, then they are perpendicular. |
| **16.** $l_1 \parallel l_2$. | **16.** If two lines are perpendicular to a third line, then they are parallel lines.  □ |

Certain other pairs of angles formed by a transversal can be used to show that lines are parallel, as the next theorems indicate.

---

**Theorem 3.2**  If two lines are cut by a transversal so that a pair of corresponding angles are congruent, then the lines are parallel.

---

*Given*:  Line $l_3$ is a transversal intersecting $l_1$ and $l_2$.
    $\angle 1 \cong \angle 2$.

*Prove*:  $l_1 \parallel l_2$.

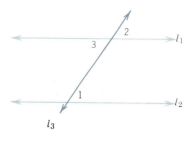

*Proof*:

| STATEMENTS | REASONS |
|---|---|
| **1.** Line $l_3$ is a transversal intersecting $l_1$ and $l_2$. | **1.** Given. |
| **2.** $\angle 1 \cong \angle 2$. | **2.** Given. |
| **3.** $\angle 3$ and $\angle 2$ are vertical angles. | **3.** Nonadjacent angles formed by two intersecting lines are vertical angles. |
| **4.** $\angle 3 \cong \angle 2$. | **4.** Vertical angles are congruent. |
| **5.** $\angle 1 \cong \angle 3$. | **5.** Transitive law of congruence. |
| **6.** $l_1 \parallel l_2$. | **6.** If two lines are cut by a transversal so that a pair of alternate interior angles are congruent, then the lines are parallel.  □ |

> **Theorem 3.3** If two lines are cut by a transversal so that two interior angles on the same side of the transversal are supplementary, then the lines are parallel.

The proof of Theorem 3.3 is left as an exercise.

Theorem 3.1 enables us to construct a line parallel to a given line through a point not on that given line. For example, suppose that we wish to construct a line parallel to a given line *l* that passes through a given point *P* shown in Figure 3.4. First we draw a line through the given point *P* that intersects the given line *l*. This line will be a transversal. We then copy ∠1 in the position of ∠2 using point *P* as vertex. Because the alternate interior angles are congruent by the construction, Theorem 3.1 guarantees the second side of ∠2, line *r*, will be parallel to the given line *l*. This construction demonstrates the existence of a parallel to a line through a point that is not on that line.

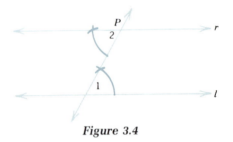

*Figure 3.4*

We shall assume there is only one such line.

> **Postulate 3.2** Given a line *l* and a point *P* not on line *l*, there is only one line through point *P* that is parallel to line *l*.

Postulate 3.2 is called the **parallel postulate**. Because it seems to be more complicated than many of the other postulates, mathematicians in the past spent a great deal of time trying to prove it as a theorem. It was a breakthrough in mathematical history for mathematicians to discover that this postulate is independent from the others, that is, it cannot be proved by using the other postulates. This discovery lead to the development of the non-Euclidean geometries discussed later in this book.

Postulate 3.2 is the key to the proof of the next theorem.

> **Theorem 3.4** If a line is perpendicular to one of two parallel lines, then it is also perpendicular to the other line.

*Given*: $l_1 \parallel l_2$.
$\qquad l_3 \perp l_2$.

*Prove*: $l_3 \perp l_1$.

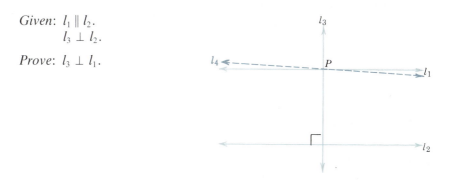

*Plan*: Construct $l_4 \perp l_3$ at point $P$ and show that $l_4$ coincides with $l_1$.

*Proof*:

| STATEMENTS | REASONS |
|---|---|
| **1.** Construct $l_4 \perp l_3$ at point $P$. | **1.** A perpendicular to $l_3$ passing through point $P$ does exist. |
| **2.** $l_3 \perp l_2$. | **2.** Given. |
| **3.** $l_4 \parallel l_2$. | **3.** Two lines perpendicular to a third line are parallel to each other. |
| **4.** $l_1 \parallel l_2$. | **4.** Given. |
| **5.** $l_1$ coincides with $l_4$. | **5.** Only one parallel to $l_2$ can exist through point $P$. |
| **6.** Since $l_3 \perp l_4$, then $l_3 \perp l_1$. | **6.** $l_4$ and $l_1$ are different names for the same line.    □ |

## EXERCISE 3.1

In Exercises 1–4, refer to Illustration 1 in which $l_1 \parallel l_2$.

**1.** Name two pairs of alternate interior angles.
**2.** Name four pairs of corresponding angles.
**3.** Name two pairs of alternate exterior angles.
**4.** Name four angles that are supplementary to $\angle a$.

*Illustration 1*

In Exercises 5–8, refer to Illustration 2.

**5.** If $\angle 1 \cong \angle 2$, is $l_1 \parallel l_2$? Explain.
**6.** If $\angle 1 \cong \angle 3$, is $l_1 \parallel l_2$? Explain.
**7.** If $\angle 4 \cong \angle 5$, is $l_1 \parallel l_2$? Explain.
**8.** If $\angle 5 \cong \angle 2$, is $l_1 \parallel l_2$? Explain.

*Illustration 2*

In Exercises 9–14, refer to Illustration 3.

**9.** If $\angle 1 \cong \angle 5$, can line $\overleftrightarrow{ED}$ intersect line $\overleftrightarrow{AB}$? Explain.
**10.** If $\angle 1 \cong \angle 2$, can line $\overleftrightarrow{ED}$ intersect line $\overleftrightarrow{AB}$? Explain.
**11.** If $\angle 1 \cong \angle 3$, can line $\overleftrightarrow{ED}$ intersect line $\overleftrightarrow{AB}$? Explain.
**12.** If $\angle 1 \cong \angle 3$, must line $\overleftrightarrow{ED}$ intersect line $\overleftrightarrow{AB}$? Explain.
**13.** If $\angle 1 \cong \angle 4$, can line $\overleftrightarrow{ED}$ intersect line $\overleftrightarrow{AB}$? Explain.
**14.** If $\angle 1 \cong \angle 4$, must line $\overleftrightarrow{ED}$ intersect line $\overleftrightarrow{AB}$? Explain.

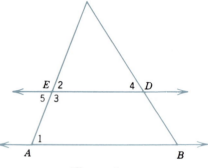

*Illustration 3*

In Exercises 15–18, refer to the four-sided figure in Illustration 4.

**15.** If $\angle 1$ is supplementary to $\angle 4$, must $\overline{AD} \parallel \overline{BC}$? Explain.
**16.** If $\angle 1$ is supplementary to $\angle 2$, must $\overline{AD} \parallel \overline{BC}$? Explain.
**17.** If $\angle 4 \cong \angle 3$, must $\overline{AD} \parallel \overline{BC}$? Explain.
**18.** If $\angle 4 \cong \angle 2$, must $\overline{AB} \parallel \overline{DC}$? Explain.

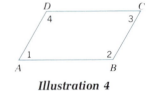

*Illustration 4*

**19.** *Prove Theorem 3.3*: If two lines are cut by a transversal so that two interior angles on the same side of the transversal are supplementary, then the lines are parallel.

**20.** *Prove*: If two lines are cut by a transversal so that alternate exterior angles are congruent, then the lines are parallel.

**21.** Refer to Illustration 5.

    *Given*: $\triangle ABC$ is isosceles with $\overline{AC} \cong \overline{BC}$.
          $\angle 1 \cong \angle 4$.

    *Prove*: $\overline{ED} \parallel \overline{AB}$.

**22.** Refer to Illustration 6.

    *Given*: $\angle 2$ is supplementary to $\angle 3$.

    *Prove*: $l_1 \parallel l_2$.

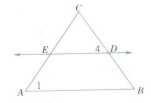

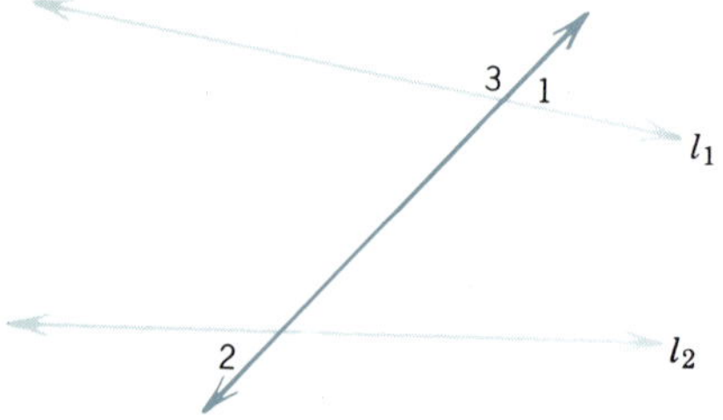

*Illustration 5*

*Illustration 6*

**23.** Refer to Illustration 7.

*Given*: ∠1 and ∠2 are supplementary.

*Prove*: $l_1 \parallel l_2$.

**24.** Refer to Illustration 8.

*Given*: △*EDC* is isosceles with base $\overline{ED}$.
∠*A* ≅ ∠1.

*Prove*: $\overline{AB} \parallel \overline{ED}$.

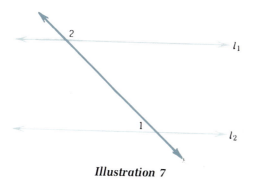

**Illustration 7**

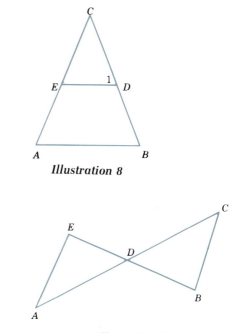

**Illustration 8**

**25.** Refer to Illustration 9.

*Given*: Segments $\overline{AC}$ and $\overline{EB}$ bisect
each other at point *D*.

*Prove*: $\overline{AE} \parallel \overline{BC}$.

**Illustration 9**

**26.** On a piece of paper, draw a line about 2 in. long. Mark a point *P* about 1 in. from the line. Through point *P*, construct a parallel to the line. Can you devise a method of constructing the required parallel by using Postulate 3.1?

## 3.2 INTRODUCTION TO INDIRECT PROOF

Statements such as

*That child is sick.*

and

*That child is not sick.*

are called **negations** of each other. If one of these statements is true, then the other is false. Statements that are negations of each other can never both be true, and they can never both be false. Should two statements that are negations ever appear to be true simultaneously, an impossible situation occurs. In such a situation, we say that a contradiction exists.

In the previous theorems and exercises we have always reasoned directly from whatever information was given to the required conclusion. There is another method of reasoning called **indirect proof**. This method of proof is illustrated by a simple example.

When John registered for the early morning geometry class, he knew that he would have one of three possible instructors: Professors Schultz, Koenig, or Yarwood. Because he was curious, he made some inquiries to learn his instructor's name. He found out that Professor Schultz teaches only in the afternoon and that Professor Koenig only teaches in the evenings. John concluded that the only remaining possibility was Professor Yarwood and, therefore, Professor Yarwood would be his instructor. By eliminating all other possibilities, John had reached his conclusion indirectly.

In *The Casebook of Sherlock Holmes*, the famous detective explains that his methods involve indirect proof:

"That process," said Holmes, "starts upon the supposition that when you have eliminated all which is impossible, then whatever remains, however improbable, must be the truth."

To eliminate possibilities we make use of the fact that a statement and its negation cannot both be true at the same time. If they were, we would have a contradiction. In any indirect proof, we will assume to be true the negation of what we are trying to prove. From this assumption, the given information and previous theorems, we will try to find a contradiction. If one does develop, we know that the assumption is false and must be rejected. This leaves the statement we are trying to prove as the only remaining possibility.

If we wish to prove that two angles are congruent by the indirect method, we would first assume that they are not congruent and then show that this assumption leads to a contradiction. Similarly, if we wish to prove that two lines are parallel, we would first assume that the lines are not parallel, and then show that this assumption leads to a contradiction. If, in each case, we can eliminate the unwanted possibilities, the desired conclusion is the only remaining choice.

We will use the method of indirect proof in the proof of Theorem 3.5, which is the converse of Theorem 3.1. The symbol $\nparallel$ is read as "is not parallel to." The symbol $\neq$ is read as "is not congruent to."

---

**Theorem 3.5** If two parallel lines are cut by a transversal, then all pairs of alternate interior angles are congruent.

*Given*: $l_1 \parallel l_2$.

*Prove*: $\angle 1 \cong \angle 2$.

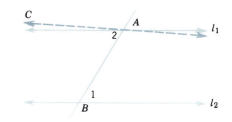

*Plan*: Use an indirect proof. Either $\angle 1 \cong \angle 2$ or $\angle 1 \not\cong \angle 2$. Assume that $\angle 1 \not\cong \angle 2$ and show that this leads to a contradiction. Then $\angle 1 \cong \angle 2$ must be true.

*Proof*:

| STATEMENTS | REASONS |
|---|---|
| **1.** Assume $\angle 1 \not\cong \angle 2$. | **1.** Possible assumption. |
| **2.** At point $A$ construct $\angle BAC \cong \angle 1$. | **2.** Possible construction. |
| **3.** $l_2 \parallel \overline{AC}$. | **3.** If two lines are cut by a transversal so that a pair of alternate interior angles are congruent, then the lines are parallel. |
| **4.** $l_1 \parallel l_2$. | **4.** Given. |
| **5.** $\overleftrightarrow{AC}$ and $l_1$ are both parallel to $l_2$ through point $A$. | **5.** Steps 3 and 4. |
| **6.** Statement 5 is impossible. | **6.** There exists only one parallel to $l_2$ through point $A$. |
| **7.** $\angle 1 \not\cong \angle 2$ is false. | **7.** It leads to a contradiction. |
| **8.** $\angle 1 \cong \angle 2$. | **8.** It is the only remaining possibility. |

The other pair of alternate interior angles can be proved to be congruent with a similar argument.  □

We will encounter indirect proofs again in the next two sections. You will be asked to do indirect proofs in the exercises for those sections.

---

***Theorem 3.6*** If two parallel lines are cut by a transversal, then all pairs of corresponding angles are congruent.

*Given*: $l_1 \parallel l_2$.

*Prove*: $\angle 1 \cong \angle 2$.

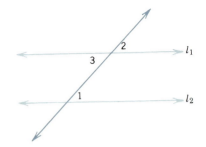

*Proof*:

| STATEMENTS | REASONS |
|---|---|
| **1.** $l_1 \parallel l_2$. | **1.** Given. |
| **2.** $\angle 1 \cong \angle 3$. | **2.** If two parallel lines are cut by a transversal, then alternate interior angles are congruent. |
| **3.** $\angle 3$ and $\angle 2$ are vertical angles. | **3.** Nonadjacent angles formed by two intersecting lines are vertical angles. |
| **4.** $\angle 3 \cong \angle 2$. | **4.** Vertical angles are congruent. |
| **5.** $\angle 1 \cong \angle 2$. | **5.** Transitive law of congruence. |

☐

---

**Theorem 3.7** If two parallel lines are cut by a transversal, then all pairs of interior angles on the same side of the transversal are supplementary.

---

*Given*: $l_1 \parallel l_2$.

*Prove*: $\angle 1$ is supplementary to $\angle 2$.

*Proof*:

| STATEMENTS | REASONS |
|---|---|
| **1.** $l_1 \parallel l_2$. | **1.** Given. |
| **2.** $\angle 1 \cong \angle 3$. | **2.** If two parallel lines are cut by a transversal, then alternate interior angles are congruent. |
| **3.** $m(\angle 1) = m(\angle 3)$. | **3.** If two angles are congruent, then they have equal measures. |
| **4.** $\angle 3$ is supplementary to $\angle 2$. | **4.** Their sum is a straight angle. |
| **5.** $m(\angle 3) + m(\angle 2) = 180°$. | **5.** If two angles are supplementary, then the sum of their measures is 180°. |
| **6.** $m(\angle 1) + m(\angle 2) = 180°$. | **6.** Substitution. |
| **7.** $\angle 1$ is supplementary to $\angle 2$. | **7.** If the sum of the measures of two angles is 180°, then the angles are supplementary. ☐ |

**EXAMPLE 1**

In Figure 3.5, $m(\angle A) = 60°$, $\overline{DC} \cong \overline{EC}$, and $\overline{DE} \parallel \overline{AB}$. Find $m(\angle 2)$.

Solution

Because $\angle 1$ and $\angle A$ are corresponding angles formed by a transversal and two parallel lines, we know that $m(\angle 1) = 60°$. Because $\overline{DC} \cong \overline{EC}$, we know that $\angle 1 \cong \angle 2$, or that $m(\angle 1) = m(\angle 2)$. By substitution, we conclude that $m(\angle 2) = 60°$.

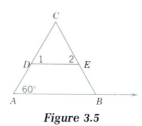

**Figure 3.5**    ■

## EXERCISE 3.2

In Exercises 1–4, refer to Illustration 1 in which $l_1 \parallel l_2$ and $m(\angle 1) = 30°$.

**1.** Find $m(\angle 2)$.
**2.** Find $m(\angle 3)$.
**3.** Find $m(\angle 4)$.
**4.** Find $m(\angle 5)$.

**Illustration 1**

In Exercises 5–8, refer to Illustration 2 in which $l_1 \parallel l_2$. Find the measure of each angle.

**5.** m($\angle 1$)

**6.** m($\angle 2$)

**7.** m($\angle 3$)

**8.** m($\angle 4$)

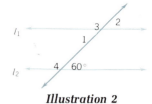

*Illustration 2*

In Exercises 9–12, refer to Illustration 3 in which $\overline{AD} \parallel \overline{BC}, \overline{AB} \parallel \overline{DC}, \overline{ED} \cong \overline{DA}$, and m($\angle B$) = 70°.

**9.** Find m($\angle C$).

**10.** Find m($\angle 1$).

**11.** Find m($\angle 2$).

**12.** Find m($\angle E$).

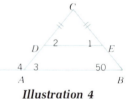

*Illustration 3*

In Exercises 13–16, refer to Illustration 4 in which $\overline{DE} \parallel \overline{AB}$ and $\overline{DC} \cong \overline{EC}$. Find the measure of each angle.

**13.** m($\angle 1$)

**14.** m($\angle 2$)

**15.** m($\angle 3$)

**16.** m($\angle 4$)

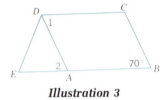

*Illustration 4*

In Exercises 17–20, refer to Illustration 5 in which $\overline{AB} \parallel \overline{DC}, \overline{AD} \parallel \overline{BC}, \overline{DB} \cong \overline{BC}$, and m($\angle ABD$) = 50°.

**17.** Find m($\angle 1$).

**18.** Find m($\angle 2$).

**19.** Find m($\angle 4$).

**20.** Find m($\angle 3$).

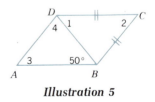

*Illustration 5*

In Exercises 21–24, refer to Illustration 6 in which $\overline{AB} \parallel \overline{DC}$, m($\angle A$) = 90°, m($\angle ABC$) = 120°, and m($\angle BCE$) = 30°.

**21.** Find m(∠1).

**22.** Find m(∠2).

**23.** Find m(∠3).

**24.** Find m(∠4).

*Illustration 6*

**25.** Refer to Illustration 7.

If $l_1 \parallel l_2$, prove that ∠1 is supplementary to ∠2.

**26.** Refer to Illustration 8.

*Given*: $\overline{AB} \parallel \overline{DC}$.
$\overline{AB} \cong \overline{CD}$.

*Prove*: ∠A ≅ ∠C.

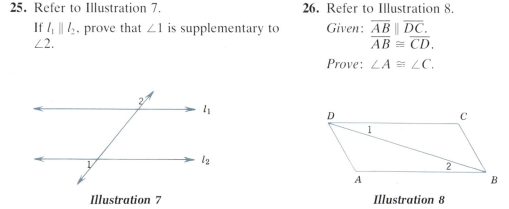

*Illustration 7*           *Illustration 8*

**27.** Refer to Illustration 9. If line $\overleftrightarrow{AB}$ is parallel to line $\overleftrightarrow{CD}$ and line $\overleftrightarrow{EF}$ is parallel to line $\overleftrightarrow{GH}$, prove that ∠1 ≅ ∠2.

**28.** Refer to Illustration 10. If line $\overleftrightarrow{AB}$ is parallel to line $\overleftrightarrow{CD}$ and $\overline{BC} \cong \overline{DC}$, prove that line $\overleftrightarrow{BD}$ bisects ∠CBA.

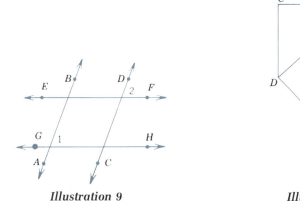

*Illustration 9*

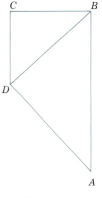

*Illustration 10*

**29.** Refer to Illustration 11.

*Given*: Lines $\overleftrightarrow{AD}$ and $\overleftrightarrow{BC}$ intersect at $E$.
  $\overline{AB} \parallel \overline{CD}$.
  $\overline{CE} \cong \overline{DE}$.

*Prove*: $\overline{AE} \cong \overline{BE}$.

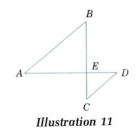

*Illustration 11*

**30.** Refer to Illustration 12.

*Given*: $\triangle ABE$
  $\overline{AB} \parallel \overline{CD}$.
  $\overline{CE} \cong \overline{DE}$.

*Prove*: $\overline{AE} \cong \overline{BE}$.

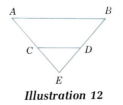

*Illustration 12*

**31.** Refer to Illustration 13.

*Given*: $\overline{AD} \parallel \overline{BE}$.
  $\overline{BD} \parallel \overline{CE}$.
  $B$ is the midpoint of $\overline{AC}$.

*Prove*: $\overline{BE} \cong \overline{AD}$.

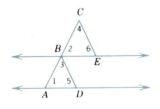

*Illustration 13*

**32.** *Prove*: If two parallel lines are cut by a transversal, then all pairs of alternate exterior angles are congruent.

# 3.3 THE SUM OF THE ANGLES IN A TRIANGLE

We can now prove one of the most important theorems discussed in this course and many **corollaries** to this theorem. A corollary is an easy theorem to prove because it is an immediate consequence of a major theorem.

> *Theorem 3.8* The sum of the measures of the interior angles of a triangle is 180°.

*Given*: $\triangle ABC$.

*Prove*: $m(\angle A) + m(\angle 2) + m(\angle B) = 180°$.

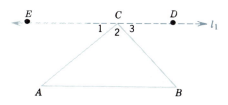

*Construction*: Construct line $l_1$ through point $C$ and parallel to line $\overleftrightarrow{AB}$.

*Proof*:

| STATEMENTS | REASONS |
|---|---|
| 1. Line $l_1$ passes through point $C$ and is parallel to $\overline{AB}$. | 1. By construction. |
| 2. $\angle DCE$ is a straight angle. | 2. An angle whose sides form a line is a straight angle. |
| 3. $m(\angle DCE) = 180°$. | 3. The measure of a straight angle is 180°. |
| 4. $m(\angle 1) + m(\angle 2) + m(\angle 3)$ $= m(\angle DCE)$. | 4. Postulate 1.11. |
| 5. $m(\angle 1) + m(\angle 2) + m(\angle 3)$ $= 180°$. | 5. Substitution. |
| 6. $\angle 1 \cong \angle A$. | 6. If two parallel lines are cut by a transversal, then the alternate interior angles are congruent. |
| 7. $m(\angle 1) = m(\angle A)$. | 7. If two angles are congruent, then they have equal measures. |
| 8. $\angle 3 \cong \angle B$. | 8. If two parallel lines are cut by a transversal, then the alternate interior angles are congruent. |
| 9. $m(\angle 3) = m(\angle B)$. | 9. If two angles are congruent, then they have equal measures. |
| 10. $m(\angle A) + m(\angle 2) + m(\angle B)$ $= 180°$. | 10. Substitution. |

> **Corollary 3.9** A triangle can have at most one right angle. A triangle can have at most one obtuse angle.

The proof is indirect. Assume that a triangle has more than one right angle. Then the sum of its angles must be greater than 180°. This contradicts Theorem 3.8. You will be asked to write out this proof in a two column format in the exercises. The case for the obtuse angle is also left as an exercise.

> **Corollary 3.10** The acute angles of a right triangle are complementary.

The proof of Corollary 3.10 is left as an exercise.

> **Corollary 3.11** If two angles of one triangle are congruent, respectively, to two angles of a second triangle, then their third angles are congruent.

*Given*: $\angle A \cong \angle D$.
$\qquad \angle C \cong \angle F$.

*Prove*: $\angle B \cong \angle E$.

*Proof*:

| STATEMENTS | REASONS |
|---|---|
| **1.** $\angle A \cong \angle D$. | **1.** Given. |
| **2.** $m(\angle A) = m(\angle D)$. | **2.** Congruent angles have equal measures. |
| **3.** $\angle C \cong \angle F$. | **3.** Given. |
| **4.** $m(\angle C) = m(\angle F)$. | **4.** Congruent angles have equal measures. |
| **5.** $m(\angle A) + m(\angle B) + m(\angle C) = 180°$. | **5.** The sum of the measures of the angles of a triangle is 180°. |
| **6.** $m(\angle D) + m(\angle E) + m(\angle F) = 180°$. | **6.** The sum of the measures of the angles of a triangle is 180°. |

| STATEMENTS | REASONS |
|---|---|
| **7.** $m(\angle A) + m(\angle B) + m(\angle C)$ $= m(\angle D) + m(\angle E) + m(\angle F)$. | 7. Substitution. |
| **8.** $m(\angle B) = m(\angle E)$. | 8. Equals subtracted from equals are equal.  □ |

**Corollary 3.12** If two angles and any side of one triangle are congruent, respectively, to two angles and the corresponding side of a second triangle, then the triangles are congruent ($AAS \cong AAS$).

The proof of Corollary 3.12 is left as an exercise. Because of Corollary 3.12, we now have five ways to prove two triangles congruent.

$SSS \cong SSS$, $SAS \cong SAS$, $ASA \cong ASA$, $hl \cong hl$, and $AAS \cong AAS$

**Definition 3.5** If one of the sides of a triangle is extended, then the smaller angle formed that is adjacent to an angle of a triangle is called an **exterior angle** of the triangle.

In Figure 3.6, $\angle a$, $\angle b$, and $\angle c$ are exterior angles of the triangle.

*Figure 3.6*

**Corollary 3.13** The measure of an exterior angle of a triangle is equal to the sum of the measures of the nonadjacent interior angles.

*Given:* △*ABC* with ∠*A*, ∠*C*,
and ∠3 as interior angles.

*Prove:* m(∠1) = m(∠*A*) + m(∠*C*).

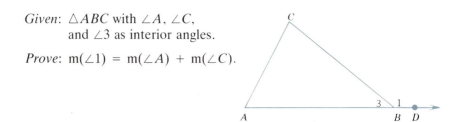

*Proof:*

| STATEMENTS | REASONS |
| --- | --- |
| **1.** △*ABC* with ∠*A*, ∠*C*, and ∠3 as interior angles. | **1.** Given. |
| **2.** m(∠*A*) + m(∠*C*) + m(∠3) = 180°. | **2.** The sum of the measures of the angles of a triangle is 180°. |
| **3.** ∠*ABD* is a straight angle. | **3.** If the sides of an angle form a line, the angle is a straight angle. |
| **4.** m(∠*ABD*) = 180°. | **4.** The measure of a straight angle is 180°. |
| **5.** m(∠1) + m(∠3) = m(∠*ABD*). | **5.** Postulate 1.11. |
| **6.** m(∠1) + m(∠3) = 180°. | **6.** Substitution. |
| **7.** m(∠*A*) + m(∠3) + m(∠*C*) = m(∠1) + m(∠3). | **7.** Substitution. |
| **8.** m(∠*A*) + m(∠*C*) = m(∠1). | **8.** Equals subtracted from equals are equal.   □ |

> **Theorem 3.14** If the hypotenuse and an acute angle of one right triangle are congruent, respectively, to the hypotenuse and an acute angle of a second triangle, then the triangles are congruent.

The proof of Theorem 3.14 is left as an exercise. We now have six ways of proving right triangles congruent. Theorem 3.14 is often abbreviated as *ha* ≅ *ha*.

**EXAMPLE 1**    In Figure 3.7, $\overline{AC} \cong \overline{BC}$, $\overline{DC} \cong \overline{EC}$, m(∠*A*) = 60°, and m(∠*CED*) = 70°. Find (a) m(∠2), (b) m(∠1), and (c) m(∠3).

Solution

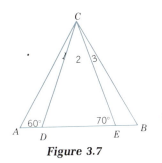

**Figure 3.7**

(a) Because $\overline{DC} \cong \overline{EC}$, it follows that $\angle CDE \cong \angle CED$. Because m($\angle CED$) = 70°, it follows that m($\angle CDE$) = 70°. Because the sum of the measures of the angles of $\triangle DEC$ must be 180°, it follows that

$$m(\angle 2) + 70° + 70° = 180°$$
$$m(\angle 2) + 140° = 180° \qquad \text{Combine terms.}$$
$$m(\angle 2) = 40° \qquad \text{Add } -140° \text{ to both sides.}$$

(b) Because $\angle ADC$ and $\angle CDE$ are supplementary and m($\angle CDE$) = 70°, it follows that m($\angle ADC$) = 110°. Because the sum of the angles of $\triangle ADC$ must be 180°, it follows that

$$m(\angle 1) + 60° + 110° = 180°$$
$$m(\angle 1) + 170° = 180° \qquad \text{Combine terms.}$$
$$m(\angle 1) = 10° \qquad \text{Add } -170° \text{ to both sides.}$$

(c) Because $\angle BEC$ and $\angle CED$ are supplementary and m($\angle CED$) = 70°, it follows that m($\angle BEC$) = 110°. Because $\overline{AC} \cong \overline{BC}$ and m($\angle A$) = 60°, it follows that m($\angle B$) = 60°. Because the sum of the measures of the angles of $\triangle BEC$ must be 180°, it follows that

$$m(\angle 3) + 60° + 110° = 180°$$
$$m(\angle 3) + 170° = 180° \qquad \text{Combine terms.}$$
$$m(\angle 3) = 10° \qquad \text{Add } -170° \text{ to both sides.} \quad \blacksquare$$

## EXERCISE 3.3

In Exercises 1–4, refer to Illustration 1 in which $\overline{AB} \parallel \overline{DC}$ and $\overline{AD} \parallel \overline{BC}$.

**1.** Find m($\angle B$).
**2.** Find m($\angle 1$).
**3.** Find m($\angle D$).
**4.** Find m($\angle 2$).

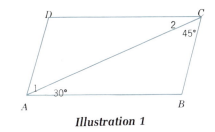

**Illustration 1**

In Exercises 5–8, refer to Illustration 2 in which $\overline{BD} \perp \overline{AC}$, $\overline{AB} \perp \overline{BC}$, and m($\angle A$) = 50°.

**5.** Find m($\angle 1$).
**6.** Find m($\angle 2$).
**7.** Find m($\angle 5$).
**8.** Find m($\angle 6$).

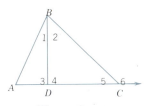

**Illustration 2**

In Exercises 9–12, refer to Illustration 3 in which $\overline{AC} \perp \overline{BD}$, m($\angle 5$) = 70°, m($\angle DAB$) = 110°, $\overline{AD} \cong \overline{DC}$, and $\overline{AB} \cong \overline{BC}$.

**9.** Find m($\angle 10$).

**10.** Find m($\angle 12$).

**11.** Find m($\angle ADC$).

**12.** Find m($\angle ABC$).

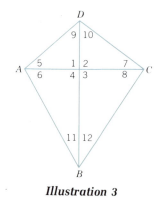

**Illustration 3**

In Exercises 13–16, refer to Illustration 4 in which $\overline{AB} \parallel \overline{DC}$, m($\angle 1$) = 40°, m($\angle ADC$) = 120°, and $\overline{BD}$ bisects $\angle ABC$.

**13.** Find m($\angle 3$).

**14.** Find m($\angle 4$).

**15.** Find m($\angle A$).

**16.** Find m($\angle C$).

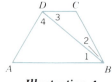

**Illustration 4**

**17.** Find the measure of each angle of an equilateral triangle.

**18.** Find the measure of each angle of an isosceles-right triangle.

**19.** Refer to Illustration 5.

Given: △ABC.
  $\overline{AC} \cong \overline{BC}$.
  $\angle 1 \cong \angle 2$.

Prove: $\angle 3 \cong \angle 4$.

**20.** Refer to Illustration 6.

Given: Lines $\overline{AC}$ and $\overline{ED}$ intersect at B.
  $\overline{AE} \perp \overline{AC}$.
  $\overline{CD} \perp \overline{AC}$.
  $\overline{BE} \cong \overline{BD}$.

Prove: $\overline{AB} \cong \overline{CB}$.

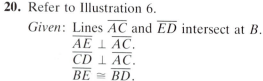

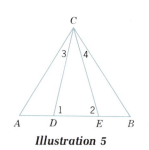

**Illustration 5**

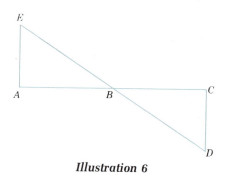

**Illustration 6**

**21.** Refer to Illustration 7.

*Given*: $\overline{AC} \perp \overline{BD}$.
$\angle 1 \cong \angle 2$.

*Prove*: $\overline{AC}$ bisects $\overline{BD}$.

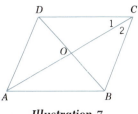

***Illustration 7***

**22.** *Prove*: A triangle can have at most one right angle.

**23.** *Prove*: A triangle can have at most one obtuse angle.

**24.** *Prove Corollary 3.10*: The acute angles of a right triangle are complementary.

**25.** *Prove Corollary 3.12*: If two angles and any side of one triangle are congruent, respectively, to two angles and the corresponding side of a second triangle, then the triangles are congruent.

**26.** *Prove Theorem 3.14*: If the hypotenuse and an acute angle of one right triangle are congruent, respectively, to the hypotenuse and an acute angle of a second triangle, then the triangles are congruent.

**27.** *Prove*: If two exterior angles of a triangle are congruent, then the triangle is isosceles.

**28.** *Prove*: If the sum of the measures of two angles of a triangle equals the measure of the third angle, then the triangle is a right triangle.

**29.** *Prove*: The sum of the measures of the exterior angles of a triangle is 360°.

**30.** *Prove*: The sum of the measures of the interior angles and the exterior angles of a triangle is 540°.

# 3.4 THE SUM OF THE ANGLES OF A POLYGON

A figure is called a **simple figure** if no pair of nonadjacent sides intersect. See Figure 3.8. A **diagonal** of a simple figure is a line segment that connects two nonadjacent vertices of the figure. If any diagonal of a simple figure is partly on the exterior of the figure, the figure is called **concave**. Otherwise, the figure is called **convex**.

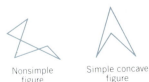

Nonsimple figure

Simple concave figure

Simple convex figure

Simple concave figure with diagonals $AC$ and $BD$

**Figure 3.8**

> **Definition 3.6** A **polygon** is a simple figure with many angles.

In this course the discussion will be limited to convex polygons.

> **Postulate 3.3** A polygon has the same number of sides as it has angles.

There are some useful theorems about the angles of a polygon. Because their proofs do not lend themselves to a two-column format, we will write the proofs in a paragraph style.

> **Theorem 3.15** The sum of the measures of the interior angles of a polygon is given by the formula
>
> $$S = (n - 2)180°$$
>
> where $n$ is the number of sides of the polygon.

*Given*: A polygon with $n$ sides.

*Prove*: The sum of the measures of the interior angles equals $(n - 2)180°$.

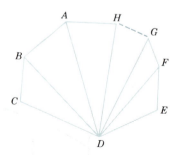

*Proof*: Pick some vertex, say vertex $D$. Draw all possible diagonals in the polygon from vertex $D$. Because vertex $C$ and vertex $E$ are adjacent to vertex $D$, it is impossible to draw diagonals to vertex $C$ or vertex $E$. We can draw $n - 3$ diagonals, one to each vertex except vertices $C$, $D$, and $E$. Diagonal $\overline{DB}$ forms $\triangle BDC$, diagonal $\overline{DA}$ forms $\triangle ADB$, diagonal $\overline{DH}$ forms $\triangle HDA$, and so on until the last diagonal $\overline{DF}$ forms *two* triangles: $\triangle FDG$ and $\triangle FDE$. Therefore, $n - 2$ triangles are formed. The sum of the measures of the interior angles of each of these triangles is 180°. Thus, the sum of the measures of the angles of the $n - 2$ triangles is $(n - 2)180°$. Because the sum of the measures of the interior angles of the $n - 2$ triangles is equal to the sum of the measures of the interior angles of the polygon, the sum of the measures of the interior angles of the polygon is $(n - 2)180°$.  □

The reference polygon in the preceding proof did not have a particular number of sides. Thus, the proof is general and applies to all polygons, regardless of the number of sides.

---

**Theorem 3.16** The sum of the measures of the exterior angles of a polygon is 360°.

---

*Given*: A polygon with $n$ sides.

*Prove*: The sum of the measures of the exterior angles is 360°.

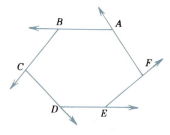

*Proof*: There are $n$ straight angles through the $n$ vertices of the polygon. Each of these $n$ straight angles is the sum of one interior angle and one exterior angle of the polygon. The measure of the $n$ straight angles is $n \cdot 180°$, and the sum of the measures of the interior angles is $(n - 2)180°$. Thus, if we call the sum of the measures of the exterior angles $E$, we have

$$E + (n - 2)180° = n \cdot 180°$$
$$E + n \cdot 180° - 360° = n \cdot 180° \qquad \text{Remove parentheses.}$$
$$E - 360° = 0 \qquad \text{Add } -n \cdot 180° \text{ to both sides.}$$
$$E = 360° \qquad \text{Add } 360° \text{ to both sides.} \quad \square$$

---

**Definition 3.7** A polygon is called an **equilateral polygon** if, and only if, all of its sides are congruent.

---

**Definition 3.8** A polygon is called an **equiangular polygon** if, and only if, all of its angles are congruent.

---

**Definition 3.9** A **regular polygon** is a polygon that is both equilateral and equiangular.

> **Theorem 3.17** The measure of an interior angle $a$ of a regular polygon is given by the formula
>
> $$a = \frac{(n - 2)180°}{n}$$
>
> where $n$ is the number of sides of the polygon.

*Proof*: The sum of the measures of the interior angles of any polygon is given by the formula

$$S = (n - 2)180°$$

and there are $n$ equal angles in an $n$-sided regular polygon. Thus, the measure of each angle in the regular polygon is equal to the sum of the measures of the angles of the polygon divided by $n$. Hence, we have

$$S = \frac{S}{n} = \frac{(n - 2)180°}{n}$$

□

**EXAMPLE 1**

Find the sum of the measures of the interior angles of a polygon with seven sides.

Solution

Substitute 7 for $n$ in the formula for the sum of the measures of the interior angles of a polygon and simplify.

$$\begin{aligned}
S &= (n - 2)180° \\
&= (7 - 2)180° \\
&= 5 \cdot 180° \\
&= 900°
\end{aligned}$$

The sum of the measures of the interior angles of a polygon with seven sides is 900°.    ∎

**EXAMPLE 2**

Find the measure of one interior angle of a regular hexagon (a six-sided polygon).

Solution

The measure $a$ of an interior angle of a regular hexagon is the sum of the measures of the six interior angles divided by 6, the number of interior angles. This is expressed by the formula

$$a = \frac{(n - 2)180°}{n}$$

where $n = 6$. Substitute 6 for $n$ and simplify.

$$a = \frac{(6 - 2)180°}{}$$

$$= 4(30°)$$
$$= 120°$$

The measure of one interior angle of a regular hexagon is 120°.   ■

**EXAMPLE 3**   Can the sum of the measures of the interior angles of a polygon be 1440°?

*Solution*   Substitute 1440° for $S$ in the formula for the sum of the measures of the interior angles of a polygon and solve for $n$.

$$S = (n - 2)180°$$
$$1440° = (n - 2)180°$$
$$1440° = n \cdot 180° - 360° \qquad \text{Remove parentheses.}$$
$$1800° = n \cdot 180° \qquad \text{Add 360° to both sides.}$$
$$10 = n \qquad \text{Divide both sides by 180°.}$$

Because $n$ is a whole number, a polygon does exist with the sum of its interior angles equal to 1440°. The polygon has 10 sides.

If $n$ had not been a whole number, then no such polygon would exist.   ■

## EXERCISE 3.4

In Exercises 1–6, find the sum of the measures of the interior angles of the given polygon.

**1.** A square

**2.** A pentagon (a 5-sided polygon)

**3.** A hexagon (a 6-sided polygon)

**4.** An octagon (an 8-sided polygon)

**5.** A decagon (a 10-sided polygon)

**6.** A dodecagon (a 12-sided polygon)

In Exercises 7–12, find the measure of one interior angle of each regular polygon.

**7.** A square

**8.** A regular pentagon

**9.** A regular hexagon

**10.** A regular octagon

**11.** A regular nonogon (a 9-sided polygon)

**12.** A regular decagon

In Exercises 13–20, each quantity might be the sum of the measures of the interior angles of a polygon. Find, if possible, the number of sides of the polygon. If no such polygon exists, so indicate.

**13.** 2880°

**14.** 2700°

**15.** 2000°

**16.** 1480°

**17.** 3960°

**18.** 1620°

**19.** 3600°

**20.** 2120°

In Exercises 21–28, each quantity might be the measure of one interior angle of a regular polygon. Find, if possible, the number of sides of the polygon. If no such polygon exists, so indicate.

**21.** 150°        **22.** 135°        **23.** 140°        **24.** 148°

**25.** 115°        **26.** 162°        **27.** 156°        **28.** $157\frac{1}{2}°$

**29.** Find the sum of the measures of the exterior angles of a hexagon.

**30.** Can the measure of an exterior angle of any regular polygon equal 15°?

**31.** Find the measure of an exterior angle of a regular polygon with 15 sides.

**32.** Show that the formula

$$a = 180° - \frac{360°}{n}$$

is equivalent to the formula given in Theorem 3.17.

**33.** In a certain regular polygon, twice the measure of an exterior angle equals the measure of one interior angle. How many sides does the polygon have?

**34.** The number of degrees in an exterior angle of a regular polygon is 10 times the number of sides. How many sides does the polygon have?

**35.** If the number of sides of a polygon were doubled, the sum of the measures of the interior angles would increase by 900°. How many sides does the polygon have?

**36.** *Prove*: If the sum of the measures of the exterior angles of a polygon equals the sum of the measures of the interior angles, then the polygon is a four-sided polygon.

## 3.5 SOME INDIRECT PROOFS

In this section, we will present several theorems that can be proved by the indirect method. You will be asked to do several indirect proofs in the exercises. The symbol $\nparallel$ is read as "is not parallel to."

---

**Theorem 3.18** If two lines are cut by a transversal so that alternate interior angles are not congruent, then the lines are not parallel.[1]

---

[1] Theorem 3.18 has a special relationship to Theorem 3.5 and it is called the **contrapositive** of Theorem 3.5. Because of this relationship, the two theorems are equivalent and Theorem 3.18 is not a new result. Contrapositives of statements are discussed in more detail in Chapter 11. The proof of Theorem 3.18 is included only as an example of indirect proof.

*Given*: $\angle 1 \not\equiv \angle 2$.

*Prove*: $l_1 \nparallel l_2$.

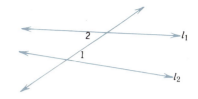

*Plan*: Use an indirect proof and assume $l \parallel l_2$.
Show that this leads to a contradiction.

*Proof*:

| STATEMENTS | REASONS |
|---|---|
| **1.** $l_1 \parallel l_2$ | **1.** Possible assumption. |
| **2.** $\angle 1 \cong \angle 2$. | **2.** If two parallel lines are cut by a transversal, then alternate interior angles are congruent. |
| **3.** $\angle 1 \not\equiv \angle 2$. | **3.** Given. |
| **4.** $l_1 \parallel l_2$ is false. | **4.** It leads to a contradiction in Statements 2 and 3. |
| **5.** $l_1 \nparallel l_2$. | **5.** It is the only remaining possibility. □ |

---

**Theorem 3.19** If two lines are cut by a transversal so that corresponding angles are not congruent, then the lines are not parallel.

---

*Given*: $\angle 1 \not\equiv \angle 2$.

*Prove*: $l_1 \nparallel l_2$.

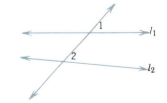

*Plan*: Use an indirect proof and assume $l \parallel l_2$. Show that this leads to a contradiction.

*Proof*:

| STATEMENTS | REASONS |
|---|---|
| 1. $l_1 \parallel l_2$. | **1.** Possible assumption. |

| STATEMENTS | REASONS |
|---|---|
| **2.** $\angle 1 \cong \angle 2$. | **2.** If two parallel lines are cut by a transversal, then corresponding angles are congruent. |
| **3.** $\angle 1 \not\cong \angle 2$. | **3.** Given. |
| **4.** $l_1 \parallel l_2$ is false. | **4.** It leads to a contradiction in Statements 2 and 3. |
| **5.** $l_1 \not\parallel l_2$. | **5.** It is the only remaining possibility.    □ |

> **Theorem 3.20** If two lines are cut by a transversal so that two interior angles on the same side of the transversal are not supplementary, then the lines are not parallel.

The proof of Theorem 3.20 is left as an exercise.

> **Theorem 3.21** If two nonparallel lines are cut by a transversal, then alternate interior angles are not congruent.

The proof of Theorem 3.21 is left as an exercise.

> **Theorem 3.22** Two lines can intersect in at most one point.

*Given*: Lines $l_1$ and $l_2$.

*Prove*: Lines $l_1$ and $l_2$ can intersect in at most one point.

*Plan*: Use an indirect proof. Assume that lines $l_1$ and $l_2$ intersect in more than one point, perhaps points $A$ and $B$. Show that this assumption leads to a contradiction.

| STATEMENTS | REASONS |
|---|---|
| **1.** Assume that $l_1$ and $l_2$ intersect at more than one point, say points $A$ and $B$. | **1.** Possible assumption. |
| **2.** Points $A$ and $B$ determine two lines. | **2.** Because of the assumption, two lines pass through $A$ and $B$. |

| STATEMENTS | REASONS |
| --- | --- |
| **3.** Points $A$ and $B$ determine one and only one line. | **3.** Postulate 1.1. |
| **4.** The assumption that $l_1$ and $l_2$ intersect in more than one point is false. | **4.** It leads to a contradiction in Statement 2 and 3. |
| **5.** Two lines can intersect in at most one point. | **5.** It is the only remaining possibility.     □ |

Theorem 3.22 implies that if two lines intersect, they intersect in exactly one point.

**Theorem 3.23** If two different lines are parallel to a third line, then they are parallel to each other.

You will be asked to prove Theorem 3.23 in the exercises both directly and indirectly.

## EXERCISE 3.5

1. *Prove*: If two sides of a triangle are not congruent, then the angles opposite those sides are not congruent.
2. *Prove*: If two angles of a triangle are not congruent, then the sides opposite those angles are not congruent.
3. *Prove*: If a triangle has two noncongruent sides, then it is not an equilateral triangle.
4. *Prove Theorem 3.20*: If two lines are cut by a transversal so that two interior angles on the same side of the transversal are not supplementary, then the lines are not parallel.
5. *Prove Theorem 3.21*: If two nonparallel lines are cut by a transversal, then alternate interior angles are not congruent.
6. *Prove*: If two nonparallel lines are cut by a transversal, then pairs of corresponding angles are not congruent.
7. *Prove*: If two nonparallel lines are cut by a transversal, then angles on the same side of the transversal and on the interior of the lines are not supplementary.
8. *Prove*: If the sum of the measures of the angles of a polygon is not 180°, then the polygon is not a triangle.
9. Prove Theorem 3.23 by using direct reasoning.
10. Prove Theorem 3.23 by using indirect reasoning.

## CHAPTER THREE REVIEW EXERCISES

In Review Exercises 1–6, refer to Illustration 1 in which transversal $l$ intersects parallel lines $l_1$ and $l_2$.

**1.** Name each pair of alternate interior angles.

**2.** Name each pair of corresponding angles.

**3.** Name each pair of alternate exterior angles.

**4.** Name each angle that is congruent to $\angle 1$.

**5.** Name each angle that is congruent to $\angle 2$.

**6.** Name each angle that is supplementary to $\angle 3$.

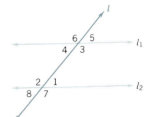

*Illustration 1*

In Review Exercises 7–10, refer to Illustration 2 in which transversal $l$ intersects lines $l_1$ and $l_2$.

**7.** If $\angle 1 \cong \angle 4$, is $l_1 \parallel l_2$? Explain.

**8.** If $\angle 8 \cong \angle 5$, is $l_1 \parallel l_2$? Explain.

**9.** If $\angle 3 \cong \angle 1$, is $l_1 \parallel l_2$? Explain.

**10.** If $\angle 1$ is supplementary to $\angle 6$, is $l_1 \parallel l_2$? Explain.

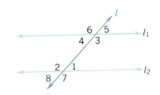

*Illustration 2*

In Review Exercises 11–14, refer to Illustration 3 in which $\overline{ED} \parallel \overline{AC}$, $\overline{BD} \cong \overline{CD}$, and $\overline{AE} \parallel \overline{BD}$.

**11.** Find m($\angle 1$).

**12.** Find m($\angle A$).

**13.** Find m($\angle E$).

**14.** Find m($\angle 2$).

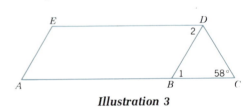

*Illustration 3*

**15.** Refer to Illustration 4.

*Given*: $\triangle DEC$ is isosceles with $\overline{DC} \cong \overline{EC}$.

　　　$l_1 \parallel l_2$.

*Prove*: $\triangle ABC$ is isosceles.

**16.** Refer to Illustration 5.

*Given*: $\overline{AC}$ and $\overline{ED}$ bisect each other.

*Prove*: $\overline{AE} \parallel \overline{CD}$.

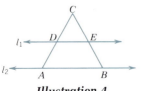

*Illustration 4*

*Illustration 5*

In Review Exercises 17–20, refer to Illustration 6.

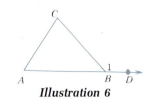

**17.** If m($\angle A$) = 70° and m($\angle B$) = 60°, find m($\angle C$).

**18.** If m($\angle A$) = 70° and m($\angle C$) = 60°, find m($\angle 1$).

**19.** If m($\angle ABC$) = 50° and $\overline{CB} \cong \overline{AB}$, find m($\angle A$).

**20.** If m($\angle 1$) = 135° and $\overline{AC} \cong \overline{BC}$, find m($\angle C$).

*Illustration 6*

In Review Exercises 21–24, refer to Illustration 7 in which $\overline{AC} \cong \overline{BC}$ and $\overline{DC} \cong \overline{EC}$.

**21.** Find m($\angle 2$).

**22.** Find m($\angle ACB$).

**23.** Find m($\angle 1$).

**24.** Find m($\angle 3$).

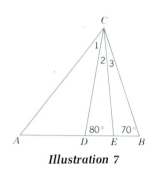

*Illustration 7*

**25.** Find the sum of the measures of the interior angles of a 15-sided polygon.

**26.** If the sum of the measures of the interior angles of a polygon is 1440°, how many sides does the polygon have?

**27.** Find the measure of one of the interior angles of a regular hexagon (a 6-sided figure).

**28.** The measure of one interior angle of a regular polygon is 108°. How many sides does the polygon have?

**29.** The sum of the measures of the interior angles of a polygon is equal to the sum of the measures of its exterior angles. How many sides does the polygon have?

**30.** *Prove*: If the sum of the measures of the interior angles of a polygon is not 360°, then the polygon is not a quadrilateral.

## CHAPTER THREE TEST

In Exercises 1–25, classify each statement as true or false.

**1.** If two lines are cut by a transversal, then all pairs of alternate interior angles are congruent.

**2.** Two lines perpendicular to the same line are perpendicular to each other.

**3.** If a line is perpendicular to one of two perpendicular lines, then it is parallel to the other.

**4.** Interior angles on the same side of the transversal can never be complementary.

**5.** The sum of the measures of the interior angles of a quadrilateral equals the sum of the measures of the exterior angles of an octagon.

**6.** If two angles of one triangle are congruent to two angles of a second triangle, then their third angles are congruent.

7. An exterior angle of a triangle could be supplementary to the sum of the measures of the nonadjacent interior angles.

8. Through a point not on a line, there exists exactly one parallel to that line.

9. If two lines are cut by a transversal so that two interior angles on the same side of the transversal are supplementary, then the lines are parallel.

10. The negation of the statement *The lines are parallel* is *The lines are not parallel.*

11. In an indirect proof we always begin by assuming the negation of what we are trying to prove.

12. The sum of the measures of the angles of a triangle is the same as the sum of the measures of two right angles.

13. The acute angles of a right triangle are supplementary.

14. If two sides of one triangle are congruent to two sides of a second triangle, then their third sides are congruent.

15. A triangle can have both a right angle and an obtuse angle.

16. An exterior angle of an equilateral triangle is greater than one of its interior angles.

17. The measure of an exterior angle of a polygon is always greater than the measure of its corresponding interior angle.

18. A triangle must have at least two acute angles.

19. The sum of the measures of the interior angles of a polygon with seven sides is $900°$.

20. If the sum of the measures of the interior angles of a polygon equals $1800°$, the polygon has twelve sides.

21. The sum of the measures of the interior angles of a quadrilateral equals the sum of the measures of the exterior angles of a decagon.

22. An exterior angle of an equilateral triangle is congruent to an interior angle of a regular hexagon.

23. The measure of one interior angle of a regular decagon (10-sided figure) is $144°$.

24. If the measure of one interior angle of a regular polygon is $135°$, the polygon is an octagon (8-sided figure).

25. If the number of sides of a polygon were doubled, the sum of its interior angles would increase by $540°$. The polygon has three sides.

# 4

*Hypatia (370 A.D. to 415 A.D.). Hypatia is the earliest known woman in the history of mathematics. She was a professor at the University of Alexandria. Because of her scientific beliefs, she was considered to be a heretic. At the age of 45, she was murdered for these beliefs.*

# Parallelograms and Trapezoids

U ntil now we have dealt principally with lines and triangles and their properties. In this chapter, we shall broaden the discussion by considering other important geometric figures.

## 4.1 PARALLELOGRAMS

A common geometric figure is a polygon with four sides.

*Definition 4.1* A **quadrilateral** is a polygon with four sides.

An important quadrilateral is the parallelogram.

> **Definition 4.2** A **parallelogram** is a quadrilateral whose opposite sides are parallel.

A convenient symbol for parallelogram is ▱. Three parallelograms are shown in Figure 4.1.

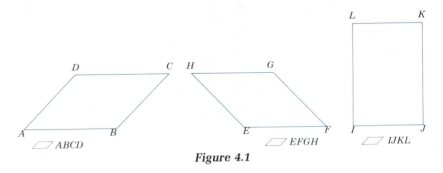

**Figure 4.1**

The following theorem is the key to finding many of the properties of parallelograms.

> **Theorem 4.1** A diagonal of a parallelogram divides the parallelogram into two congruent triangles.

*Given:* $ABCD$ is a ▱ with diagonal $\overline{DB}$.

*Prove:* $\triangle ABD \cong \triangle CDB$.

*Proof:*

| STATEMENTS | REASONS |
| --- | --- |
| 1. $ABCD$ is a ▱ with diagonal $\overline{BD}$. | 1. Given. |
| 2. $\overline{AB} \parallel \overline{DC}$; $\overline{AD} \parallel \overline{BC}$. | 2. The opposite sides of a parallelogram are parallel. |

| STATEMENTS | REASONS |
|---|---|
| **3.** $\angle 1 \cong \angle 2$; $\angle 3 \cong \angle 4$. | **3.** If two parallel lines are cut by a transversal, then alternate interior angles are congruent. |
| **4.** $\overline{BD} \cong \overline{DB}$. | **4.** Reflexive law. |
| **5.** $\triangle ABD \cong \triangle CDB$. | **5.** $ASA \cong ASA$.    □ |

---

**Corollary 4.2** Opposite sides of a parallelogram are congruent.

---

The proof of Corollary 4.2 is left as an exercise.

---

**Corollary 4.3** Nonconsecutive angles of a parallelogram are congruent.

---

The proof of Corollary 4.3 is left as an exercise.

---

**Theorem 4.4** Consecutive angles of a parallelogram are supplementary.

---

The proof of Theorem 4.4 is left as an exercise.

---

**Definition 4.3** The **distance from a point to a line** is the length of the perpendicular segment drawn from the point to the line.

---

In Figure 4.2, the distance from point $P$ to line $\overleftrightarrow{AB}$ is the measure of line segment $\overline{PD}$.

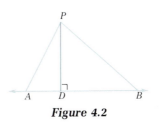

*Figure 4.2*

> **Theorem 4.5** Parallel lines are always the same distance apart.

*Given*: $l_1 \parallel l_2$.

*Prove*: $l_1$ and $l_2$ are always the same distance apart.

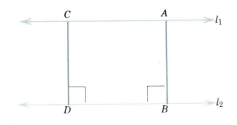

*Construction*: $\overline{AB} \perp l_2$.
$\overline{CD} \perp l_2$.

*Plan*: Show that two arbitrary points $A$ and $C$ on line $l_1$ are equidistant from line $l_2$. Since the points are arbitrary, all points on line $l_1$ are equidistant from line $l_2$. Hence, $l_1$ and $l_2$ are always the same distance apart.

*Proof*:

| STATEMENTS | REASONS |
|---|---|
| **1.** $l_1 \parallel l_2$. | **1.** Given. |
| **2.** $\overline{AB} \perp l_2$; $\overline{CD} \perp l_2$. | **2.** Construction. |
| **3.** m($\overline{AB}$) is the distance from point $A$ to line $l_2$; m($\overline{CD}$) is the distance from point $C$ to line $l_2$. | **3.** Definition of the distance from a point to a line. |
| **4.** $\overline{CD} \parallel \overline{AB}$. | **4.** Two lines perpendicular to a third line are parallel. |
| **5.** Quadrilateral $CDBA$ is a $\square$. | **5.** If the opposite sides of a quadrilateral are parallel, then the quadrilateral is a parallelogram. |
| **6.** $\overline{CD} \cong \overline{AB}$. | **6.** Opposite sides of a parallelogram are congruent. |
| **7.** m($\overline{CD}$) = m($\overline{AB}$). | **7.** Congruent line segments have equal measures. |

| STATEMENTS | REASONS |
|---|---|
| **8.** $l_1$ and $l_2$ are always the same distance apart. | **8.** Because $A$ and $C$ are arbitrary points on $l_1$ and they are equidistant from $l_2$, all points on $l_1$ are equidistant from $l_2$.   □ |

---

**Theorem 4.6** If both pairs of opposite sides of a quadrilateral are congruent, then the quadrilateral is a parallelogram.

---

*Given:* $\overline{AB} \cong \overline{CD}$.
$\overline{CB} \cong \overline{AD}$.

*Prove:* $ABCD$ is a □.

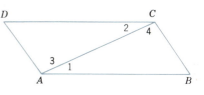

*Proof:*

| STATEMENTS | REASONS |
|---|---|
| **1.** $\overline{AB} \cong \overline{CD}$. | **1.** Given. |
| **2.** $\overline{CB} \cong \overline{AD}$. | **2.** Given. |
| **3.** $\overline{CA} \cong \overline{AC}$. | **3.** Reflexive law. |
| **4.** $\triangle ABC \cong \triangle CDA$. | **4.** $SSS \cong SSS$. |
| **5.** $\angle 1 \cong \angle 2$. | **5.** cpctc. |
| **6.** $\angle 3 \cong \angle 4$. | **6.** cpctc. |
| **7.** $\overline{AB} \parallel \overline{DC}$. | **7.** If two lines are cut by a transversal so the alternate interior angles are congruent, then the lines are parallel. |
| **8.** $\overline{AD} \parallel \overline{BC}$. | **8.** If two lines are cut by a transversal so that alternate interior angles are congruent, then the lines are parallel. |
| **9.** $ABCD$ is a □. | **9.** If a quadrilateral has both pairs of opposite sides parallel, then it is a parallelogram.   □ |

---

**Theorem 4.7** If two opposite sides of a quadrilateral are both parallel and congruent, then the quadrilateral is a parallelogram.

---

The proof of Theorem 4.7 is left as an exercise.

---

**Theorem 4.8** The diagonals of a parallelogram bisect each other.

---

*Given*: $ABCD$ is a $\square$ with diagonals $\overline{AC}$ and $\overline{BD}$.

*Prove*: $\overline{AC}$ and $\overline{BD}$ bisect each other.

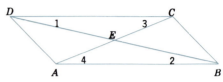

*Proof*:

| STATEMENTS | REASONS |
| --- | --- |
| 1. $ABCD$ is a $\square$ with diagonals $\overline{AC}$ and $\overline{BD}$. | 1. Given. |
| 2. $\overline{DC} \parallel \overline{AB}$. | 2. Opposite sides of a parallelogram are parallel. |
| 3. $\angle 1 \cong \angle 2$. | 3. If parallel lines are cut by a transversal, then alternate interior angles are congruent. |
| 4. $\angle 3 \cong \angle 4$. | 4. If parallel lines are cut by a transversal, then alternate interior angles are congruent. |
| 5. $\overline{DC} \cong \overline{BA}$. | 5. Opposite sides of a parallelogram are congruent. |
| 6. $\triangle DCE \cong \triangle BAE$. | 6. ASA $\cong$ ASA. |
| 7. $\overline{CE} \cong \overline{AE}$. | 7. cpctc. |
| 8. $\overline{DE} \cong \overline{BE}$. | 8. cpctc. |
| 9. $\overline{AC}$ and $\overline{BD}$ bisect each other. | 9. If two line segments divide each other into congruent segments, then they bisect each other. |

---

**Theorem 4.9** If the diagonals of a quadrilateral bisect each other, then the quadrilateral is a parallelogram.

*Given*: In quadrilateral $ABCD$, $\overline{DB}$ and $\overline{AC}$ bisect each other at point $E$.

*Prove*: $ABCD$ is a $\square$.

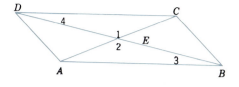

*Proof*:

| STATEMENTS | REASONS |
|---|---|
| **1.** In quadrilateral $ABCD$, $\overline{DB}$ and $\overline{AC}$ bisect each other at point $E$. | **1.** Given. |
| **2.** $\overline{DE} \cong \overline{BE}$. | **2.** A bisected line segment is divided into two congruent segments. |
| **3.** $\overline{CE} \cong \overline{AE}$. | **3.** A bisected line segment is divided into two congruent segments. |
| **4.** $\angle 1$ and $\angle 2$ are vertical angles. | **4.** Nonadjacent angles formed by two intersecting line segments are vertical angles. |
| **5.** $\angle 1 \cong \angle 2$. | **5.** Vertical angles are congruent. |
| **6.** $\triangle DCE \cong \triangle BAE$. | **6.** $SAS \cong SAS$. |
| **7.** $\angle 3 \cong \angle 4$. | **7.** cpctc. |
| **8.** $\overline{DC} \parallel \overline{BA}$. | **8.** If two lines are cut by a transversal so that alternate interior angles are congruent, then the lines are parallel. |
| **9.** $\overline{DC} \cong \overline{BA}$. | **9.** cpctc. |
| **10.** $ABCD$ is a $\square$. | **10.** If two opposite sides of a quadrilateral are both parallel and congruent, then the quadrilateral is a parallelogram. $\square$ |

## EXERCISE 4.1

In Exercises 1–4, refer to Illustration 1 in which $\angle 1$ corresponds to $\angle 2$.

**1.** If $\angle 1 \cong \angle 2$, must quadrilateral $ABCD$ be a parallelogram? Explain.

**2.** If $\triangle AOB \cong \triangle COD$, must quadrilateral $ABCD$ be a parallelogram? Explain.

**3.** If ∠1 ≅ ∠2 and $\overline{AD} ≅ \overline{CB}$, must quadrilateral *ABCD* be a parallelogram? Explain.

**4.** If ∠1 ≅ ∠2 and $\overline{AB} ≅ \overline{CD}$, must quadrilateral *ABCD* be a parallelogram? Explain.

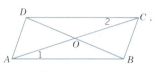

*Illustration 1*

In Exercises 5–8, refer to the parallelogram shown in Illustration 2.

**5.** Find m(∠1).

**6.** Find m(∠C).

**7.** Find m(∠2).

**8.** Find m(∠3).

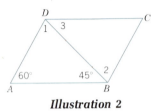

*Illustration 2*

In Exercises 9–16, refer to the parallelogram shown in Illustration 3.

**9.** Find m(∠1).     **10.** Find m(∠2).

**11.** Find m(∠DAB).  **12.** Find m(∠3).

**13.** Find m(∠4).     **14.** Find m(∠5).

**15.** Find m(∠6).     **16.** Find m($\overline{AD}$).

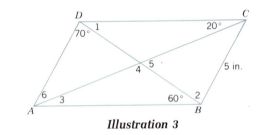

*Illustration 3*

**17.** *Prove Corollary 4.2:*   Opposite sides of a parallelogram are congruent.

**18.** *Prove Corollary 4.3:*   Nonconsecutive angles of a parallelogram are congruent.

**19.** *Prove Theorem 4.4:*   Consecutive angles of a parallelogram are supplementary.

**20.** *Prove Theorem 4.7:*   If two opposite sides of a quadrilateral are both parallel and congruent, then the quadrilateral is a parallelogram.

**21.** *Prove:*   If all pairs of consecutive angles of a quadrilateral are supplementary, then the quadrilateral is a parallelogram.

**22.** If the measure of one angle of a parallelogram is twice another, find the measure of each of its angles.

**23.** In Illustration 4, *ABCD* is a parallelogram with points *E* and *F* midpoints of $\overline{AB}$ and $\overline{DC}$, respectively. Prove that *EBFD* is a parallelogram.

**24.** In Illustration 5, $ABCD$ is a parallelogram in which $\overline{LN}$ bisects diagonal $\overline{AC}$ at point $M$. Prove that $\overline{AC}$ bisects $\overline{LN}$.

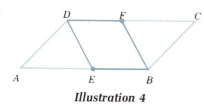

*Illustration 4*

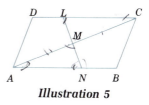

*Illustration 5*

**25.** Refer to Illustration 6.

*Given*: Quadrilateral $ABCD$.
$\overline{AB} \parallel \overline{DC}$.
$\angle A \cong \angle C$.
*Prove*: $ABCD$ is a parallelogram.

**26.** Refer to Illustration 7.

*Given*: Quadrilateral $ABCD$, with its diagonals intersecting at point $M$.
$\angle 1 \cong \angle 2$.
$\overline{AB} \cong \overline{CD}$.
*Prove*: $\overline{MD} \cong \overline{MB}$.

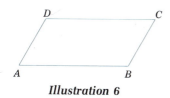

*Illustration 6*

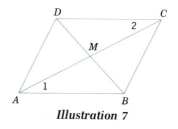

*Illustration 7*

**27.** Refer to Illustration 8.

*Given*: $\square ABCD$ with diagonal $\overline{AC}$.
$\overline{AM} \cong \overline{CN}$.
*Prove*: $MBND$ is a $\square$.

**28.** Refer to Illustration 9.

*Given*: $\square ABCD$.
$\overline{AM} \perp \overline{DB}$.
$\overline{CN} \perp \overline{DB}$.
*Prove*: $AMCN$ is a $\square$.

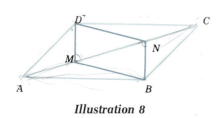

*Illustration 8*

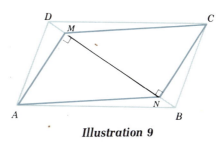

*Illustration 9*

# 4.2 RECTANGLES AND RHOMBUSES

In this section we shall discuss two special parallelograms called **rectangles** and **rhombuses**.

---

**Definition 4.3** A **rectangle** is a parallelogram with one right angle.

---

**Theorem 4.10** All angles of a rectangle are right angles.

---

The proof of Theorem 4.10 is left as an exercise.

---

**Theorem 4.11** The diagonals of a rectangle are congruent.

---

*Given*: Quadrilateral $ABCD$ is a rectangle with diagonals $\overline{DB}$ and $\overline{AC}$.

*Prove*: $\overline{DB} \cong \overline{AC}$.

*Proof*:

| STATEMENTS | REASONS |
|---|---|
| **1.** Quadrilateral $ABCD$ is a rectangle with diagonals $\overline{DB}$ and $\overline{AC}$. | **1.** Given. |
| **2.** $\overline{AB} \cong \overline{DC}$. | **2.** Because a rectangle is a parallelogram, its opposite sides are congruent. |
| **3.** $\angle BAD$ and $\angle CDA$ are right angles. | **3.** All angles of a rectangle are right angles. |
| **4.** $\angle BAD \cong \angle CDA$. | **4.** All right angles are congruent. |
| **5.** $\overline{AD} \cong \overline{DA}$. | **5.** Reflexive law. |

| STATEMENTS | REASONS |
|---|---|
| **6.** $\triangle BAD \cong \triangle CDA$. | **6.** $SAS \cong SAS$. |
| **7.** $\overline{DB} \cong \overline{AC}$. | **7.** cpctc.                            ☐ |

---

**Theorem 4.12** If the diagonals of a parallelogram are congruent, then the parallelogram is a rectangle.

---

*Given*: $ABCD$ is a ☐ with diagonals $\overline{BD}$ and $\overline{CA}$. $\overline{BD} \cong \overline{CA}$.

*Prove*: $ABCD$ is a rectangle.

*Proof*:

| STATEMENTS | REASONS |
|---|---|
| **1.** $ABCD$ is a ☐ with diagonals $\overline{BD}$ and $\overline{CA}$. | **1.** Given. |
| **2.** $\overline{AB} \cong \overline{DC}$. | **2.** Opposite sides of a parallelogram are congruent. |
| **3.** $\overline{AD} \cong \overline{DA}$. | **3.** Reflexive law. |
| **4.** $\overline{BD} \cong \overline{CA}$. | **4.** Given. |
| **5.** $\triangle ABD \cong \triangle DCA$. | **5.** $SSS \cong SSS$. |
| **6.** $\angle BAD \cong \angle CDA$. | **6.** cpctc. |
| **7.** $\angle BAD$ and $\angle CDA$ are supplementary. | **7.** Consecutive angles in a parallelogram are supplementary. |
| **8.** $\angle BAD$ and $\angle CDA$ are right angles. | **8.** If two angles are supplementary and congruent, then they are right angles. |
| **9.** $ABCD$ is a rectangle. | **9.** If a parallelogram has a right angle, then it is a rectangle.        ☐ |

---

**Definition 4.4** A **rhombus** is a parallelogram with two adjacent sides that are congruent.

---

> **Theorem 4.13** All sides of a rhombus are congruent.

The proof of Theorem 4.13 is left as an exercise.

> **Theorem 4.14** The diagonals of a rhombus are perpendicular.

*Given*: $ABCD$ is a rhombus with diagonals $\overline{AC}$ and $\overline{BD}$.

*Prove*: $\overline{AC} \perp \overline{BD}$.

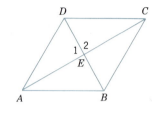

*Proof*:

| STATEMENTS | REASONS |
| --- | --- |
| **1.** $ABCD$ is a rhombus with diagonals $\overline{AC}$ and $\overline{BD}$. | **1.** Given. |
| **2.** $\overline{AE} \cong \overline{CE}$. | **2.** Because a rhombus is a parallelogram, its diagonals bisect each other. |
| **3.** $\overline{ED} \cong \overline{ED}$. | **3.** Reflexive law. |
| **4.** $\overline{AD} \cong \overline{CD}$. | **4.** Adjacent sides of a rhombus are congruent. |
| **5.** $\triangle AED \cong \triangle CED$. | **5.** $SSS \cong SSS$. |
| **6.** $\angle 1 \cong \angle 2$. | **6.** cpctc. |
| **7.** $\angle 1$ is adjacent to $\angle 2$. | **7.** Two angles with a common vertex and a common side between them are adjacent angles. |
| **8.** $\overline{AC} \perp \overline{BD}$. | **8.** If two line segments meet and form congruent adjacent angles, then they are perpendicular. □ |

> **Theorem 4.15** If the diagonals of a parallelogram are perpendicular, then the parallelogram is a rhombus.

*Given:* *ABCD* is a $\square$ with
diagonals $\overline{BD}$ and $\overline{AC}$.
$\overline{BD} \perp \overline{AC}$.

*Prove:* *ABCD* is a rhombus.

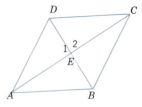

*Proof*:

| STATEMENTS | REASONS |
|---|---|
| **1.** $\overline{BD} \perp \overline{AC}$. | **1.** Given. |
| **2.** $\angle 1 \cong \angle 2$. | **2.** Perpendicular line segments meet and form congruent adjacent angles. |
| **3.** $\overline{ED} \cong \overline{ED}$. | **3.** Reflexive law. |
| **4.** *ABCD* is a $\square$ with diagonals $\overline{BD}$ and $\overline{AC}$. | **4.** Given. |
| **5.** $\overline{AE} \cong \overline{CE}$. | **5.** The diagonals of a parallelogram bisect each other. |
| **6.** $\triangle AED \cong \triangle CED$. | **6.** $SAS \cong SAS$. |
| **7.** $\overline{AD} \cong \overline{CD}$. | **7.** cpctc. |
| **8.** *ABCD* is a rhombus. | **8.** If two adjacent sides of a parallelogram are congruent, then the parallelogram is a rhombus. |

$\square$

---

**Theorem 4.16** The diagonals of a rhombus bisect the angles of the rhombus.

---

The proof of Theorem 4.16 is left as an excercise.

---

**Definition 4.5** A **square** is a rhombus with a right angle.

---

Because a square is both a rhombus and a rectangle, it has all the properties of both a rhombus and a rectangle.

> **Theorem 4.17** If a line joins the midpoints of two sides of a triangle, then the line is parallel to the third side and is half as long as the third side.

*Given*: $\triangle ABC$ with $D$ and $E$ midpoints of $\overline{AC}$ and $\overline{BC}$, respectively.

*Prove*: $\overline{DE} \parallel \overline{AB}$.
  $m(\overline{DE}) = \frac{1}{2}m(\overline{AB})$.

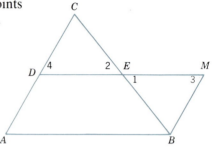

*Construction*: Continue line segment $\overline{DE}$ through point $E$ to point $M$ so that $\overline{DE} \cong \overline{ME}$.

*Proof*:

| STATEMENTS | REASONS |
|---|---|
| **1.** $\triangle ABC$ with $E$ as the midpoint of $\overline{BC}$ and $D$ as the midpoint of $\overline{AC}$. | **1.** Given. |
| **2.** $\overline{CE} \cong \overline{BE}$. | **2.** A midpoint divides a line segment into two congruent line segments. |
| **3.** $\angle 1$ and $\angle 2$ are vertical angles. | **3.** Nonadjacent angles formed by two intersecting line segments are vertical angles. |
| **4.** $\angle 1 \cong \angle 2$. | **4.** Vertical angles are congruent. |
| **5.** $\overline{DE} \cong \overline{ME}$. | **5.** By construction. |
| **6.** $\triangle DEC \cong \triangle MEB$. | **6.** $SAS \cong SAS$. |
| **7.** $\overline{DC} \cong \overline{MB}$. | **7.** cpctc. |
| **8.** $\overline{DC} \cong \overline{DA}$. | **8.** A midpoint divides a line segment into two congruent line segments. |
| **9.** $\overline{MB} \cong \overline{DA}$. | **9.** If two line segments are congruent to the same line segment, then they are congruent to each other. |

| STATEMENTS | REASONS |
|---|---|
| **10.** $\angle 3 \cong \angle 4$. | **10.** cpctc. |
| **11.** $\overline{MB} \parallel \overline{DA}$. | **11.** If two lines are cut by a transversal so that alternate interior angles are congruent, then the line are parallel. |
| **12.** $ABMD$ is a $\square$. | **12.** If opposite sides of a quadrilateral are both congruent and parallel, then the quadrilateral is a parallelogram. |
| **13.** $\overline{DE} \parallel \overline{AB}$. | **13.** Opposite sides of a parallelogram are parallel. |
| **14.** $\overline{DM} \cong \overline{AB}$. | **14.** Opposite sides of a parallelogram are congruent. |
| **15.** $m(\overline{DM}) = m(\overline{AB})$. | **15.** Congruent line segments have equal measures. |
| **16.** $m(\overline{DE}) = \frac{1}{2}m(\overline{DM})$. | **16.** $E$ is the midpoint of $\overline{DM}$ and divides $\overline{DM}$ into two congruent parts each of which is one-half as long as the whole. |
| **17.** $m(\overline{DE}) = \frac{1}{2}m(\overline{AB})$. | **17.** Substitution.     □ |

> **Theorem 4.18** If line segments are drawn connecting the midpoints of the consecutive sides of a quadrilateral, then those segments form a parallelogram.

The proof of Theorem 4.18 is left as an exercise.

## EXERCISE 4.2

In Exercises 1–8, refer to Illustration 1 in which quadrilateral $ABCD$ is a parallelogram.

**1.** If $m(\angle BAD) = 90°$ and $m(\overline{DB}) = 10$ in., find $m(\overline{AC})$.
**2.** If $\overline{AD} \cong \overline{DC}$ and $m(\angle 1) = 50°$, find $m(\angle 2)$.
**3.** If $\overline{AC} \cong \overline{BD}$ and $m(\angle 1) = 40°$, find $m(\angle 3)$.
**4.** If $m(\angle A) = 90°$ and $m(\overline{OB}) = 7$ cm, find $m(\overline{AC})$.
**5.** If $\overline{DB} \perp \overline{AC}$ and $m(\overline{BC}) = 5$ cm, find $m(\overline{DC})$.
**6.** If $\overline{DB} \perp \overline{AC}$ and $m(\angle 1) = 45°$, is $ABCD$ a rectangle? Explain.

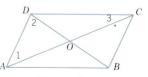

*Illustration 1*

**7.** If $\overline{DB} \perp \overline{AC}$ and m($\angle 1$) = 30°, find m($\angle ADC$).

**8.** If $\overline{AC} \cong \overline{BD}$ and $\overline{AC} \perp \overline{BD}$, is $ABCD$ a square? Explain.

In Exercises 9–12, refer to the triangle in Illustration 2 in which $D$ and $F$ are midpoints of $\overline{AC}$ and $\overline{BC}$.

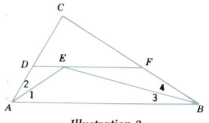

**9.** If m($\overline{DF}$) = 10 m, find m($\overline{AB}$).

**10.** If m($\overline{AD}$) = 5 in. and $\angle 1 \cong \angle 2$, find m($\overline{DE}$).

**11.** If $\angle 1 \cong \angle 2$, $\angle 3 \cong \angle 4$, m($\overline{AD}$) = 2 cm,
and m($\overline{BF}$) = 3 cm, find m($\overline{AB}$).

**12.** If $\overline{AE}$ bisects $\angle A$, $\overline{BE}$ bisects $\angle B$, m($\overline{AD}$) = 3 yd,
and m($\overline{AB}$) = 16 yards, find m($\overline{BF}$).

*Illustration 2*

**13.** *Prove*: A square is a rectangle.

**14.** *Prove Theorem 4.10*: All angles of a rectangle are right angles.

**15.** *Prove Theorem 4.13*: All sides of a rhombus are congruent.

**16.** *Prove Theorem 4.16*: The diagonals of a rhombus bisect the angles of the rhombus.

**17.** *Prove*: A rectangle with two adjacent sides congruent is a square.

**18.** Use the following information and Illustration 3 to prove Theorem 4.18.

> *Given*: $ABCD$ is a quadrilateral with $E$, $F$,
> $G$, and $H$ as midpoints of their respective
> sides.
> *Prove*: $EFGH$ is a $\square$.
> (*Hint*: Draw $\overline{DB}$ and apply Theorem 4.17.)

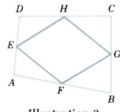

*Illustration 3*

**19.** *Prove*: The bisectors of two consecutive angles of a parallelogram are perpendicular.

**20.** *Prove*: The line segments connecting the midpoints of the sides of an equilateral triangle form another equilateral triangle.

**21.** *Prove*: If two nonadjacent angles of a quadrilateral are right angles, then the bisectors of the other two angles either coincide or are parallel.

**22.** *Prove*: Lines connecting the midpoints of adjacent sides of a rhombus form a rectangle.

**23.** Refer to Illustration 4.

> *Given*: $\triangle ABC$ is an equilateral
> triangle.
> $L$, $M$, and $N$ are the midpoints
> of $\overline{AC}$, $\overline{CB}$, and $\overline{AB}$, respectively.
> *Prove*: $LMBN$ is a rhombus.

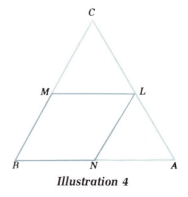

*Illustration 4*

**24.** Refer to Illustration 5.

> *Given*: △*ABC* is a right
> triangle with m(∠*B*) = 90°.
> *L* and *M* are midpoints of its
> two legs.
> *N* is the midpoint of the
> hypotenuse.
> *Prove*: *LBMN* is a rectangle.

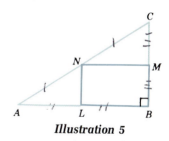

*Illustration 5*

In Exercises 25–26, a carpenter wants to build a square shed. To make sure that the foundation is a square, he makes the given measurements.

**25.** The carpenter measures each side of the foundation and determines that they are of equal length. Can he be sure the foundation is a square? Explain.

**26.** The carpenter measures each diagonal of the foundation and determines that they are of equal length and that they bisect each other. Can he be sure the foundation is a square? Explain.

In Exercises 27–30, a carpenter wants to build a rectangular shed. To make sure that the foundation is "square," she makes the given measurements.

**27.** The carpenter determines that the diagonals of the foundation bisect each other. Can she be sure the foundation is square? Explain.

**28.** The carpenter determines that its opposite sides of the foundation are of equal length and that the diagonals are of equal length. Can she be sure the foundation is square? Explain.

**29.** The carpenter measures each diagonal of the foundation and finds that they are of equal length and that they bisect each other. Can she be sure the foundation is square? Explain.

**30.** The carpenter determines that the diagonals of the foundation bisect each other and that one pair of opposite sides are of equal length. Can she be sure the foundation is square? Explain.

**31.** *Prove*: Lines connecting the midpoints of adjacent sides of a rectangle form a rhombus.

## 4.3 TRAPEZOIDS

In this section we shall discuss a special quadrilateral with exactly two sides parallel.

> **Definition 4.6** A **trapezoid** is a quadrilateral with two, and only two, sides parallel.
> The parallel sides are called **bases** and the nonparallel sides are called **legs**.

Two trapezoids are shown in Figure 4.3.

> **Definition 4.7** An **isosceles trapezoid** is a trapezoid whose legs are congruent.
>    A pair of angles including one of the parallel sides of a trapezoid is called a pair of **base angles**.

An isosceles trapezoid with base angles $\angle 1$ and $\angle 2$ and base angles $\angle 3$ and $\angle 4$ is shown in Figure 4.3.

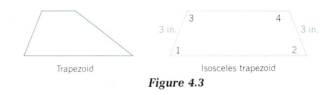

Trapezoid            Isosceles trapezoid

**Figure 4.3**

> **Theorem 4.19** The base angles of an isosceles trapezoid are congruent.

*Given*: Isosceles trapezoid $ABCD$ with $\overline{AB} \parallel \overline{DC}$.

*Prove*: $\angle A \cong \angle B$.

*Construction*: Construct perpendiculars from points $D$ and $C$ to line segment $\overline{AB}$.

*Proof*:

| STATEMENTS | REASONS |
|---|---|
| **1.** $\overline{DE} \perp \overline{AB}$; $\overline{CF} \perp \overline{AB}$ forming right triangles $AED$ and $BFC$. | **1.** By construction. |
| **2.** $ABCD$ is an isosceles. trapezoid with $\overline{AB} \parallel \overline{DC}$. | **2.** Given. |
| **3.** $\overline{AD} \cong \overline{BC}$. | **3.** The nonparallel sides of an isosceles trapezoid are congruent. |

| STATEMENTS | REASONS |
|---|---|
| **4.** $\overline{DE} \cong \overline{CF}$. | **4.** Parallel lines are always the same distance apart. |
| **5.** $\triangle ADE \cong \triangle BCF$. | **5.** $hl \cong hl$. |
| **6.** $\angle A \cong \angle B$. | **6.** cpctc. ☐ |

---

**Definition 4.8** The **median of a trapezoid** is the line segment connecting the midpoints of its legs.

---

**Theorem 4.20** The median of a trapezoid is parallel to the bases and its measure is one-half the sum of the measures of the bases.

*Given*: Trapezoid *ABCD* with median $\overline{EF}$.
$\overline{DC} \parallel \overline{AB}$.

*Prove*: $\overline{EF} \parallel \overline{AB}$.
$\mathrm{m}(\overline{EF}) = \frac{1}{2}[\mathrm{m}(\overline{AB}) + \mathrm{m}(\overline{DC})]$.

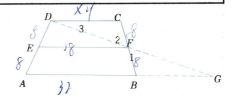

*Construction*: Construct line $\overleftrightarrow{DF}$. Because $\overleftrightarrow{DF}$ cannot be parallel to line $\overleftrightarrow{AB}$, it will intersect the extension of segment $\overline{AB}$ at point *G*.

*Proof*:

| STATEMENTS | REASONS |
|---|---|
| **1.** $\overleftrightarrow{DF}$ intersects line $\overleftrightarrow{AB}$ at point *G*. | **1.** By construction. |
| **2.** *ABCD* is a trapezoid with $\overline{EF}$ as median. | **2.** Given. |
| **3.** *E* is the midpoint of $\overline{AD}$; *F* is the midpoint of $\overline{BC}$. | **3.** Definition of a median of a trapezoid. |
| **4.** $\overline{CF} \cong \overline{BF}$; $\overline{DE} \cong \overline{AE}$. | **4.** A midpoint divides a line segment into two congruent segments. |
| **5.** $\angle 1$ and $\angle 2$ are vertical angles. | **5.** Nonadjacent angles formed by two intersecting line segments are vertical angles. |
| **6.** $\angle 1 \cong \angle 2$. | **6.** Vertical angles are congruent. |

| STATEMENTS | REASONS |
|---|---|
| **7.** $\overline{DC} \parallel \overline{AB}$. | **7.** Given. |
| **8.** $\angle 3 \cong \angle G$. | **8.** If two parallel lines are cut by a transversal, then alternate interior angles are congruent. |
| **9.** $\triangle DCF \cong \triangle GBF$. | **9.** $AAS \cong AAS$. |
| **10.** $\overline{DF} \cong \overline{GF}$. | **10.** cpctc. |
| **11.** $\overline{DC} \cong \overline{GB}$. | **11.** cpctc. |
| **12.** m($\overline{DC}$) = m($\overline{GB}$). | **12.** Congruent line segments have equal measures. |
| **13.** m($\overline{EF}$) = $\frac{1}{2}$m($\overline{AG}$). | **13.** A line connecting the midpoints of two sides of a triangle is one-half the measure of the third side. |
| **14.** m($\overline{AG}$) = m($\overline{AB}$) + m($\overline{GB}$). | **14.** Postulate 1.10. |
| **15.** m($\overline{AG}$) = m($\overline{AB}$) + m($\overline{DC}$). | **15.** Substitution. |
| **16.** m($\overline{EF}$) = $\frac{1}{2}$[m($\overline{AB}$) + m($\overline{DC}$)]. | **16.** Substitution. |
| **17.** $\overline{EF} \parallel \overline{AB}$. | **17.** A line connecting the midpoints of two sides of a triangle is parallel to the third side. $\square$ |

---

**Theorem 4.21** If three or more parallel lines cut off congruent segments on one transversal, then they cut off congruent segments on all transversals.

---

*Given:* $l_1 \parallel l_2 \parallel l_3$.
   $\overline{AB} \cong \overline{BC}$.

*Prove:* $\overline{DE} \cong \overline{EF}$.

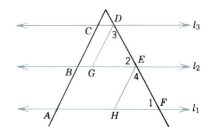

*Construction:* Construct lines parallel to $\overline{AC}$ passing through points *D and E.*

*Proof*:

| STATEMENTS | REASONS |
|---|---|
| **1.** $l_1 \parallel l_2 \parallel l_3$; $\overline{AB} \cong \overline{BC}$. | **1.** Given. |
| **2.** $\overline{GD} \parallel \overline{BC}$; $\overline{HE} \parallel \overline{AB}$. | **2.** By construction. |
| **3.** *BCDG* and *ABEH* are parallelograms. | **3.** If opposite sides of a quadrilateral are parallel, then the quadrilateral is a parallelogram. |
| **4.** $\overline{BC} \cong \overline{GD}$; $\overline{AB} \cong \overline{HE}$. | **4.** Opposite sides of a parallelogram are congruent. |
| **5.** $\overline{GD} \cong \overline{HE}$. | **5.** If two line segments are congruent to congruent line segments, then they are congruent to each other. |
| **6.** $\overline{GD} \parallel \overline{HE}$. | **6.** Two lines parallel to a third line are parallel to each other. |
| **7.** $\angle 1 \cong \angle 2$. | **7.** If two parallel lines are cut by a transversal, then corresponding angles are congruent. |
| **8.** $\angle 3 \cong \angle 4$. | **8.** If two parallel lines are cut by a transversal, then corresponding angles are congruent. |
| **9.** $\triangle GED \cong \triangle HFE$. | **9.** $AAS \cong AAS$. |
| **10.** $\overline{DE} \cong \overline{EF}$. | **10.** cpctc. □ |

Theorem 4.21 provides a method for dividing a line segment into any number of congruent parts. Suppose, for example, that we wish to divide a line segment $\overline{AB}$ into five congruent parts. See Figure 4.4. We first draw any ray $\overrightarrow{AC}$, originating at $A$. We fix a compass at a convenient setting, and copy five congruent segments on the ray $\overrightarrow{AC}$, originating at the initial point $A$. Thus,

$$\overline{AP} \cong \overline{PQ} \cong \overline{QR} \cong \overline{RS} \cong \overline{ST}$$

Draw line $\overleftrightarrow{TB}$ and then construct other lines, parallel to $\overleftrightarrow{TB}$, passing through points $P$, $Q$, $R$, and $S$. The intersection of these lines with the segment $\overline{AB}$ divides $\overline{AB}$ into five congruent parts.

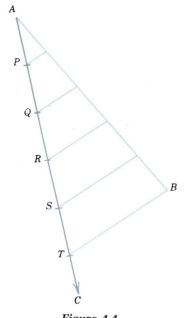

**Figure 4.4**

## EXERCISE 4.3

In Exercises 1–6, refer to Illustration 1 in which $\overline{EF}$ is the median of trapezoid $ABCD$.

**1.** If m($\angle A$) = 60°, find m($\angle D$).

**2.** If m($\angle C$) = 110° and $\overline{AD} \cong \overline{BC}$, find m($\angle A$).

**3.** If m($\overline{AB}$) = 20 cm and m($\overline{DC}$) = 8 cm, find m($\overline{EF}$).

**4.** If m($\overline{DC}$) = 12 m and m($\overline{EF}$) = 20 m, find m($\overline{AB}$).

**5.** If m($\overline{AB}$) = 30 in. and m($\overline{EF}$) = 20 in., find m($\overline{DC}$).

**6.** If m($\overline{DC}$) = $\frac{1}{2}$ m($\overline{AB}$) and m($\overline{EF}$) = 12 cm, find m($\overline{AB}$).

**Illustration 1**

In Exercises 7–12, refer to Illustration 2 in which $\overline{EF}$ is the median of trapezoid $ABCD$.

**7.** If m($\angle D$) = 70°, find m($\angle 1$).

**8.** If m($\angle C$) = 80°, find m($\angle 2$).

**9.** If m($\overline{AB}$) = 8 cm, find m($\overline{EF}$).

**10.** If m($\overline{EF}$) = 13 cm, find m($\overline{AB}$).

**11.** If m($\angle D$) = 75°, find m($\angle 3$).

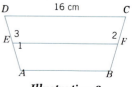

**Illustration 2**

**12.** If m($\angle C$) = 85° and $\overline{AD} \cong \overline{BC}$, find m($\angle A$).

**13.** *Prove*: If the base angles of a trapezoid are congruent, then the trapezoid is an isosceles trapezoid.

**14.** *Prove*: The line segment connecting the midpoints of the bases of an isosceles trapezoid is perpendicular to the bases.

**15.** *Prove*: The figure formed by connecting the midpoints of the sides of an isosceles trapezoid is a rhombus.

**16.** Refer to Illustration 3.

*Given*: Quadrilateral $ABCD$ is a $\square$. Points $F$, $G$, $H$, and $I$ are the midpoints of $\overline{DE}$, $\overline{AE}$, $\overline{BE}$, and $\overline{CE}$, respectively.

*Prove*: Quadrilateral $FGHI$ is a $\square$.

**17.** Refer to Illustration 4.

*Given*: $\triangle ABC$ is an equilateral triangle. Points $E$ and $F$ are midpoints of $\overline{AC}$ and $\overline{BC}$, respectively. $G$ is the midpoint of $\overline{AE}$. $H$ is the midpoint of $\overline{BF}$.

*Prove*: m($\overline{GH}$) = $\frac{3}{4}$m($\overline{AB}$).

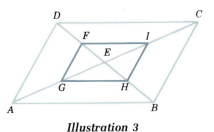

**Illustration 3**

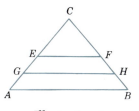

**Illustration 4**

**18.** Draw a line segment about 3 in. long. Using only a compass and a straightedge, divide the segment into seven congruent parts.

## CHAPTER FOUR REVIEW EXERCISES

In Review Exercises 1–4, refer to the quadrilateral shown in Illustration 1.

**1.** If m($\overline{DC}$) = 8 cm, must $ABCD$ be a parallelogram? Explain.

**2.** If $\overline{AD} \parallel \overline{BC}$, must $ABCD$ be a parallelogram? Explain.

**3.** If $\angle 1 \cong \angle 2$, find m($\overline{DE}$).

**4.** If $\angle 1 \cong \angle 2$ and $\overline{DB} \perp \overline{AC}$, find m($\overline{BC}$).

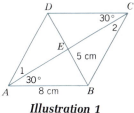

**Illustration 1**

In Review Exercises 5–8, refer to the parallelogram shown in Illustration 2.

**5.** Find m($\angle 1$).

**6.** Find m($\angle 2$).

**7.** Find m($\angle D$).

**8.** Find m($\angle 3$).

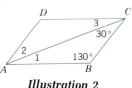

**Illustration 2**

**9.** Refer to Illustration 3.

*Given*: □*ABCD*.
    $\overline{DF} \cong \overline{BE}$.
*Prove*: Quadrilateral *AECF* is a □.

**10.** Refer to Illustration 4.

*Given*: □ *ABCD* with *E*, *F*, *G*,
    and *H* as midpoints of
    their respective sides.
*Prove*: $\overline{HG} \cong \overline{FE}$.

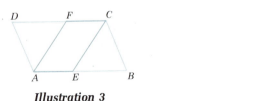

***Illustration 3***

***Illustration 4***

In Review Exercises 11–18, refer to the parallelogram shown in Illustration 5, and tell whether each statement is true. If a statement is false, explain why.

**11.** If $\overline{DB} \cong \overline{AC}$, then parallelogram *ABCD* is a rectangle.
**12.** If $\angle 1 \cong \angle 2$, then parallelogram *ABCD* is a rectangle.
**13.** If $\angle 1 \cong \angle 3$, then parallelogram *ABCD* is a rectangle.
**14.** If $\angle BAD \cong \angle CDA$, then parallelogram *ABCD* is a rectangle.
**15.** If $\overline{DB} \perp \overline{AC}$, then parallelogram *ABCD* is a rhombus.
**16.** If $\angle 1 \cong \angle 2$, then parallelogram *ABCD* is a rhombus.
**17.** If $\overline{AB} \cong \overline{BC}$, then parallelogram *ABCD* is a square.
**18.** If $\overline{DB} \perp \overline{AC}$ and $\angle 1 \cong \angle 3$, then parallelogram *ABCD* must be a square.

***Illustration 5***

In Review Exercises 19–22, refer to Illustration 6 in which *E* is the midpoint of $\overline{AC}$ and *D* is the midpoint of $\overline{BC}$.

**19.** Find m($\overline{ED}$).
**20.** If m($\angle A$) = 65°, find m($\angle 1$).
**21.** Find the perimeter of $\triangle ABC$.
**22.** If m($\angle 2$) = 130°, find m($\angle C$).

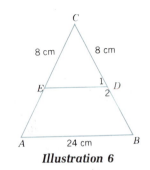

***Illustration 6***

In Review Exercises 23–26, refer to the trapezoid shown in Illustration 7, in which $\overline{EF}$ is the median of the trapezoid.

**23.** If m($\overline{AB}$) = 12 m and m($\overline{DC}$) = 8 m, find m($\overline{EF}$).

**24.** If m($\overline{AB}$) = 22 ft and m($\overline{EF}$) = 16 ft, find m($\overline{DC}$).

**25.** If m($\angle C$) = 120° and $\overline{AD} \cong \overline{BC}$, find m($\angle A$).

**26.** If m($\overline{AB}$) = 32 in., m($\overline{EF}$) = 18 in., m($\overline{AE}$) = 8 in., and $\angle A \cong \angle B$, find the perimeter of the trapezoid.

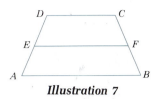

***Illustration 7***

In Review Exercises 27–30, refer to Illustration 8, in which $l_1 \parallel l_2 \parallel l_3$.

**27.** If m($\overline{AB}$) = m($\overline{BC}$) = 6 cm and m($\overline{FE}$) = 7 cm, find m($\overline{ED}$).

**28.** If m($\overline{AC}$) = 12 ft and $\overline{FE} \cong \overline{ED}$, find m($\overline{AB}$).

**29.** If quadrilateral $ABEF$ is an isosceles trapezoid and m($\angle 1$) = 70°, find m($\angle 2$).

**30.** If quadrilateral $ABEF$ is an isosceles trapezoid and m($\angle 1$) = 70°, find m($\angle 3$).

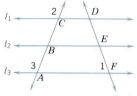

***Illustration 8***

**31.** On a piece of paper draw a line segment about 3 in. long. Then use a compass and a straightedge to divide it into five congruent pieces.

## CHAPTER FOUR TEST

In Exercises 1–20, classify each statement as true or false.

**1.** A parallelogram is always a quadrilateral.

**2.** The diagonal of a quadrilateral divides the quadrilateral into two congruent triangles.

**3.** Consecutive angles in a parallelogram are congruent.

**4.** If one pair of sides of a quadrilateral are congruent and the other pair are parallel, then the quadrilateral is a parallelogram.

**5.** A parallelogram must have two obtuse angles.

**6.** A parallelogram is a special case of a trapezoid.

**7.** The diagonals of a parallelogram bisect each other.

**8.** The diagonals of a rhombus are perpendicular.

**9.** The diagonals of a rhombus are congruent.

**10.** If the diagonals of a quadrilateral bisect each other, the quadrilateral is a trapezoid.

**11.** If one pair of angles in a trapezoid are congruent, then the other pair of angles are congruent.

**12.** The median of a trapezoid with bases of length 10 and 22 in. has a length of 16 in.

**13.** If a trapezoid has a base 15 in. long and a median that is 23 in. long, the length of its other base is 19 in.

**14.** The line connecting the midpoints of the bases of a trapezoid is perpendicular to each base.

**15.** If the diagonals of a quadrilateral are congruent, then it is a rectangle.

**16.** A rhombus cannot be a trapezoid.

**17.** If all of the angles of a rhombus are congruent, the rhombus is a square.

**18.** Line segments connecting the midpoints of the opposite sides of a quadrilateral bisect each other.

**19.** The line segment connecting the midpoints of the parallel sides of a trapezoid is perpendicular to the median of the trapezoid.

**20.** If a line segment $\overline{AB}$ joins the midpoints of two sides of a triangle with a base 25 m long, the length of $\overline{AB}$ is also 25 m.

# 5

*Pythagoras (about 584–495 B.C.). Pythagoras was a Greek philosopher and mathematician, and the leader of a secret society that believed in numerology. Although Pythagoras is most famous for the theorem that bears his name, he is sometimes called the "father of music." The Pythagoreans are credited with the discovery of the fundamentals of musical harmony. They realized the mathematical relationship between the length of a vibrating string and its pitch. They are also credited with the discovery of irrational numbers.*

# Areas and Related Topics

We now apply the results from previous chapters to develop the basic formulas for finding areas of certain polygons. Technically, a polygon does not have an area, but a polygon does enclose an area. However, it is common to speak of the area of a polygon, and we will do so in this chapter.

## 5.1 AREAS OF POLYGONS

The **area of a polygon** is a measure of the amount of surface that is enclosed by the polygon. Knowing how to find the area of a rectangle, for example, would be helpful in estimating the cost of carpeting a rectangular floor. Knowing how to find the area of a triangle would enable a sailboat manufacturer to estimate the amount of sailcloth needed for its sail.

A scale drawing can illustrate the area of a rectangle. For example, each square in Figure 5.1 represents an area of one square foot—a surface enclosed by a square one foot on each side. It is easy to count the number of squares bounded by the larger rectangle: There are 7 rows with 10 squares in each row, for a total of 70. The area of the rectangle in Figure 5.1 is 70 square feet. This is often written as 70 ft², read as 70 square feet.

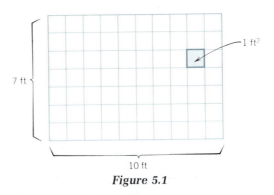

**Figure 5.1**

Because of the slanting sides of the triangle in Figure 5.2, it is more difficult to find its area by counting squares. If we count the squares that are complete and then approximate the area of the fractional squares, we can estimate the total area to be about 35 ft².

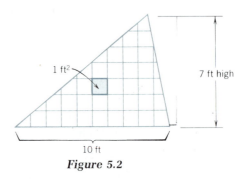

**Figure 5.2**

Fortunately, we do not need to find areas by counting squares because there are several formulas that enable us to calculate areas easily. Several of them are developed in this chapter.

Definition 1.20 explained what is meant by the *altitude* of a triangle. There are similar definitions for the altitudes of many other polygons.

> ***Definition 5.1*** An **altitude of a parallelogram** is a line segment drawn from a vertex of the parallelogram perpendicular to a nonadjacent side, or its extension.
>
> The length of an altitude is called a **height**, and the side to which it is drawn is called a **base**.

Three altitudes of $\square ABCD$ appear in Figure 5.3. They are line segments $\overline{DQ}$, $\overline{CP}$, and $\overline{DR}$. Altitude $\overline{DQ}$ is drawn to base $\overline{AB}$, altitude $\overline{CP}$ is drawn to the extension of $\overline{AB}$, and altitude $\overline{DR}$ is drawn to the extension of $\overline{BC}$.

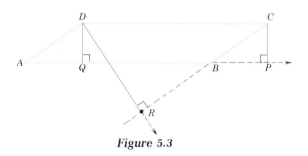

***Figure 5.3***

Refer to the rectangle of Figure 5.1. Recall that it has an area of 70 ft² because there are 7 rows of 10 squares each, and 7 times 10 is 70. This fact suggests a formula for finding the area of a rectangle.

> ***Postulate 5.1*** The **area, *A*, of a rectangle** is given by the formula
>
> $$A = bh$$
>
> where *b* is the length of the base and *h* is the height of the rectangle.

**EXAMPLE 1**    Find the area of rectangle $ABCD$ shown in Figure 5.4.

Solution    From the figure, we see that the length *b* of the base $\overline{AB}$ is 12 units and the height *h* (the length of the altitude $\overline{DA}$) is 8 units. The area is found by using the formula $A = bh$ and proceding as follows:

$$A = bh \qquad \text{The formula for the area of a rectangle.}$$
$$A = 12 \cdot 8 \qquad \text{Substitute 12 for } b \text{ and 8 for } h.$$
$$= 96 \qquad \text{Simplify.}$$

The area of the rectangle is 96 square units.

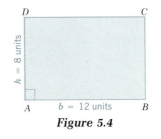

**Figure 5.4**

∎

**EXAMPLE 2**

The combined living-room/dining room area of a house has the floor plan shown in Figure 5.5. At $29 per yd², how much will it cost to carpet the area? Assume there is no waste and that all walls meet at right angles.

Solution

To find the cost of carpeting, we must first find the area of the room. Although the room is not a rectangle, the line segment $\overline{CF}$ divides the room into two rectangles. The area of the living room is the product of the length of its base 4 and its height 7. Thus, we have

Area of living room $= b \cdot h$
$= 4 \cdot 7$
$= 28$

The area of the living room is 28 yd².

The area of the dining room is the product of the length of its base and its height. The base $\overline{FE}$ is 9 yards minus 4 yards, which is 5 yards. The height $\overline{DE}$ is 4 yards. Thus, we have

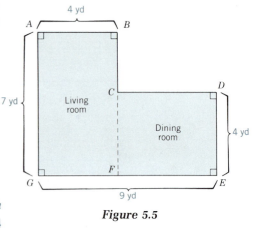

**Figure 5.5**

Area of dining room $= b \cdot h$
$= 5 \cdot 4$
$= 20$

The area of the dining room is 20 yd². The total area to be carpeted is 28 yd² + 20 yd², or 48 yd². At $29 for each square yard, it will cost 48 · $29, or $1392 to carpet the entire area. ∎

To simplify writing proofs of area formulas for other polygons, we shall use the notation $A(\triangle ABC)$ to represent the area of triangle $ABC$. Similarly, $A(\square ABCD)$ represents the area of parallelogram $ABCD$. In general, the notation $A(\quad)$ represents the area of the polygon that is named within the parentheses. Three more ideas are required before we proceed.

---

**Postulate 5.2**  Any two congruent figures have the same area.

**Postulate 5.3  The distributive law.**   If $a$, $b$, and $c$ are real numbers, then

$$a(b + c) = ab + ac$$

**Postulate 5.4  The additive properties of areas.**   If line segments divide a given area into smaller areas, then the given area is equal to the sum of the smaller areas.

**EXAMPLE 3**

The additive property of areas points out that a large area is the sum of all of the smaller areas that it might contain. In Figure 5.6, the area of quadrilateral $ABCD$ is the sum of the areas of the two triangles, $\triangle ABC$ and $\triangle ADC$. Thus,

$$A(ABCD) = A(\triangle ABC) + A(\triangle ADC)$$

Similarly, in Figure 5.7, the area of the trapezoid $ABCD$ is the sum of the areas of three polygons—$\triangle ABM$, $\triangle DCN$, and rectangle $AMND$.

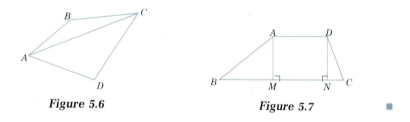

**Figure 5.6**                    **Figure 5.7**          ■

**Theorem 5.1**  The **area**, $A$, **of a parallelogram** is given by the formula

$$A = bh$$

where $b$ is the length of a base and $h$ is the corresponding height of the parallelogram.

*Given*: $ABCD$ is a parallelogram.

*Prove*: $A(\square\,ABCD) = bh$.

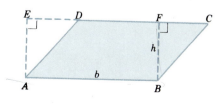

*Construction*: Construct a perpendicular from point $A$ to line segment $\overline{DC}$ extended and a perpendicular from point $B$ to line segment $\overline{DC}$.

*Proof*:

| STATEMENTS | REASONS |
|---|---|
| 1. $ABCD$ is a $\square$. | 1. Given. |
| 2. $\overline{AD} \cong \overline{BC}$. | 2. Opposite sides of a parallelogram are congruent. |
| 3. $\overline{AB} \parallel \overline{DC}$. | 3. Opposite sides of a parallelogram are parallel. |
| 4. $\overline{AE} \perp \overline{DC}$; $\overline{BF} \perp \overline{DC}$; triangles $AED$ and $BFC$ are right triangles. | 4. By construction. |
| 5. $\overline{AE} \cong \overline{BF}$. | 5. Parallel lines are always the same distance apart. |
| 6. $\triangle ADE \cong \triangle BCF$. | 6. $hl \cong hl$. |
| 7. $A(\square\,ABCD) = A(\triangle BCF) + A(ABFD)$. | 7. The additive property of areas. |
| 8. $A(\triangle BCF) = A(\triangle ADE)$. | 8. Congruent figures have equal areas. |
| 9. $A(\square\,ABCD) = A(\triangle ADE) + A(ABFD)$. | 9. Substitution. |
| 10. $A(ABFE) = A(\triangle ADE) + A(ABFD)$. | 10. The additive property of areas. |
| 11. $A(\square\,ABCD) = A(ABFE)$. | 11. If two quantities are equal to the same quantity, then they are equal to each other. |
| 12. $\overline{AE} \parallel \overline{BF}$. | 12. Because $\overline{AE}$ and $\overline{BE}$ are each perpendicular to $\overline{DC}$, they are parallel. |
| 13. $ABFE$ is a parallelogram. | 13. If opposite sides of a quadrilateral are parallel, then the quadrilateral is a parallelogram. |
| 14. $ABFE$ is a rectangle. | 14. A parallelogram with one right angle is a rectangle. |
| 15. $A(ABFE) = bh$. | 15. Postulate 5.1. |
| 16. $A(\square\,ABCD) = bh$. | 16. Substitution.     □ |

*EXAMPLE 4*

*Solution*

Find the area of the parallelogram shown in Figure 5.8.

Do not be misled by the 18-ft measure of side $\overline{DA}$. The area of the parallelogram is the product of the length of its base and its height. The height is the length of the altitude $\overline{AE}$, given in the figure as 15 ft. The length of the base $\overline{DC}$ is given as 30 ft.

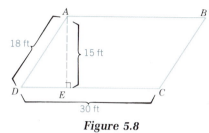

**Figure 5.8**

$$A = bh \qquad \text{The formula for the area of a } \square.$$
$$A = 30 \cdot 15 \qquad \text{Substitute 30 for } b \text{ and 15 for } h.$$
$$\phantom{A} = 450$$

The area of the parallelogram is 450 ft².                 ∎

Theorem 5.2 provides a formula for finding the area of a triangle.

---

**Theorem 5.2** The **area**, *A*, **of a triangle** is given by the formula

$$A = \frac{1}{2}bh$$

where *b* is the length of a base and *h* is the corresponding height of the triangle.

---

*Given*: $\triangle ABC$.

*Prove*: $A(\triangle ABC) = \dfrac{1}{2}bh.$

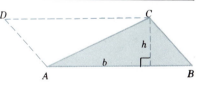

*Construction*: Construct a line through point *C* parallel to line segment $\overline{AB}$ and construct a line through point *A* parallel to line segment $\overline{BC}$.

*Proof*:

| STATEMENTS | REASONS |
|---|---|
| **1.** $\overline{AB} \parallel \overline{DC}$; $\overline{AD} \parallel \overline{BC}$. | **1.** By construction. |
| **2.** $ABCD$ is a $\square$. | **2.** If opposite sides of a quadrilateral are parallel, then the quadrilateral is a parallelogram. |

| STATEMENTS | REASONS |
|---|---|
| **3.** $\triangle ABC \cong \triangle CDA$. | **3.** The diagonal of a parallelogram divides it into two congruent triangles. |
| **4.** $A(\triangle ABC) = A(\triangle CDA)$. | **4.** Congruent figures have equal areas. |
| **5.** $A(\square\ ABCD) = A(\triangle ABC) + A(\triangle CDA)$. | **5.** The additive property of areas. |
| **6.** $A(\square\ ABCD) = A(\triangle ABC) + A(\triangle ABC) = 2A(\triangle ABC)$. | **6.** Substitution and simplification. |
| **7.** $A(\square\ ABCD) = bh$. | **7.** Theorem 5.1. |
| **8.** $2A(\triangle ABC) = bh$. | **8.** Substitution. |
| **9.** $A(\triangle ABC) = \dfrac{1}{2}bh$. | **9.** If equal quantities are divided by equal quantities, the results are equal.    □ |

**EXAMPLE 5**

Find the area of a triangle with a base of 5 in. and a height of 8 in. See Figure 5.9.

Solution

By Theorem 5.2, the area $A$ is found by the formula

$$A = \frac{1}{2}bh$$

In the triangle, $b = 5$ and $h = 8$. Thus,

$$A = \frac{1}{2}bh$$

$$= \frac{1}{2}\cdot 5 \cdot 8$$

$$= 20$$

Thus, the area of the triangle is 20 in².

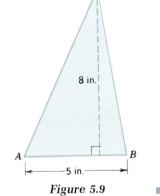

*Figure 5.9*    ■

**EXAMPLE 6**

The area of a certain triangle is 52 ft² and the height is 13 ft. Find the length of the base of a triangle. See Figure 5.10.

Solution

Since we wish to find the length of the base $b$, let us solve the equation

$$A = \frac{1}{2}bh$$

for *b*. By multiplying both sides of the equation by 2 and dividing by *h*, we have

$$A = \frac{1}{2}bh$$

$$2A = bh$$

(1) $$\frac{2A}{h} = b$$

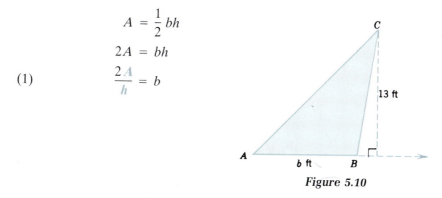

**Figure 5.10**

Because we are given that $A = 52$ and $h = 13$, we substitute these values into Equation 1 to obtain

$$\frac{(2)(52)}{13} = b$$

$$8 = b$$

Thus, the measure of the base of the triangle is 8 ft. ∎

An **altitude of a trapezoid** is a line segment drawn from a vertex of the trapezoid to the nonadjacent parallel side or its extension. The length of an altitude is called the **height of the trapezoid**.

---

**Theorem 5.3** The **area, *A*, of a trapezoid** is given by the formula

$$A = \frac{1}{2}h(b + b')$$

where *h* is the height and *b* and *b'* are the lengths of the bases of the trapezoid.

---

*Given*: Trapezoid *ABCD*.

*Prove*: $A(ABCD) = \frac{1}{2}h(b + b')$.

*Construction*: Draw diagonal $\overline{DB}$.

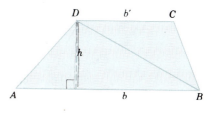

*Proof*:

| STATEMENTS | REASONS |
|---|---|
| **1.** *ABCD* is a trapezoid. | **1.** Given. |
| **2.** $\overline{DB}$ is a diagonal of trapezoid *ABCD*. | **2.** By construction. |
| **3.** $A(ABCD) = A(\triangle ABD) + A(\triangle BCD)$. | **3.** By the additive property of areas. |
| **4.** $A(\triangle ABD) = \dfrac{1}{2} bh$. | **4.** Theorem 5.2. |
| **5.** $A(\triangle BCD) = \dfrac{1}{2} b'h$. | **5.** Theorem 5.2. |
| **6.** $A(ABCD) = \dfrac{1}{2} bh + \dfrac{1}{2} b'h$. | **6.** The additive property of areas. |
| **7.** $A(ABCD) = \dfrac{1}{2} h(b + b')$. | **7.** Use the distributive property to factor out $\frac{1}{2}h$.     ☐ |

**EXAMPLE 7**

The trapezoid shown in Figure 5.11 has bases measuring 6 and 10 in., and a height of 1 ft. Find its area.

**Solution**

By Theorem 5.3, the area of a trapezoid is given by the formula

$$A = \frac{1}{2} h(b + b')$$

In this example, $b = 6$ and $b' = 10$. It would be incorrect to say that $h = 1$ because the height of 1 must be expressed in inches to be consistent with the units of the bases. Thus, $h = 12$. Substituting in the above formula, we have

$$A = \frac{1}{2} h(b + b')$$

$$A = \frac{1}{2} \ 12)(6 + 10)$$

$$= \frac{1}{2} (12)(16)$$

$$= 6(16)$$

$$= 96$$

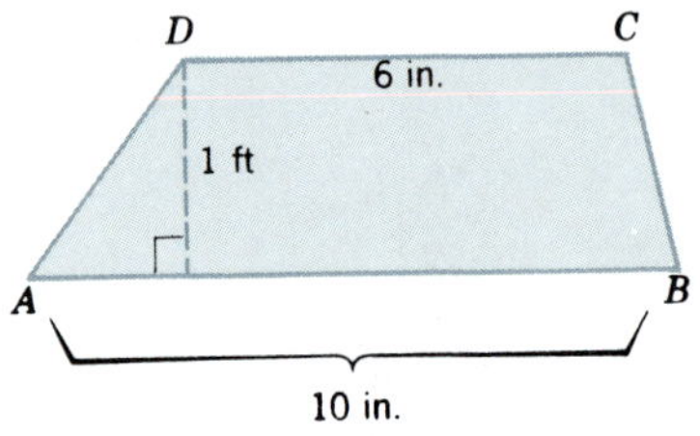

*Figure 5.11*

The area of the trapezoid is 96 in². ■

> **Theorem 5.4** The **area of a rhombus** is equal to one-half the product of the measures of its diagonals.

*Given*: $ABCD$ is a rhombus with diagonals $\overline{AC}$ and $\overline{BD}$.

*Prove*: $A(ABCD) = \dfrac{1}{2}[m(\overline{BD})m(\overline{AC})]$.

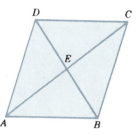

*Proof*:

| STATEMENTS | REASONS |
|---|---|
| 1. $ABCD$ is a rhombus with diagonals $\overline{AC}$ and $\overline{BD}$. | 1. Given. |
| 2. $\overline{AC} \perp \overline{BD}$. | 2. The diagonals of a rhombus are perpendicular. |
| 3. $A(ABCD) = A(\triangle ABD) + A(\triangle CBD)$. | 3. The additive property of areas. |
| 4. $A(\triangle ABD) = \frac{1}{2}[m(\overline{BD})m(\overline{AE})]$. | 4. Theorem 5.2. |
| 5. $A(\triangle CBD) = \frac{1}{2}[m(\overline{BD})m(\overline{EC})]$. | 5. Theorem 5.2. |
| 6. $A(ABCD) = \frac{1}{2}[m(\overline{BD})m(\overline{AE})] + \frac{1}{2}[m(\overline{BD})m(\overline{EC})]$. | 6. Substitution. |
| 7. $A(ABCD) = \frac{1}{2}[m(\overline{BD})][m(\overline{AE} + \overline{EC})]$. | 7. Use the distributive property to factor out $\frac{1}{2}[m(\overline{BD})]$. |
| 8. $m(\overline{AE}) + m(\overline{EC}) = m(\overline{AC})$. | 8. Postulate 1.10. |
| 9. $A(ABCD) = \dfrac{1}{2}[m(\overline{BD})m(\overline{AC})]$. | 9. Substitution. |

□

**EXAMPLE 8**    The wall at one end of an attic room has the shape of a trapezoid because of a slanted ceiling. The wall is 8 ft high at one end, 10 ft wide, and 3 ft high at the other end. Find the area of the wall.

Solution    The wall forms the shape of a trapezoid with bases of 8 ft and 3 ft, and a height of 10 ft. See Figure 5.12. Substituting into the formula for the area of a trapezoid, we have

$$A = \frac{1}{2} h(b + b')$$

$$= \frac{1}{2} (10)(8 + 3)$$

$$= \frac{1}{2} (10)(11)$$

$$= 55$$

The area of the wall is 55 ft².

8 ft

3 ft

10 ft

Figure 5.12

## EXERCISE 5.1

In Exercises 1–20, find the shaded area of each figure.

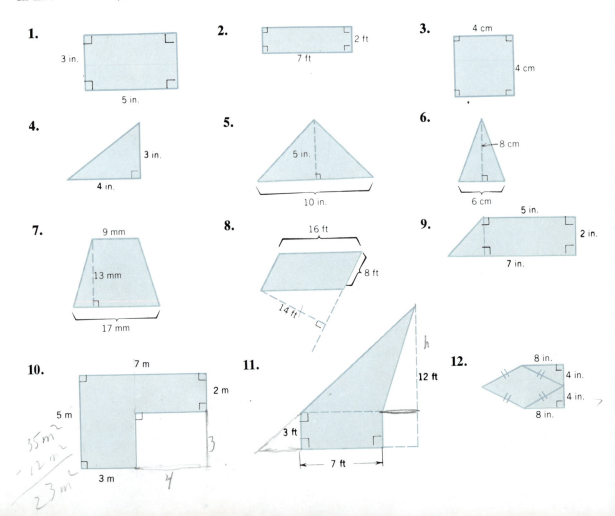

1.    3 in.    5 in.

2.    2 ft    7 ft

3.    4 cm    4 cm

4.    3 in.    4 in.

5.    5 in.    10 in.

6.    8 cm    6 cm

7.    9 mm    13 mm    17 mm

8.    16 ft    8 ft    14 ft

9.    5 in.    2 in.    7 in.

10.    7 m    2 m    5 m    3 m

35m²
12 m
3 m

11.    12 ft    h    3 ft    7 ft

12.    8 in.    4 in.    4 in.    8 in.

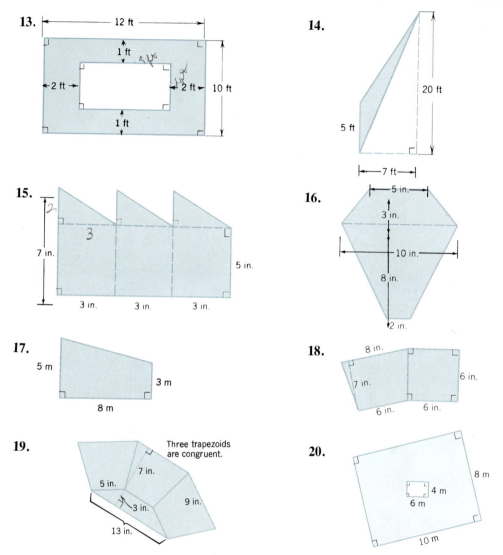

**21.** Each square in Illustration 1 has a side that is one-half as long as the side of the next larger square. Let $s$ represent the length of the smallest side. Find a formula for the total area.

**22.** In Illustration 2, the largest of the three isosceles triangles has a base of length $s$ cm. The congruent sides of each triangle are $\frac{2}{3}$ as long as the length of the triangle's base. Find the perimeter of polygon $ABCDE$.

*Illustration 1*                    *Illustration 2*

**23.** If the side of a square is doubled, what change occurs in its area?

**24.** How many altitudes does a parallelogram have? How many heights? How many altitudes does a trapezoid have? How many heights?

**25.** A rhombus has diagonals measuring 16 in. and 18 in. Find its area.

**26.** A square has a diagonal measuring 7 in. Find its area.

**27.** A square has a diagonal of $3\sqrt{2}$ cm. Find its perimeter.

**28.** A square has a perimeter of $12\sqrt{5}$ m. Find its area.

**29.** A living room has the shape of a rectangle measuring 6 yd by 4 yd. At \$42 per $yd^2$, how much will it cost to carpet the room? Neglect any waste.

**30.** A basement recreation room is in the form of a rectangle measuring 14 ft by 20 ft. Vinyl floor tiles are 1 $ft^2$ and cost \$1.29 each. How much will the tile cost to cover the floor? Neglect any waste.

**31.** The north wall of a barn is in the shape of a rectangle 23 ft high and 72 ft long. There are five windows in the wall, each one in the shape of a rectangle with dimensions of 4 ft by 6 ft. If a gallon of paint will cover 300 $ft^2$ and the wall will need two coats, find how many gallons of paint will be needed to paint the wall.

**32.** A picture window is in the shape of a triangle with a base 5 ft long and with a height of 65 in. Find the area of the window.

**33.** The gable end of a warehouse is in the shape of an isosceles triangle with a height of 4 yd and a base of 23 yd. It will require one coat of primer and one coat of paint to paint the triangular surface. Primer costs \$17 per gallon and paint costs \$23 per gallon. Find the cost of primer and paint needed to paint the gable end of the warehouse. Assume that one gallon covers 300 $ft^2$.

**34.** A swimming pool is in the shape of a trapezoid with the dimensions shown in Illustration 3. Find how many square meters of plastic sheeting are needed to cover the pool. Assume no waste.

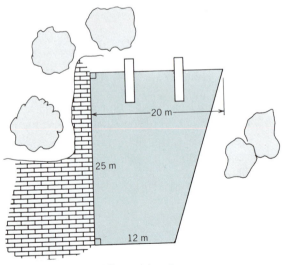

*Illustration 3*

**35.** The pattern of a tiled kitchen floor contains the eight congruent trapezoids with dimensions as shown in Illustration 4. If the total area is 162 in.², find the value of $x$.

**36.** The isosceles trapezoid shown in Illustration 5(a) is folded to form the rectangle shown in Illustration 5b. Find the area of rectangle $ABCD$.

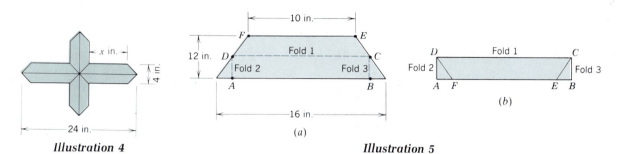

*Illustration 4*

*Illustration 5*

**37.** Refer to Illustration 6.

    *Given*: $\triangle ABC$ with $\overline{DG}$ intersecting $\overline{AB}$ and $\overline{CB}$ at midpoints $E$ and $F$. $DACG$ is a rectangle.

    *Prove*: $A(DACG) = A(\triangle ABC)$.

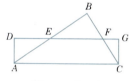

*Illustration 6*

**38.** Refer to Illustration 7.

    *Given*: $\square ABCD$.
              Point $E$ is between $B$ and $C$.

    *Prove*: $A(\triangle AED) = \dfrac{1}{2} A(\square ABCD)$.

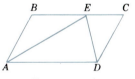

*Illustration 7*

**39.** *Given*: $\triangle ABC$ and $\triangle ABD$ shown in Illustration 8 have equal areas.
      *Prove*: $\overline{CD}$ is bisected by $\overline{AB}$.

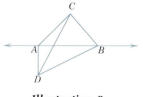

*Illustration 8*

**40.** One base of a trapezoid is 5 in. longer than the other base. The height of the trapezoid is 8 in., and the area of the trapezoid is 100 in². Find the length of the shorter base.

**41.** The length of one diagonal of a rhombus exceeds the other by 8 ft, and the area is 24 ft². Find the sum of the lengths of the diagonals.

**42.** *Prove*: The area of an isosceles right triangle is one-fourth the square of the length of the hypotenuse.

**43.** *Prove*: The median of a triangle divides the triangle into two triangles with equal areas.

**44.** *Prove*: The diagonals of a parallelogram divide it into four nonoverlapping triangles with equal areas.

**45.** *Prove*: The area of a trapezoid is given by the formula $A = mh$, where $h$ is the height of the trapezoid and $m$ is the length of the median.

**46.** The area of a right triangle is one-half the product of the lengths of its legs.

## 5.2 AREAS OF CIRCLES

In this section we shall discuss how to find areas of circles.

> ***Definition 5.2*** A **circle** is the set of points in the same plane that are equidistant from a fixed point called its **center**.

> ***Definition 5.3*** A **diameter of a circle** is a line segment that passes through the center of the circle and with its endpoints on the circle.
>    A **radius of a circle** is a line segment drawn from the center of the circle to a point on the circle.

In the circle shown in Figure 5.13, $O$ is the center of the circle, $\overline{AB}$ is a diameter, and $\overline{OC}$, $\overline{OA}$, and $\overline{OB}$ are radii (plural of radius) of the circle. From Figure 5.13, we see that a diameter of a circle is twice as long as a radius.

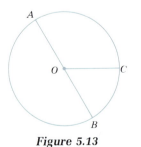

*Figure 5.13*

---

*Definition 5.4* The **circumference of a circle** is the distance around the circle.

---

Since early history people knew that the ratio of the circumference of a circle to the length of its diameter is always the same number. This number is **pi** $(\pi)$, a constant with a value of approximately 3.14. Computers have computed $\pi$ to millions of decimal places. Its actual value is

$$\pi = 3.141592653589\ldots$$

The three dots following the digits, called the **ellipsis**, indicate that $\pi$ is an unending decimal.

Because $\pi$ is the ratio of the circumference $C$ of a circle to the length of its diameter $d$, we have

(1)
$$\pi = \frac{C}{d}$$

If we multiply both sides of Equation 1 by $d$, we have a formula for finding the circumference $C$ of a circle:

$$C = \pi d$$

Because a diameter of a circle is twice as long as its radius, this formula is often written as

$$C = 2\pi r$$

where $r$ is the length of a radius of the circle.

The constant $\pi$ also appears in the formula for the area of a circle. Although the following argument is not a proof, it does make the formula for the area of a circle plausible.

---

*Theorem 5.5* The **area**, $A$, **of a circle** is given by the formula

$$A = \pi r^2$$

where $r$ is the radius of the circle.

---

Consider the circle $O$ in Figure 5.14a. If we divide the circle into an even number of congruent pie-shaped pieces and regroup them as shown in Figure 5.14b, we obtain a figure that resembles a parallelogram. Its base is approximately one-half of the circumference of the circle and its height is approximately equal to the radius of the circle.

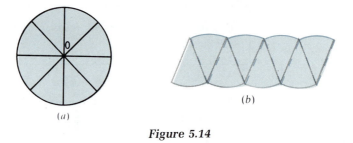

*(a)*                 *(b)*

**Figure 5.14**

As the number of pie-shaped pieces increases, Figure 5.14*b* will look even more like a parallelogram. The length of its base will be much closer to one-half the circumference of the circle, and the height will be much closer to the length of a radius of the circle.

We now use the formula for the area of a parallelogram

$$A = bh$$

If we let the number of pie-shaped pieces become very large, *b* becomes very close to $\frac{1}{2}C$ and *h* becomes very close to *r*. Thus, we have

$$
\begin{aligned}
A &= bh \\
  &= \tfrac{1}{2}Cr && \text{Substitute } \tfrac{1}{2}C \text{ for } b \text{ and } r \text{ for } h. \\
  &= \tfrac{1}{2}(2\pi r)r && \text{Substitute } 2\pi r \text{ for } C. \\
  &= \pi r^2
\end{aligned}
$$

Since the area of the pie-shaped pieces do not change when we rearrange them, the area of the circle is given by the formula

$$A = \pi r^2 \qquad \qquad \square$$

**EXAMPLE 1**

Find the area of a circle with a diameter of 10 cm.

*Solution*

Because the diameter of the circle is 10 cm, the radius is 5 cm. We substitute 5 for *r* in the formula for the area of a circle and proceed as follows:

$$
\begin{aligned}
A &= \pi r^2 \\
  &= \pi(5^2) && \text{Substitute 5 for } r. \\
  &= 25\pi
\end{aligned}
$$

The area of a circle with a diameter of 10 cm is $25\pi$ cm$^2$.   ∎

**EXAMPLE 2**

Orange paint is available only in gallons that cost \$19 each and cover 375 ft$^2$. Find how much it will cost to paint a helicopter pad 60 feet in diameter.

*Solution*

First calculate the area of the helicopter pad. Then determine how many gallons of paint will be needed to cover the area, and, finally, calculate the cost of the paint.

$$A = \pi r^2$$
$$= \pi(30^2) \qquad \text{Substitute } \tfrac{1}{2}(60) \text{ for } r.$$
$$= 900\pi$$
$$\approx 2826 \qquad \text{Let } \pi = 3.14 \text{ and multiply.}$$

The area of the helicopter pad is approximately 2826 ft². Because each gallon covers 375 ft², the number of gallons needed is founded by dividing 2826 by 375.

$$\text{number of gallons} = \frac{2826}{375} = 7.5$$

The painters will need approximately 7.5 gal. Because they will need to buy 8 gal and the cost of each gal is $19, the paint will cost

$$8 \cdot \$19 = \$152 \qquad \blacksquare$$

**EXAMPLE 3**

Find the length of one side of a square that has the same area as a circle with a diameter that is 6 in. long.

Solution

We can let $x$ represent the length in inches of one side of the square. Then the area of the square is $x^2$ in². Because a radius of a circle is half as long as its diameter, the radius of the circle is $\tfrac{6}{2}$ or 3 in. Thus, the number of square inches in the area of the circle is

$$A = \pi r^2 = \pi(3^2) = 9\pi$$

After setting the area of the square equal to the area of the circle and solving for $x$, we have

$$x^2 = 9\pi$$
$$x = \sqrt{9\pi} \qquad \text{Take the square root of both sides.}$$
$$= 3\sqrt{\pi} \qquad \sqrt{9\pi} = \sqrt{9}\sqrt{\pi} = 3\sqrt{\pi}.$$

The length of one side of the square is $3\sqrt{\pi}$ in. $\qquad \blacksquare$

## EXERCISE 5.2

In Exercises 1–14, find the shaded area in each figure.

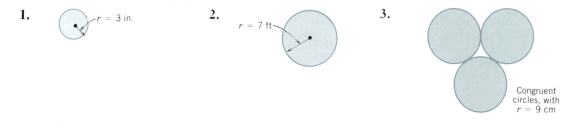

**1.** $r = 3$ in.

**2.** $r = 7$ ft

**3.** Congruent circles, with $r = 9$ cm

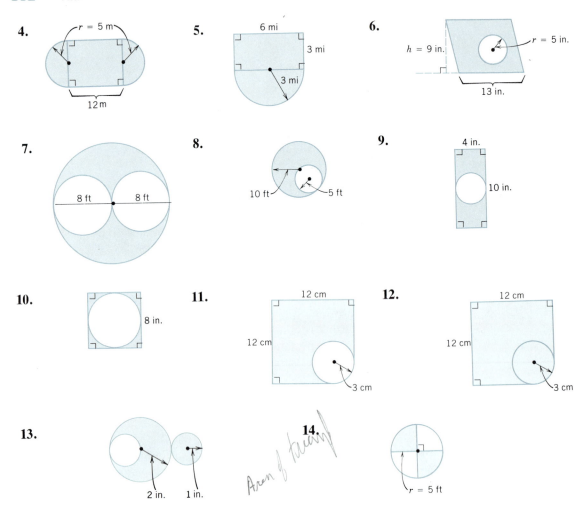

**4.** $r = 5$ m  12 m

**5.** 6 mi  3 mi  3 mi

**6.** $h = 9$ in.  $r = 5$ in.  13 in.

**7.** 8 ft  8 ft

**8.** 10 ft  5 ft

**9.** 4 in.  10 in.

**10.** 8 in.

**11.** 12 cm  12 cm  3 cm

**12.** 12 cm  12 cm  3 cm

**13.** 2 in.  1 in.

**14.** *Area of twenty*  $r = 5$ ft

15. Round Lake has a circular shoreline 2 miles in diameter. Find the area of the lake.

16. A Girl Scout is planning to hike completely around the shore of Round Lake. Approximately how far will the girl walk if the lake is 2 miles in diameter?

17. A circular track has a diameter of $\frac{1}{4}$ mile. If a runner wishes to jog for 10 miles, how many times must the runner circle the track?

18. The rotunda of a state capitol building is circular in shape with a diameter of 100 ft. The state legislature wishes to appropriate money to have the rotunda carpeted. The lowest bid is $83 per yd$^2$, including installation. How much will the state legislature spend? (Use $\pi = 3.14$.)

19. A steel band is drawn tightly around the earth's equator. (Assume the earth is a smooth ball.) The band is then loosened by increasing its length a mere 10 ft, and the resulting slack is distributed evenly along the band's entire length. How far above the earth's surface will the band be? (*Hint:* You don't need to know the radius of the earth.)

**20.** The radius of a circle is *a* units long. A square has the same area as the circle. Find the perimeter of the square.

**21.** A square has a side of length *s* units. A circle has the same area as the square. Find the length of the radius of the circle.

**22.** If the radius of a circle is doubled in length, by what factor is the area increased?

**23.** If the diameter of a circle is tripled in length, by what factor is the area increased?

**24.** If the radius of a circle is doubled in length, by what factor is the circumference increased?

# 5.3 THE PYTHAGOREAN THEOREM

Of the many theorems of geometry, one stands above the others in its usefulness and importance. It has applications in many branches of mathematics as well as in many of the sciences. This theorem is called **Pythagorean theorem**. It was first proved by Pythagoras, a Greek mathematician who lived about 550 B.C., some 200 years before Euclid.

Because the proof of the Pythagorean theorem does not lend itself to a two-column format, we shall present the proof in paragraph style.

> **Theorem 5.6** In a right triangle, the square of the length of the hypotenuse is equal to the sum of the squares of the lengths of the other two sides.

*Given*: $\triangle ABC$ is a right triangle with $m(\angle C) = 90°$.

*Prove*: $a^2 + b^2 = c^2$.

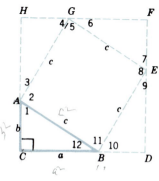

*Proof*: Construct a square $AGEB$ on the hypotenuse, $\overline{AB}$, of the right triangle $ABC$. By this construction, $\angle 2$, $\angle 5$, $\angle 8$, and $\angle 11$ are right angles, and sides $\overline{AG}$, $\overline{GE}$, $\overline{EB}$, and $\overline{BA}$ are congruent. Then, through point $G$, construct a line parallel to $\overline{CB}$ and through point $E$, construct a line parallel to $\overline{AC}$.

Since $\angle C$ is a right angle, $\overline{HF} \parallel \overline{CD}$ and $\overline{HC} \parallel \overline{FD}$, we know that quadrilateral $CHFD$ is a rectangle. Hence, $\angle H$, $\angle F$, and $\angle D$ are right angles. We can now show that the four triangles in the drawing are congruent.

Angle $GAC$ is an exterior angle of $\triangle AGH$ and is, therefore, congruent to the sum of the measures of the nonadjacent interior angles. Thus,

$$m(\angle GAC) = m(\angle H) + m(\angle 4)$$

Because

$$m(\angle GAC) = m(\angle 1) + m(\angle 2)$$

we can substitute to obtain

$$m(\angle 1) + m(\angle 2) = m(\angle 4) + m(\angle H)$$

Because $\angle 2$ and $\angle H$ are right angles, we have

$$m(\angle 1) + 90° = m(\angle 4) + 90°$$

or, if we subtract equal quantities from equal quantities,

$$m(\angle 1) = m(\angle 4)$$

Since $\overline{BA}$ and $\overline{AG}$ are congruent, we know by $ha \cong ha$ that $\triangle ABC \cong \triangle GAH$. In a similar way, we can show that the other triangles are congruent.

Because each side of the larger square has a length of $a + b$, we know its area is $(a + b)^2$, or $a^2 + 2ab + b^2$. Because each side of the smaller square has a length of $c$, we know that its area is $c^2$. The area of one triangle is $\frac{1}{2}ab$, and since the triangles are all congruent, the area of the four triangles is four times as big as the area of one triangle. Thus,

$$\text{Area of the four triangles} = 4\left(\frac{1}{2}ab\right) = 2ab$$

By the additive property of areas, we know that the area of the large square, $CHFD$, equals the sum of the areas of the small square, $AGEB$, and the four triangles. Therefore, we know that

$$a^2 + 2ab + b^2 = 2ab + c^2$$

or, by subtracting $2ab$ from both sides,

$$a^2 + b^2 = c^2 \qquad \square$$

**EXAMPLE 1**    The legs of a right triangle shown in Figure 5.15 are 3 ft and 4 ft in length. Find the length of the hypotenuse of the triangle.

Solution

By the Pythagorean theorem, the sum of the squares of the lengths of the two legs, $a$ and $b$, of a right triangle is equal to the square of the length of the hypotenuse, $c$. Thus,

$$a^2 + b^2 = c^2$$

We are given the lengths of two legs: $a = 3$ and $b = 4$. Substitute these values into the above formula and simplify.

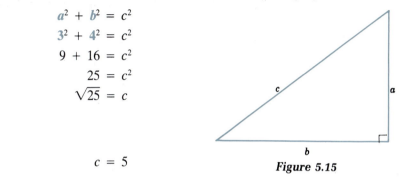

$$a^2 + b^2 = c^2$$
$$3^2 + 4^2 = c^2$$
$$9 + 16 = c^2$$
$$25 = c^2$$
$$\sqrt{25} = c$$

or

$$c = 5$$

**Figure 5.15**

The length of the hypotenuse is 5 ft.    ■

**EXAMPLE 2**

The isosceles-right triangle shown in Figure 5.16 has dimensions as indicated. Find the length of its hypotenuse.

Solution

Let $x$ represent the length of the hypotenuse of the right triangle. By the Pythagorean theorem, the sum of the squares of the lengths of the two legs is equal to the square of the length of the hypotenuse. Because the length of each leg is 8 units and the length of the hypotenuse is $x$ units, we have

$$8^2 + 8^2 = x^2$$
$$64 + 64 = x^2$$
$$64 \cdot 2 = x^2$$
$$\sqrt{64 \cdot 2} = \sqrt{x^2} \qquad \text{Take the positive square root of both sides.}$$
$$8\sqrt{2} = x \qquad \text{Simplify.}$$

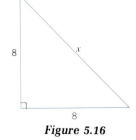

**Figure 5.16**

The length of the hypotenuse of the isosceles right triangle is $8\sqrt{2}$.    ■

**EXAMPLE 3**    The legs of a certain right triangle are congruent and the hypotenuse is $2\sqrt{2}$ units long. Find the length of each leg of the triangle.

Solution    We refer to the triangle shown in Figure 5.17 in which $x$ represents the length of each leg. Because of the Pythagorean theorem, we have

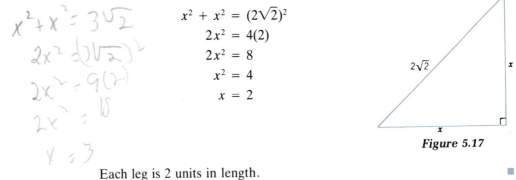

$$x^2 + x^2 = (2\sqrt{2})^2$$
$$2x^2 = 4(2)$$
$$2x^2 = 8$$
$$x^2 = 4$$
$$x = 2$$

*Figure 5.17*

Each leg is 2 units in length.    ∎

**EXAMPLE 4**    The base of a 32-ft ladder is 7 ft from the wall shown in Figure 5.18. The top of the ladder just reaches the bottom of a window. Find how far the bottom of the window is above the ground.

Solution    The ground, wall, and ladder form the right triangle shown in the figure. Because $x$ represents the height of the window, we can use the Pythagorean theorem to obtain

$$x^2 + 7^2 = 32^2$$
$$x^2 + 49 = 1024$$
$$x^2 = 975 \qquad \text{Subtract 49 from both sides.}$$
$$x = \sqrt{975} \qquad \text{Take the positive square root of both sides.}$$
$$= 31.2 \qquad \text{Use a calculator to find } \sqrt{975}.$$

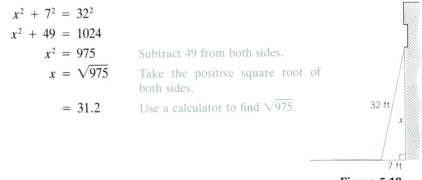

*Figure 5.18*

The window is approximately 31 ft above the ground.    ∎

> **Theorem 5.7**  If a triangle has sides of length $a$, $b$ and $c$, and $c^2 = a^2 + b^2$, then the triangle is a right triangle.

*Given*: $\triangle ABC$ with $c^2 = a^2 + b^2$.

*Prove*: $\triangle ABC$ is a right triangle.

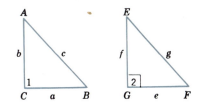

*Plan*: Construct a right triangle $EFG$ with $e = a$, $f = b$, and m($\angle 2$) = 90°. We will prove that $\triangle ABC \cong \triangle EFG$ and, thereby, show that m($\angle 1$) = m($\angle 2$) = 90°.

*Proof*:

| STATEMENTS | REASONS |
|---|---|
| **1.** $\triangle ABC$ with $c^2 = a^2 + b^2$. | **1.** Given. |
| **2.** $\triangle EFG$ with $e = a$, $f = b$, and m($\angle 2$) = 90°. | **2.** By construction. |
| **3.** $g^2 = e^2 + f^2$. | **3.** Pythagorean theorem. |
| **4.** $e^2 = a^2$, $f^2 = b^2$. | **4.** Square both sides of the first two equations in Step 2. |
| **5.** $g^2 = a^2 + b^2$. | **5.** Substitution. |
| **6.** $g^2 = c^2$. | **6.** If two quantities are equal to the same quantity, then they are equal to each other. |
| **7.** $g = c$. | **7.** The square roots of equal positive quantities are equal. |
| **8.** $\triangle ABC \cong \triangle EFG$. | **8.** $SSS \cong SSS$. |
| **9.** $\angle 1 \cong \angle 2$. | **9.** cpctc. |
| **10.** m($\angle 1$) = m($\angle 2$). | **10.** If two angles are congruent, then they have equal measures. |
| **11.** m($\angle 1$) = 90°. | **11.** Substitution. |
| **12.** $\angle 1$ is a right angle. | **12.** If an angle measures 90°, then it is a right angle. |
| **13.** $\triangle ABC$ is a right triangle. | **13.** A triangle with a right angle is a right triangle.    □ |

**EXAMPLE 5**    A triangle has sides that measure 20, 21, and 29 cm. Determine whether the triangle is a right triangle.

*Solution*    By Theorem 5.7, the triangle must be a right triangle if we can show that the square of the length of one side is equal to the sum of the squares of the lengths of the other two sides. Because we know the values of $a$, $b$, and $c$, we can compute their squares:

$$a^2 = 20^2 = 400$$
$$b^2 = 21^2 = 441$$
$$c^2 = 29^2 = 841$$

Because $400 + 441 = 841$, it follows that $a^2 + b^2 = c^2$ and the triangle is a right triangle.    ■

**EXAMPLE 6**    Is a triangle with sides of 3, 7, and 11 meters a right triangle?

*Solution*    We compute the squares of the values of $a$, $b$, and $c$.

$$a^2 = 9$$
$$b^2 = 49$$
$$c^2 = 121$$

Because $9 + 49 \neq 121$, the triangle is not a right triangle.    ■

## EXERCISE 5.3

In Exercises 1–6, refer to Illustration 1. Use the Pythagorean theorem and the values given to find the required dimension.

**1.** $a = 4$, $b = 3$. Find $c$.
**2.** $a = 5$, $b = 12$. Find $c$.
**3.** $b = 8$, $c = 10$. Find $a$.
**4.** $c = 17$, $b = 15$. Find $a$.
**5.** $c = 25$, $a = 7$. Find $b$.
**6.** $c = 41$, $a = 40$. Find $b$.

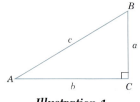

***Illustration 1***

In Exercises 7–12, determine whether a triangle with the indicated sides is a right triangle.

**7.** $a = 6$, $b = 8$, $c = 10$
**8.** $a = 5$, $b = 12$, $c = 13$
**9.** $a = 8$, $b = 15$, $c = 19$
**10.** $a = 2$, $b = 2$, $c = \sqrt{3}$
**11.** $a = 2$, $b = 5$, $c = \sqrt{29}$
**12.** $a = \sqrt{2}$, $b = \sqrt{3}$, $c = \sqrt{5}$

**13.** The sides of a square measure 2 m. Find the length of its diagonal.

**14.** A rectangle has a side 12 in. long and a diagonal 13 in. long. Find its area.

**15.** Is a triangle with sides 9, 12, and 15 cm long a right triangle?

**16.** Is a triangle with sides 6, 15, and 19 cm long a right triangle?

**17.** Is a triangle with sides 25, 60, and 65 in. long a right triangle?

**18.** The cube shown in Illustration 2 has an edge that is 1 ft long. Find the length of its diagonal.

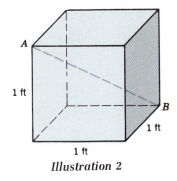

*Illustration 2*

**19.** A rectangle has sides 4 and 9 ft long. Find the dimensions of a square with the same area.

**20.** A 13-ft ladder is placed against the side of a building with its foot 5 ft from the base of the building. How far above the ground will the ladder touch the building?

**21.** A television tower on a flat roof is 150 ft tall. It is secured by three guy wires fastened to the top of the tower and to hooks 15 ft from its base. How much guy wire is used?

**22.** A rectangular garden has dimensions of 28 ft by 45 ft. Find the length of a path that extends from one corner to the opposite corner.

**23.** A woman drives 4.2 miles north and then 4.0 miles east. How far is she from her starting point?

**24.** Two sides of a triangular wall brace are 14 in. long. How long is the third side?

**25.** The gable end of the roof shown in Illustration 3 is an isosceles right triangle with a span of 32 ft. Find the distance *d* from the eave to the peak.

**26.** Find the height, *h*, of the gable end of the roof described in Exercise 25.

**27.** Illustration 4 shows a cube with edges 1 ft long. Point *S* is the midpoint of an edge. A spider at point *S* walks on the faces of the cube along the shortest path to catch a fly at point *F*. How far must the spider walk?

*Illustration 3*                    *Illustration 4*

**28.** In Exercise 27, how far would the spider have to tunnel through the cube to reach the fly?

**29.** A baseball diamond is a square with sides of 90 ft. The pitcher's mound is 60 ft, 6 in. from home plate. How far is it from second base?

**30.** A shortstop fields a grounder at a point one-third of the way from second base to third. How far must he throw the ball to make an out at first base?

**31.** Refer to Illustration 5 and find *a*.

**32.** Refer to Illustration 5 and find *b*.

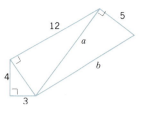

**Illustration 5**

**33.** Devise a construction technique that will multiply the length of a given line segment by $\sqrt{2}$.

**34.** Devise a construction technique that will multiply the length of a given line segment by $\sqrt{3}$.

**35.** Devise a construction technique that will multiply the length of a given line segment by $\sqrt{5}$.

**36.** Devise a construction technique that will multiply the length of a given line segment by $\sqrt{7}$.

**37.** Refer to Illustration 6.

*Given*: Squares *ABCD* and *ACEF*.
*Prove*: $A(ACEF) = 2 \cdot A(ABCD)$.

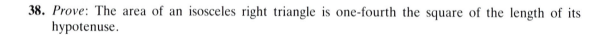

**Illustration 6**

**38.** *Prove*: The area of an isosceles right triangle is one-fourth the square of the length of its hypotenuse.

## 5.4 SOME SPECIAL RIGHT TRIANGLES

Two special right triangles are important in more advanced courses in mathematics. We now consider their properties.

---

> ***Theorem 5.8*** In a 30°-60° right triangle, the hypotenuse is twice as long as the side opposite the 30° angle.
>
> The side opposite the 60° angle is equal to the length of the side opposite the 30° angle multiplied by $\sqrt{3}$.

---

*Given*: Right $\triangle ABC$ with m($\angle A$) = 60°,
m($\angle ABC$) = 30°, and m($\angle C$) = 90°.

*Prove*: m($\overline{AB}$) = 2[m($\overline{AC}$)]; m($\overline{BC}$) = m($\overline{AC}$) $\sqrt{3}$.

*Construction*: Construct a 30° angle, $\angle CBD$ and
locate point $D$ so that point $D$ lies
on the extension of side $\overline{AC}$
forming $\triangle DBC$. This forms $\angle ABD$
which measures 60°.

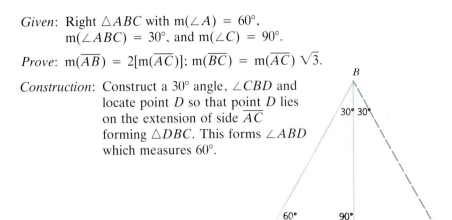

*Proof*:

| STATEMENTS | REASONS |
| --- | --- |
| **1.** Right $\triangle ABC$; m($\angle A$) = 60°, m($\angle ABC$) = 30°, and m($\angle ACB$) = 90°. | **1.** Given. |
| **2.** $\triangle DBC$, m($\angle DBC$) = 30°, m($\angle ABD$) = 60°. | **2.** By construction. |
| **3.** m($\angle A$) + m($\angle ABD$) + m($\angle D$) = 180°. | **3.** The sum of the measures of the angles of a triangle is 180°. |
| **4.** 60° + 60° + m($\angle D$) = 180°. | **4.** Substitution. |
| **5.** m($\angle D$) = 60°. | **5.** Equals subtracted from equals are equal. |
| **6.** $\angle A \cong \angle D$. | **6.** If two angles have equal measures, then they are congruent. |
| **7.** $\triangle ADB$ is an isosceles triangle with base angles $\angle A$ and $\angle D$. | **7.** If a triangle has two congruent angles, then it is an isosceles triangle. |
| **8.** $\angle ABC \cong \angle DBC$. | **8.** If two angles have the same measure, then they are congruent. |
| **9.** $\overline{BC}$ bisects $\angle ABD$. | **9.** If a line segment divides an angle into two congruent angles, then it bisects the angle. |

| STATEMENTS | REASONS |
|---|---|
| **10.** $\overline{AC} \cong \overline{CD}$. | **10.** The bisector of the vertex angle of an isosceles triangle bisects its base. |
| **11.** $m(\overline{AC}) = m(\overline{CD})$. | **11.** Congruent line segments have equal measures. |
| **12.** $m(\overline{AD}) = m(\overline{AC}) + m(\overline{CD})$. | **12.** Postulate 1.10. |
| **13.** $m(\overline{AD}) = m(\overline{AC}) + m(\overline{AC})$ $= 2\ m(\overline{AC})$. | **13.** Substitution. |
| **14.** $\overline{AD} \cong \overline{AB}$. | **14.** In a triangle if two angles are congruent, the sides opposite those angles are congruent. |
| **15.** $m(\overline{AB}) = 2m(\overline{AC})$. | **15.** Substitution. |
| **16.** $[m(\overline{AB})]^2 = [m(\overline{AC})]^2 + [m(\overline{BC})]^2$. | **16.** Because $\triangle ABC$ is a right triangle, the Pythagorean theorem applies. |
| **17.** $[2m(\overline{AC})]^2 = [m(\overline{AC})]^2 + [m(\overline{BC})]^2$. | **17.** Substitution. |
| **18.** $4[m(\overline{AC})]^2 = [m(\overline{AC})]^2 + [m(\overline{BC})]^2$. | **18.** $[2m(\overline{AC})]^2 = 4[m(\overline{AC})]^2$. |
| **19.** $3[m(\overline{AC})]^2 = [m(\overline{BC})]^2$. | **19.** Equal subtracted from equals are equal. |
| **20.** $m(\overline{AC})\ \sqrt{3} = m(\overline{BC})$. | **20.** Square roots of equal positive numbers are equal.    □ |

**EXAMPLE 1**

A young child, lying on level ground 43 ft from the base of a flagpole, sights its top at an angle of 30°. See Figure 5.19. Find the height of the flagpole.

*Solution*

The triangle in the figure is a 30°-60° right triangle. By Theorem 5.8, the side opposite the 60° angle is $\sqrt{3}$ times the length of the side opposite the 30° angle. If $h$ represents the length of the flagpole, we can proceed as follows:

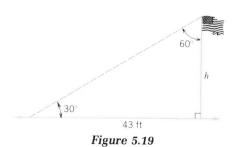

*Figure 5.19*

Side opposite the 60° angle = $\sqrt{3}$ (side opposite the 30° angle)

$$43 = \sqrt{3}\, h$$

$$\frac{43}{\sqrt{3}} = h \qquad \text{Equal quantities divided by equal quantities are equal.}$$

$$\frac{43}{1.73} = h \qquad \sqrt{3} \text{ is approximately 1.73.}$$

$$24.9 = h$$

The flagpole is approximately 25 ft tall.  ■

**EXAMPLE 2**

The equilateral triangle shown in Figure 5.20 has sides 8 in. long. Find its height.

Solution

Draw the bisector of $\angle C$ so that $\triangle ADC$ is a 30°-60° right triangle with $m(\angle A) = 60°$. By Theorem 5.8, the side opposite the 30° angle is one-half the length of the hypotenuse.

Therefore, $m(\overline{AD}) = 4$ inches. The height of the equilateral triangle is $m(\overline{CD})$, which by Theorem 5.8 is $\sqrt{3}$ times $m(\overline{AD})$. Thus, the height of the equilateral triangle is $4\sqrt{3}$ in.

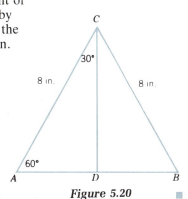

*Figure 5.20*  ■

---

**Theorem 5.9** In an isosceles right triangle, the length of the hypotenuse is equal to the length of one of the legs multiplied by $\sqrt{2}$

---

The proof of this theorem is left as an exercise.

**EXAMPLE 3**

The square shown in Figure 5.21 has sides $2\sqrt{2}$ ft long. Find the length of its diagonal.

Solution

Draw diagonal $\overline{AC}$. $\triangle ABC$ is an isosceles right triangle. By Theorem 5.9, the length of the hypotenuse $\overline{AC}$ is equal to the length of a leg multiplied by $\sqrt{2}$. Thus,

$$m(\overline{AC}) = m(\overline{DC})\sqrt{2}$$
$$= (2\sqrt{2})(\sqrt{2})$$
$$= 2(2)$$
$$= 4$$

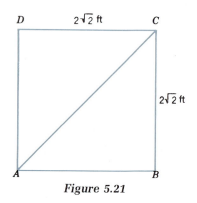

**Figure 5.21**

The diagonal of the square is 4 ft long.

*EXAMPLE 4*

The two boats shown in Figure 5.22 are the same distance from a lighthouse, one due north and one due east. The boats are 17 nautical miles from each other. How far is each boat from the lighthouse?

Solution

The boats and the lighthouse are arranged as shown in the figure. The distance of 17 miles is the length of the hypotenuse of an isosceles right triangle, and $d$ is the length of each of its congruent legs. We must find $d$. By Theorem 5.9, the length of the hypotenuse is equal to the length of one of its legs multiplied by $\sqrt{2}$. Proceed as follows:

Length of hypotenuse $= \sqrt{2} \cdot$ (length of one leg)

$$17 = d\sqrt{2}$$

$$\frac{17}{\sqrt{2}} = d \qquad \text{Equal quantities divided by equal quantities are equal.}$$

$$\frac{17}{1.4} = d \qquad \sqrt{2} \text{ is approximately 1.4.}$$

$$12.1 = d$$

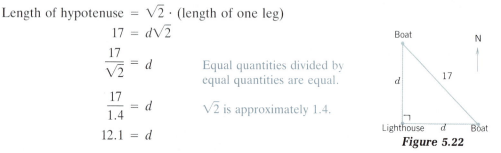

**Figure 5.22**

Each boat is approximately 12 nautical miles from the lighthouse.

## EXERCISE 5.4

In Exercises 1–10, refer to the 30°-60° right triangle shown in Illustration 1. Use the given information and find the length of the indicated side.

**1.** $a = 20$. Find $c$.

**2.** $b = 30\sqrt{3}$. Find $c$.

**3.** $a = 14$. Find $b$.

**4.** $b = 12$. Find $a$.

**5.** $a = \dfrac{\sqrt{2}}{3}$. Find $c$.

**6.** $b = \dfrac{\sqrt{3}}{2}$. Find $c$.

**7.** $c = 5$. Find $a$.

**8.** $c = 4$. Find $b$.

**9.** $c = \dfrac{2\sqrt{3}}{3}$. Find $a$

**10.** $c = \dfrac{2\sqrt{3}}{3}$. Find $b$.

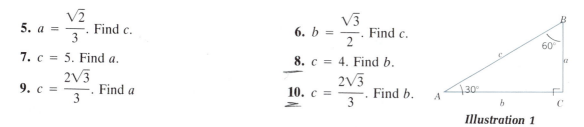

*Illustration 1*

In Exercises 11–20, refer to the isosceles right triangle shown in Illustration 2. Use the given information and find the length of the indicated side.

**11.** $a = 20$. Find $c$.

**12.** $b = 30$. Find $c$.

**13.** $a = 14$. Find $b$.

**14.** $b = 12$. Find $a$.

**15.** $a = \sqrt{3}$. Find $c$.

**16.** $b = \sqrt{2}$. Find $c$.

**17.** $c = 5$. Find $a$.

**18.** $c = 4$. Find $b$.

**19.** $c = 5\sqrt{3}$. Find $a$.

**20.** $c = 5\sqrt{3}$. Find $b$.

*Illustration 2*

**21.** Two airplanes, flying at the same altitude, are 155 miles from the airport, one due west and one due south. Find the distance between the planes.

**22.** John is at camp 220 miles due east of his home. His sister Maria is directly northeast of home and is also directly north of John. How far is Maria from home?

**23.** The guy wires to the top of a 230-ft tower makes angles of 60° with the ground. How far from the base of the tower are the wires anchored?

**24.** A circus tightrope walker ascends to a platform by walking 212 ft up a wire that makes an angle of 30° with the horizontal. Find the height of the platform.

**25.** The length of the hypotenuse of a 30°-60° right triangle is 4 dm. Find the length of each of its legs.

**26.** The length of the hypotenuse of a 30°-60° right triangle is 4 dm. Find the area of the triangle.

**27.** The area of a 30°-60° right triangle is $4\sqrt{3}$ square units. Find the length of each leg.

**28.** The area of an isosceles right triangle is 4 square units. Find the length of each leg.

**29.** The area of an isosceles right triangle is 4 square units. Find the length of its hypotenuse.

**30.** Find the altitude of an equilateral triangle with each side measuring 2 m.

**31.** Find the area of an equilateral triangle with each side measuring 2 dm.

**32.** The altitude of an equilateral triangle is 8 in. long. Find the length of a side of the triangle.

**33.** The altitude of an equilateral triangle is 8 in. long. Find the area of the triangle.

**34.** *Prove Theorem 5.9:* In an isosceles right triangle, the length of the hypotenuse is equal to the length of one of the legs multiplied by $\sqrt{2}$.

# 5.5 SURFACE AREA AND VOLUME

The geometric figures we have considered thus far have been **plane figures**—two-dimensional figures that lie in a plane. Other geometric figures are three-dimensional figures that occupy space.

Three three-dimensional figures, called **right prisms**, are shown in Figure 5.23. The **base** of the prism shown in Figure 5.23*a* is a triangle, and the base of the prism shown in Figure 5.23*b* is a rectangle. If the base of a prism is a rectangle, the prism is often called a **rectangular solid**. The rectangular solid shown in Figure 5.23*b* represents an empty box with length *l*, width *w*, and height *h*. A rectangular solid is a hollow figure that encloses

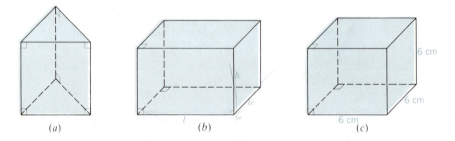

(a)　　　(b)　　　(c)

***Figure 5.23***

space. It is not like a brick that is solid all the way through. A rectangular solid whose length, width, and height are all the same length is a **cube**. Figure 5.23*c* shows a cube.

The **surface area of a rectangular solid** is the sum of the areas of the six rectangular **faces** that are its sides. The **volume of a rectangular solid** is the amount of space it encloses. We will accept the following formulas as a postulate.

---

**Postulate 5.5** The **surface area**, *A*, and the **volume**, *V*, of a **rectangular solid** are given by the formulas

$$A = 2lw + 2wh + 2lh$$
$$V = lwh$$

where *l* is its length, *w* is its width, and *h* is its height.

---

**EXAMPLE 1**

The oil storage tank shown in Figure 5.24 is a rectangular solid with dimensions of 17 by 10 by 8 ft. Find both the surface area and the volume of the tank.

*Solution*

To find the surface area, substitute 17 for $l$, 10 for $w$, and 8 for $h$ in the formula for surface area and simplify.

$$\begin{aligned} A &= 2lw + 2wh + 2lh \\ &= 2(17)(10) + 2(10)(8) + 2(17)(8) \\ &= 772 \end{aligned}$$

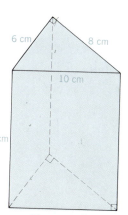

8 ft
10 ft
17 ft
**Figure 5.24**

The surface area is 772 $ft^2$.

To find the volume, substitute 17 for $l$, 10 for $w$, and 8 for $h$ in the formula for volume and simplify.

$$\begin{aligned} V &= lwh \\ &= (17)(10)(8) \\ &= 1360 \end{aligned}$$

The volume is 1360 $ft^3$. ■

To find the surface area of any right prism, we must find the sum of the areas of its faces. To find its volume, we can use the following formula.

> ***Postulate 5.6*** The **volume**, $V$, of a right prism is given by the formula
>
> $$V = Bh$$
>
> where $B$ is the area of its base and $h$ is its height.

*EXAMPLE 2* Find the (a) surface area and (b) the volume of the triangular prism shown in Figure 5.25.

*Solution*

(a) The area of each triangular base is

$$\frac{1}{2}(6)(8)\ \text{cm}^2 \qquad \text{or} \qquad 24\ \text{cm}^2$$

and the areas of the rectangular faces are

$$6(12)\ \text{cm}^2,\ 8(12)\ \text{cm}^2,\ \text{and}\ 10(12)\ \text{cm}^2$$

Thus, the surface area including the triangular ends is

$$\begin{aligned} \text{Surface area} &= 2(24) + 72 + 96 + 120 \\ &= 336 \end{aligned}$$

The surface area of the prism is 336 $cm^2$.

6 cm
8 cm
10 cm
12 cm
**Figure 5.25**

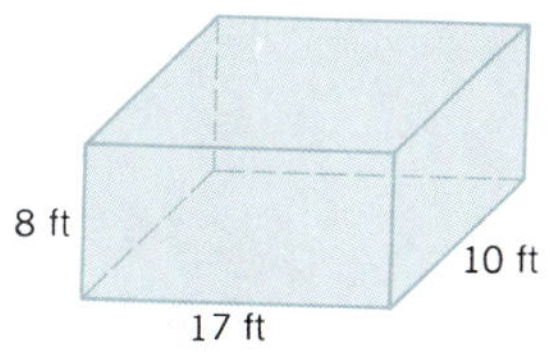

(b)  The volume of the prism is the area of its base multiplied by its height. Because the area of the triangular base is 24 cm² and the height is 12 centimeters, you have

$$A = Bh = 24(12) = 288$$

The volume of the prism is 288 cm³.                            ∎

A **sphere** is a hollow, round, ball-like *surface*. The points on the sphere shown in Figure 5.26 all lie at a fixed distance $r$ from a point called its **center**. A segment drawn from the center of the sphere to a point on the sphere is called a **radius**.

We will accept the formulas for the surface area and the volume of a sphere as postulates.

**Figure 5.26**

---

**Postulate 5.7**  The **surface area**, $A$, and the **volume**, $V$, of a sphere are given by the formulas

$$A = 4\pi r^2$$

$$V = \frac{4}{3}\pi r^3$$

where $r$ is the radius of the sphere.

---

*EXAMPLE 3*

A beachball has a diameter of 16 inches. How many square inches of plastic material were needed to make the ball? Neglect any waste.

*Solution*

Because the radius of the ball is half its diameter, the radius is 8 inches. So substitute 8 for $r$ in the formula for the surface area of a sphere and simplify.

$$A = 4\pi r^2$$
$$= 4\pi(8)^2$$
$$= 256\pi$$
$$\approx 804 \qquad \text{Substitute 3.14 for } \pi.$$

Approximately 804 in.² of material was required to make the beachball.

∎

**EXAMPLE 4**

How many cubic feet of water are needed to fill a spherical tank with a radius of 15 ft?

**Solution**

Because the tank has a radius of 15 ft, substitute 15 for $r$ in the formula for the volume of a sphere and simplify.

$$V = \frac{4}{3} \pi r^3$$

$$= \frac{4}{3} \pi (15)^3$$

$$= \frac{4}{3} \pi (3375)$$

$$= 4500\pi$$

$$\approx 14{,}130 \qquad \text{Substitute 3.14 for } \pi.$$

Approximately 14,130 ft$^3$ of water are needed to fill the tank.  ■

A **right-circular cylindrical solid** is a figure like a pop can. The **radius of a cylinder** is the radius of its circular cross section. Its height is the distance between its ends. The radius of the cylinder in Figure 5.27 is $r$ and its height is $h$.

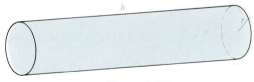

**Figure 5.27**

The surface area of a right-circular cylinder is the sum of the areas of its side and its two ends. Its volume is the amount of space enclosed by the cylinder.

---

**Postulate 5.8** The **surface area**, $A$, and the **volume**, $V$, of a right-circular cylinder are given by the formulas

$$A = 2\pi rh + 2\pi r^2 \qquad \text{(area of sides + area of ends)}$$
$$V = \pi r^2 h$$

where $r$ is the radius of the cylinder and $h$ is its height.

---

Although we accept the formula for the surface area of a right-circular cylinder as a postulate, the following reasoning makes the formula plausible. We refer to the right-circular cylinder shown in Figure 5.28 and note that the distance $C$ is the circumference of a circle with radius $r$, given by the formula $C = 2\pi r$. We imagine slicing the cylinder along line $l$ and unrolling it into a rectangle, as in Figures 5.28$b$ and 5.28$c$. The dimensions of this rectangle are $2\pi r$ by $h$, and the area of the rectangle is $A = 2\pi rh$. To obtain the formula given in Postulate 5.8, we must add the areas of the circular ends.

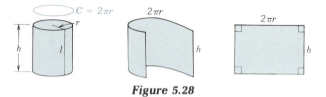

**Figure 5.28**

EXAMPLE 5

A farmer's silo is a right-circular cylinder 50 ft tall topped with a hemisphere (a half-sphere). The radius of the cylinder is 10 ft. Find (a) the surface area not including the floor and (b) the volume of the silo.

Solution

(a)  Refer to the silo shown in Figure 5.29. The total surface area is the sum of two areas. The first is the surface area of the curved side of the cylinder and the second is the surface area of the dome.

Area of the cylinder's side $= 2\pi rh$
$$= 2\pi(10)(50)$$
$$= 1000\pi$$

Area of the dome $= \dfrac{1}{2}$ (area of the sphere)

$$= \dfrac{1}{2}(4\pi r^2)$$
$$= 2\pi r^2$$
$$= 2\pi(10)^2$$
$$= 200\pi$$

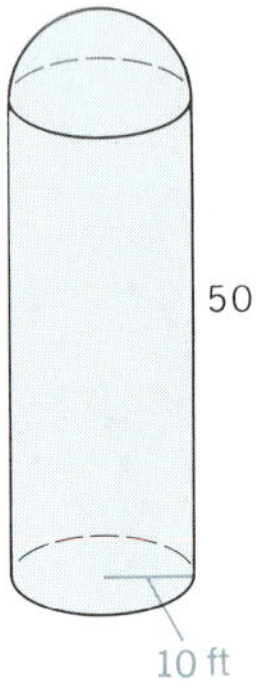

50 ft

10 ft

**Figure 5.29**

The total surface area of the silo is $1000\pi$ ft$^2$ + $200\pi$ ft$^2$, or $1200\pi$ ft$^2$, which is approximately 3768 ft$^2$.

(b)  To find the volume of the silo, find the sum of the volumes of the cylinder and the dome.

Volume of the cylinder + volume of the dome

$$= \pi r^2 h + \frac{1}{2}\left(\frac{4}{3}\pi r^3\right)$$

$$= \pi(10^2)50 + \frac{1}{2}\left(\frac{4}{3}\pi 10^3\right)$$

$$= 5000\pi + \frac{2000}{3}\pi$$

$$= \frac{17,000}{3}\pi$$

$$= 17,793 \qquad\qquad \text{Substitute 3.14 for } \pi.$$

The volume of the silo is approximately 17,793 ft³.  ∎

A **right-circular cone** is shown in Figure 5.30. The height of the cone is h and the radius of the cone is r, the radius of its circular base. The distance s measured along the side of the cone is called the **slant height**.

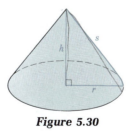

**Figure 5.30**

The solid formed by including the interior of the cone is called a **right-circular conical solid**. The formulas for its area and volume are as follows:

---

*Postulate 5.9* The **surface area**, $A$, and the **volume**, $V$, of a right-circular conical solid are given by the formulas

$$A = \pi r s + \pi r^2$$

$$V = \frac{1}{3}\pi r^2 h$$

where $r$ is the length of the radius of its base, $h$ is its height, and $s$ is its slant height.

---

**EXAMPLE 6**    Find the surface area of the conical solid shown in Figure 5.31.

Solution    Substitute 6 for $s$ and $\frac{1}{2}(8)$, or 4, for $r$ in the formula for the area of a cone and simplify.

$$\begin{aligned} A &= \pi rs + \pi r^2 \\ &= \pi(4)(6) + \pi 4^2 \\ &= 24\pi + 16\pi \\ &= 40\pi \\ &\approx 125.6 \end{aligned}$$

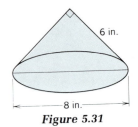

6 in.

8 in.

**Figure 5.31**

The area of the solid is approximately 126 in². ■

Two **right pyramids** are shown in Figure 5.32. The height of each pyramid is $h$. The base of the pyramid shown in Figure 5.31a is an equilateral triangle, and the base of the pyramid shown in Figure 5.31b is a square. The distance $s$ measured along a face of the pyramid is called its **slant height**.

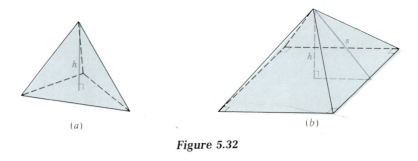

(a)          (b)

**Figure 5.32**

To find the surface area of a pyramid, we must find the sum of the areas of its faces.

---

**Postulate 5.10**  The **volume**, **$V$**, **of a pyramid** is given by the formula

$$V = \frac{1}{3}Bh$$

where $B$ is the area of its base and $h$ is its height.

---

## EXERCISE 5.5

In Exercises 1–12, find the surface area and volume of each figure.

**1.** A rectangular solid with dimensions of 3 by 4 by 5 cm.

**2.** A rectangular solid with dimensions of 5 by 8 by 10 m.

**3.** A right prism whose base is a right triangle with legs of length 3 and 4 m, whose hypotenuse is 5 m long, and whose height is 8 m.

**4.** A right prism whose base is a right triangle with legs that are 5 and 12 ft long, whose hypotenuse is 13 ft long, and whose height is 10 feet.

**5.** A sphere with a radius of 9 in.          **6.** A sphere with a diameter of 10 ft.

**7.** A right-circular cylinder with a base of radius 6 m and a height of 12 m.

**8.** A right-circular cylinder with a base of diameter 18 m and a height of 4 m.

**9.** A right-circular cone with a base of diameter 10 cm, a height of 12 cm, and a slant height of 13 cm.

**10.** A right-circular cone with a base of radius 4 in., a height of 3 in., and a slant height of 5 in.

**11.** A pyramid with a square base 10 m on each side, a height of 12 m, and a slant height of 13 m.

**12.** A pyramid with a square base 6 in. on each side, a height of 4 in., and a slant height of 5 in.

In Exercises 13–16, find the volume of each figure.

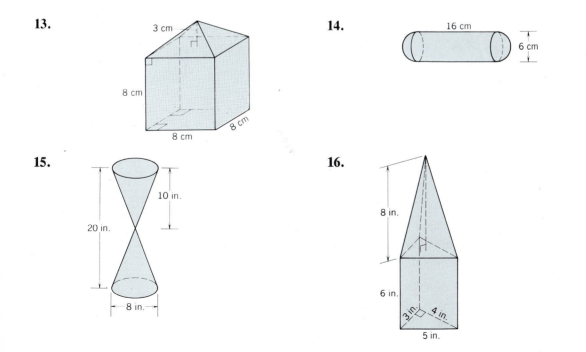

**13.**  3 cm  8 cm  8 cm  8 cm

**14.**  16 cm  6 cm

**15.**  10 in.  20 in.  8 in.

**16.**  8 in.  6 in.  3 in.  4 in.  5 in.

**17.** A sugar cube measures $\frac{1}{2}$ in. on each edge. How much space does one cube occupy?

**18.** A classroom is 40 ft long, 30 ft wide, and 9 ft high. Find the number of cubic feet contained in the room.

**19.** A cylindrical oil tank has a diameter of 6 ft and a length of 7 ft. Find the volume of the tank.

**20.** A restaurant serves pudding in a conical dish that has a diameter of 3 in. If the dish is 4 in. deep, how many cubic inches of pudding are in each dish?

**21.** How much material is needed to manufacture a spherical balloon 10 ft in diameter? Assume no waste.

**22.** How many square yards of wood is necessary to build a cubical box that measures 6 ft on each side?

**23.** The lifting power of a spherical balloon depends on its volume. How many cubic feet of gas will a balloon hold that is 40 ft in diameter?

**24.** A box of cereal measures 3 by 8 by 10 in. The manufacturer plans to market a smaller box that measures $2\frac{1}{2}$ by 7 by 8 in. By how much will the volume be reduced?

**25.** Explain why the volume of a cube can be found by using the formula

$$V = s^3$$

where $s$ is the length of one edge.

**26.** How many cubic inches are in 1 $\text{ft}^3$?

**27.** How many cubic feet are in 1 $\text{yd}^3$?

**28.** How many cubic inches are in 1 $\text{yd}^3$?

## CHAPTER FIVE REVIEW EXERCISES

**1.** Find the area of a square with sides that are 7 cm long.

**2.** Find the area of a square with sides that are 24.3 m long.

**3.** Find the area of a rectangle with dimensions of 12 by 15 ft.

**4.** Find the area of a rectangle that is 12 m long and 13 m wide.

**5.** Find the area of a parallelogram with a base that is 20 cm long and a height of 12 cm.

**6.** Find the area of a parallelogram with a height of 8 in. and a side that is 16 in. long.

**7.** Find the area of a triangle with a base 10 in. long and a height of 16 in.

**8.** Find the area of an isosceles triangle with a base that is 25 in. long and whose height is 12.2 in.

**9.** Find the area of a trapezoid with bases measuring 10 and 14 m and a height of 6 m.

**10.** Find the area of a trapezoid with a height of 8.3 cm and with bases measuring 23.3 and 43.5 cm.

**11.** Find the area of a rhombus with diagonals measuring 8 and 12 cm.

**12.** Find the area of a rhombus with diagonals measuring 6.4 and 3.2 cm.

**13.** Find how many square feet are in 1 $\text{yd}^2$.

**14.** Find how many square inches are in 1 $\text{ft}^2$.

**15.** Find the circumference of a circle that is 36 ft in diameter. ($\pi = 3.14$)

**16.** Find the radius of a circle whose circumference is $36\pi$ ft.

**17.** Find the area of a circle with a radius of 7.5 cm.

**18.** Find the diameter of a circle whose area is $25\pi$ cm².

**19.** Find the length of the hypotenuse of a right triangle with legs measuring 8 and 15 m.

**20.** Is a triangle with sides measuring 7, 24, and 25 cm a right triangle?

**21.** Three guy wires stretch from the top of a 200-ft tower to points 150 ft from its base. Find the total length of the three wires.

**22.** An express mail company will not accept any packages with a length, width, or depth that exceeds 5 ft. To ship a 6-ft length of pipe, a clever shipping clerk packages it diagonally in a cubical box. Will the package be acceptable to mail?

**23.** The leg opposite the 30° angle in a right triangle is 3 ft long. Find the length of the hypotenuse.

**24.** The leg opposite the 60° angle in a right triangle is 5 in. long. Find the length of the hypotenuse.

**25.** The leg opposite the 45° angle in a right triangle is 8 cm. Find the length of the hypotenuse.

**26.** Find the area of an equilateral triangle with a base that is 10 m long.

**27.** Find the surface area of a cube with an edge that is 15 cm long.

**28.** Find the volume of a rectangular solid with dimensions of 5 by 7 by 9 m.

**29.** Find the volume of a prism that is 12 in. tall and whose base is an equilateral triangle with a height of 6.06 in. and a perimeter of 21 in.

**30.** Find the surface area of a sphere with a diameter of 10 m.

**31.** Find the surface area of a right-circular cylinder with a radius of 1 ft and a height of 16 in.

**32.** Find the volume of the cylinder described in Exercise 31.

**33.** Find the surface area of a right-circular cone with a base with a diameter of 15 centimeters and with a height of 6 centimeters.

**34.** Find the volume of the cone described in Exercise 33.

**35.** How many cubic inches are in 2 yd³?

**36.** How many cubic feet are in 8640 in.³?

**37.** Find the volume of a pyramid with a base that is 12 in. square and a height of 12 in.

**38.** A wooden beam is 12 ft long, 6 in. wide, and 2 in. thick. If one cubic foot of wood weighs about 23 lb, how much does the entire beam weigh?

**39.** The walls, floor, and ceiling of an acoustical quiet room are covered with a special sound-absorbing material. The room is 18 by 21 ft and has 11-ft ceilings. How many square feet of material are needed?

**40.** A large sewer pipe section is a 12-ft-long tube of reinforced concrete. Its outer radius is 18 in. and its inner radius is 15 in. What volume of concrete was used in its construction?

**41.** A classroom measures 40 by 40 by 9 ft. Find the volume of the classroom in cubic yards.

**42.** A cube has a volume of 27 cm³. Find the area of one of its faces.

**43.** A right-circular cylinder encloses a sphere as in Illustration 1. Show that the volume of the sphere is two-thirds of the volume of the cylinder.

**44.** A right-circular cylinder encloses a sphere as in Illustration 1. Show that the surface area of the sphere is equal to the surface area of the cylinder, excluding the bases.

**45.** A roll of adding machine paper has dimensions as shown in Illustration 2. The package gives the length as 750 ft. How thick is the paper?

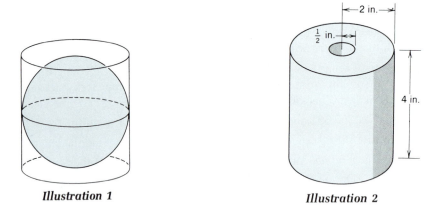

| Illustration 1 | Illustration 2 |

**46.** A cube has an edge of length $d$. Find the volume of the cube.

**47.** Jumbo gumballs are not the bargain they appear to be because they are hollow in the middle. Suppose a jumbo gumball is 1 inch in diameter and surrounds a $\frac{1}{2}$-inch-diameter bubble in its center. If the same amount of gum were used to make a solid gumball, what would be the gumball's diameter?

## CHAPTER FIVE TEST

In Exercises 1–25, classify each statement as true or false.

**1.** If two figures are congruent, then they have the same area.

**2.** If two figures have the same area, then they are congruent.

**3.** A triangle with sides measuring 3, 4, and 5 cm is a right triangle.

**4.** A triangle with sides measuring 5, 24, and 25 dm is a right triangle.

**5.** A triangle with sides measuring 3, 5, and 7 m is a right triangle.

**6.** The distance around a circle is called its **perimeter**.

**7.** The area of a parallelogram is equal to the product of the lengths of its diagonals.

**8.** A rectangle has dimensions of 6 and 8 in. A square with the same area has a side of length 7 inches.

**9.** A triangle has three altitudes.

**10.** The formula for the area of a circle is $A = \pi r^2$.

**11.** The side opposite the 30° angle in a right triangle with a hypotenuse 6 in. long is 12 in. long.

**12.** If the length of one side of a triangle is $\sqrt{2}$ times the length of another side, then the triangle is a right triangle.

**13.** The area of an equilateral triangle with a height of 6 cm is $12\sqrt{3}$ cm².

**14.** The area of a parallelogram with sides measuring 10 and 14 m is 140 m².

**15.** If two trapezoids have equal heights and medians of equal length, then they have equal areas.

**16.** A rectangle is a parallelogram.

**17.** The area of a trapezoid with bases 12 in. and 18 in. long is $15h$, where $h$ is the height of the trapezoid.

**18.** The area and circumference of a circle with a radius 2 cm long have the same number of units in their measures.

**19.** A square with sides 3 in. long has the same area as a circle with a radius 3 inches long.

**20.** A square has a diagonal that is $\sqrt{13}$ m long. The area of the square is $\dfrac{13}{2}$ m².

**21.** The volume of a sphere with a radius that is 3 units long is $36\pi$ cubic units.

**22.** The volume of a cone with a height of $h$ units and a radius of $h$ units is $\frac{1}{3}\pi h^3$.

**23.** The volume of a rectangular solid measuring 2 by 3 by 4 ft is $24\pi$ ft³.

**24.** The surface area of a cube with an edge that is $d$ units long is $6d^2$.

**25.** The surface area of a sphere with a diameter that is $d$ units long is $4\pi d^2$.

# 6

*Hipparchus (2nd century* B.C.*). Hipparchus is given credit for inventing trigonometry. He defined the trigonometric functions as ratios of a chord of a circle to the radius of the circle.*

# Ratio, Proportion, and Similarity

In this chapter we will discuss ratio, proportion, and similarity of triangles. Because these topics are algebraic in nature, we must rely on many of the results from algebra. You should review the postulates from algebra that are discussed on pages 27–28. In addition to these postulates, we will assume that you are familiar with the basic rules of fractions.

## 6.1 RATIO AND PROPORTION

An indicated quotient of two numbers is often called a **ratio**.

> *Definition 6.1* A **ratio** is the comparison of two numbers by their indicated quotient.

The previous definition implies that a ratio is a fraction. Some examples of ratios are

$$\frac{7}{8}, \quad \frac{21}{24}, \quad \text{and} \quad \frac{117}{223}$$

The fraction $\frac{7}{8}$ can be read as "the ratio of 7 to 8," the fraction $\frac{21}{24}$ can be read as "the ratio of 21 to 24," and the ratio $\frac{117}{223}$ can be read as "the ratio of 117 to 223." Because the fractions $\frac{7}{8}$ and $\frac{21}{24}$ represent equal numbers, they are called **equal ratios**.

*EXAMPLE 1*

Express each phrase as a ratio in lowest terms:

(a)   the ratio of 15 to 12.
(b)   the ratio of 3 in. to 7 in.
(c)   the ratio of 2 ft to 1 yd.
(d)   the ratio of 6 oz to 1 lb.

*Solution*

(a)   The ratio of 15 to 12 can be written as the fraction $\frac{15}{12}$. Expressed in lowest terms, it is $\frac{5}{4}$.

(b)   The ratio of 3 inches to 7 inches can be written as the fraction $\frac{3 \text{ inches}}{7 \text{ inches}}$, or just $\frac{3}{7}$.

(c)   The ratio of 2 ft to 1 yd should not be written as the fraction $\frac{2}{1}$. To express the ratio as a pure number, you must use the same units. Because there are 3 ft in 1 yd, the proper ratio is $\frac{2 \text{ ft}}{3 \text{ ft}}$, or just $\frac{2}{3}$.

(d)   The ratio 6 oz to 1 lb should not be written as the fraction $\frac{6}{1}$. To express the ratio as a pure number, you must use the same units. Because there are 16 oz in 1 lb, the proper ratio is $\frac{6 \text{ oz}}{16 \text{ oz}}$, which simplifies to $\frac{3}{8}$. ∎

---

*Definition 6.2* A **proportion** is a statement indicating that two ratios are equal.

Some examples of proportions are

$$\frac{1}{2} = \frac{3}{6}, \qquad \frac{3}{7} = \frac{9}{21}, \qquad \text{and} \qquad \frac{8}{1} = \frac{40}{5}$$

The proportion $\frac{1}{2} = \frac{3}{6}$ can be read as "1 is to 2 as 3 is to 6," the proportion $\frac{3}{7} = \frac{9}{21}$ can be read as "3 is to 7 as 9 is to 21," and the proportion $\frac{8}{1} = \frac{40}{5}$ can be read as "8 is to 1 as 40 is to 5."

A proportion can be written in general form as

$$\frac{a}{b} = \frac{c}{d}$$

where $a$, $b$, $c$, and $d$ are real numbers and $b$ and $d$ are not 0. If we write the divisions represented by the fractions in the proportion $\frac{a}{b} = \frac{c}{d}$ using the division symbol, the proportion becomes $a \div b = c \div d$. In this form, $a$ is the first term, $b$ is the second term, and so on. Also $a$ and $d$ are the extreme ends of the expression, and $b$ and $c$ are in the middle. This prompts the following definition.

---

**Definition 6.3** In the proportion

$$\frac{a}{b} = \frac{c}{d} \qquad (b \neq 0,\ d \neq 0)$$

the number $a$ is called the **first term**, $b$ is called the **second term**, $c$ is called the **third term**, and $d$ is called the **fourth term**.

The numbers $a$ and $d$ are called the **extremes**, and the numbers $b$ and $c$ are called the **means** of the proportion. The number $d$ is called the **fourth proportional**.

---

In the proportion $\frac{2}{6} = \frac{3}{9}$, the numbers 2 and 9 are the **extremes** of the proportion, and the numbers 6 and 3 are the **means**. If we find the product of the extremes and the product of the means in this proportion, we see that the products are equal:

$$2 \cdot 9 = 18 \qquad \text{and} \qquad 6 \cdot 3 = 18$$

This example illustrates the following theorem.

---

**Theorem 6.1** In any proportion, the product of the extremes is equal to the product of the means. In symbols,

$$\text{If } \frac{a}{b} = \frac{c}{d}, \text{ then } ad = bc.$$

*Given:* $\dfrac{a}{b} = \dfrac{c}{d}$.

*Prove:* $ad = bc$.

*Proof*:

| STATEMENTS | REASONS |
|---|---|
| **1.** $\dfrac{a}{b} = \dfrac{c}{d}$. | **1.** Given. |
| **2.** $bd = bd$. | **2.** A quantity is equal to itself. |
| **3.** $\dfrac{a}{b}(bd) = \dfrac{c}{d}(bd)$. | **3.** Equals multiplied by equals are equal. |
| **4.** $ad = bc$. | **4.** Each side of the equation in Statement 3 can be simplified. □ |

We can use Theorem 6.1 to decide whether two fractions are equal. For example, to decide whether the statement

$$\frac{7}{12} \overset{?}{=} \frac{21}{34}$$

is a proportion, we multiply 7 and 34 together and multiply 12 and 21 together as follows:

$$7(34) = 238$$
$$12(21) = 252$$

Only when the products are equal are the fractions equal and the statement a proportion. In this case the products are not equal. Thus, the given statement is not a proportion.

## SOLVING PROPORTIONS

Suppose that we know three numbers in the proportion

$$\frac{?}{5} = \frac{24}{20}$$

and we wish to find the missing number. To do so, we can represent the missing number by a variable, say $x$, multiply the extremes and multiply the means, set them equal, and find $x$:

$$\frac{x}{5} = \frac{24}{20}$$
$$20 \cdot x = 24 \cdot 5$$
$$20x = 120$$
$$x = 6 \qquad \text{Divide both sides of the equation by 20.}$$

The missing number is 6.

**EXAMPLE 2**    Solve the proportion $\dfrac{12}{18} = \dfrac{3}{x}$ for $x$.

Solution    Proceed as follows:

$$\frac{12}{18} = \frac{3}{x}$$

$12 \cdot x = 3 \cdot 18$    The product of the extremes equals the product of the means.

$12x = 54$    Simplify.

$x = \dfrac{54}{12}$    Divide both sides of the equation by 12.

$x = \dfrac{9}{2}$    Simplify.

Thus, $x = \dfrac{9}{2}$.    ■

**EXAMPLE 3**    Solve the proportion $\dfrac{x}{2} = \dfrac{\frac{3}{2} + x}{5}$ for $x$.

Solution    Proceed as follows:

$$\frac{x}{2} = \frac{\frac{3}{2} + x}{5}$$

$5x = 2\left(\dfrac{3}{2} + x\right)$    The product of the extremes equals the product of the means.

$5x = 3 + 2x$    Remove parentheses.

$3x = 3$    Add $-2x$ to both sides.

$x = 1$    Divide both sides by 3.

■

**EXAMPLE 4**    Solve the proportion $\dfrac{n + 3}{n} = \dfrac{4}{3}$ for $n$.

Solution    Proceed as follows:

$$\frac{n + 3}{n} = \frac{4}{3}$$

$3(n + 3) = 4n$    The product of the extremes equals the product of the means.

$3n + 9 = 4n$    Remove parentheses.

$9 = n$    Add $-3n$ to both sides.

or

$$n = 9$$  ∎

**EXAMPLE 5**

If 5 tomatoes cost \$1.15, how much will 16 tomatoes cost?

*Solution*

Let $c$ represent the cost of 16 tomatoes. The ratio of the numbers of tomatoes is the same as the ratio of their costs. Express this relationship as a proportion and find $c$.

$$\frac{5}{16} = \frac{1.15}{c}$$

$5 \cdot c = 1.15(16)$     The product of the extremes is equal to the product of the means.

$5 \cdot c = 18.4$     Do the multiplication.

$c = \dfrac{18.4}{5}$     Divide both sides by 5.

$c = 3.68$     Simplify.

Sixteen tomatoes will cost \$3.68.  ∎

A proportion such as $\frac{2}{6} = \frac{3}{9}$ can be rewritten in several different ways. For example, if each fraction is inverted, we obtain a different true proportion: $\frac{6}{2} = \frac{9}{3}$. The generalization of this property is given in Theorem 6.2.

---

**Theorem 6.2** A proportion can be written by inversion to form another proportion. In symbols,

$$\text{If } \frac{a}{b} = \frac{c}{d}, \text{ then } \frac{b}{a} = \frac{d}{c}.$$

---

*Given:* $\dfrac{a}{b} = \dfrac{c}{d}$.

*Prove:* $\dfrac{b}{a} = \dfrac{d}{c}$.

*Proof:*

| STATEMENTS | REASONS |
|---|---|
| **1.** $\dfrac{a}{b} = \dfrac{c}{d}$. | **1.** Given. |

| STATEMENTS | REASONS |
|---|---|
| **2.** $ad = bc$. | **2.** The product of the extremes is equal to the product of the means. |
| **3.** $ac = ac$. | **3.** A quantity is equal to itself. |
| **4.** $\dfrac{ad}{ac} = \dfrac{bc}{ac}$. | **4.** If equal quantities are divided by nonzero equal quantities, then the results are equal quantities. |
| **5.** $\dfrac{d}{c} = \dfrac{b}{a}$. | **5.** The fractions in Statement 4 can be simplified.    □ |

If the means in the proportion $\frac{2}{6} = \frac{3}{9}$ are interchanged, we obtain another true proportion: $\frac{2}{3} = \frac{6}{9}$. The generalization of this property is given in Theorem 6.3.

---

**Theorem 6.3** In any proportion, the means can be interchanged to form another proportion. In symbols,

$$\text{If } \frac{a}{b} = \frac{c}{d}, \text{ then } \frac{a}{c} = \frac{b}{d}.$$

---

*Given:* $\dfrac{a}{b} = \dfrac{c}{d}$.

*Prove:* $\dfrac{a}{c} = \dfrac{b}{d}$.

*Proof:*

| STATEMENTS | REASONS |
|---|---|
| **1.** $\dfrac{a}{b} = \dfrac{c}{d}$. | **1.** Given. |
| **2.** $ad = bc$. | **2.** The product of the extremes is equal to the product of the means. |
| **3.** $cd = cd$. | **3.** A quantity is equal to itself. |
| **4.** $\dfrac{ad}{cd} = \dfrac{bc}{cd}$. | **4.** If equal quantities are divided by nonzero equal quantities, then the results are equal quantities. |

| STATEMENTS | REASONS |
|---|---|
| **5.** $\dfrac{a}{c} = \dfrac{b}{d}$. | **5.** The fractions in Statement 4 can be simplified.    □ |

If the extremes in the proportion $\frac{2}{6} = \frac{3}{9}$ are interchanged, we obtain another true proportion: $\frac{9}{6} = \frac{3}{2}$. The generalization of this property is given in Theorem 6.4.

---

**Theorem 6.4** In any proportion, the extremes can be interchanged to form another proportion. In symbols,

$$\text{If } \frac{a}{b} = \frac{c}{d}, \text{ then } \frac{d}{b} = \frac{c}{a}.$$

---

The proof of Theorem 6.4 is left as an exercise.

If the denominators of the fractions in the proportion $\frac{2}{6} = \frac{3}{9}$ are added to their respective numerators, we obtain another true proportion:

$$\frac{2 + 6}{6} = \frac{3 + 9}{9} \quad \text{or} \quad \frac{8}{6} = \frac{12}{9}$$

The generalization of this property is given in Theorem 6.5.

---

**Theorem 6.5** A proportion can be written by addition to obtain another proportion. In symbols.

$$\text{If } \frac{a}{b} = \frac{c}{d}, \text{ then } \frac{a + b}{b} = \frac{c + d}{d}.$$

---

*Given:* $\dfrac{a}{b} = \dfrac{c}{d}$.

*Prove:* $\dfrac{a + b}{b} = \dfrac{c + d}{d}$.

*Proof:*

| STATEMENTS | REASONS |
|---|---|
| **1.** $\dfrac{a}{b} = \dfrac{c}{d}$. | **1.** Given. |
| **2.** $\dfrac{a}{b} + 1 = \dfrac{c}{d} + 1$. | **2.** Equal added to equals are equal. |

| STATEMENTS | REASONS |
|---|---|
| 3. $\dfrac{b}{b} = 1; \dfrac{d}{d} = 1.$ | 3. Any nonzero quantity divided by itself is 1. |
| 4. $\dfrac{a}{b} + \dfrac{b}{b} = \dfrac{c}{d} + \dfrac{d}{d}.$ | 4. Substitution. |
| 5. $\dfrac{a + b}{b} = \dfrac{c + d}{d}.$ | 5. The fractions on each side of the equation in Statement 3 can be added.    □ |

If the denominators of the fractions in the proportion $\frac{2}{6} = \frac{3}{9}$ are subtracted from their respective numerators, we obtain another true proportion:

$$\frac{2 - 6}{6} = \frac{3 - 9}{9} \qquad \text{or} \qquad \frac{-4}{6} = \frac{-6}{9}$$

The generalization of this property is given in Theorem 6.6.

---

**Theorem 6.6** A proportion can be written by subtraction to obtain another proportion. In symbols,

$$\text{If } \frac{a}{b} = \frac{c}{d}, \text{ then } \frac{a - b}{b} = \frac{c - d}{d}.$$

---

The proof of Theorem 6.6 is left as an exercise.

*EXAMPLE 6*    Write the proportion $\dfrac{2}{5} = \dfrac{6}{15}$ by (a) inversion, (b) by addition, and (c) by subtraction.

*Solution*    (a)  To write the proportion by inversion simply invert each fraction:

$$\text{If } \frac{2}{5} = \frac{6}{15}, \text{ then } \frac{5}{2} = \frac{15}{6}.$$

Note that the result $\dfrac{5}{2} = \dfrac{15}{6}$ is a proportion.

(b)  To write the proportion by addition, add each fraction's denominator to that fraction's numerator:

$$\text{If } \frac{2}{5} = \frac{6}{15}, \text{ then } \frac{2 + 5}{5} = \frac{6 + 15}{15}.$$

Note that the result is a proportion: $\dfrac{7}{5} = \dfrac{21}{15}.$

(c) To write the proportion by subtraction, subtract each fraction's denominator from that fraction's numerator:

$$\text{If } \frac{2}{5} = \frac{6}{15}, \text{ then } \frac{2 - 5}{5} = \frac{6 - 15}{15}.$$

Note that the result is a proportion: $\dfrac{-3}{5} = \dfrac{-9}{15}.$   ■

---

**Theorem 6.7** If three terms of one proportion are equal, respectively, to three terms of a second proportion, then the fourth terms are equal.

---

The proof of Theorem 6.7 is left as an exercise.

To illustrate a final theorem, we consider the equal fractions:

$$\frac{4}{5} = \frac{8}{10} = \frac{12}{15} = \frac{16}{20}$$

and note that if we add their numerators and add their denominators, we obtain an equal fraction.

$$\frac{4 + 8 + 12 + 16}{5 + 10 + 15 + 20} = \frac{40}{50} = \frac{4}{5}$$

The generalization of this property is given in Theorem 6.8.

---

**Theorem 6.8** If $a$, $b$, $c$, $d$, $e$, and $f$ are nonzero real numbers and

$$\frac{a}{b} = \frac{c}{d} = \frac{e}{f}$$

then

$$\frac{a + c + e}{b + d + f} = \frac{a}{b}$$

---

*Given:* $\dfrac{a}{b} = \dfrac{c}{d} = \dfrac{e}{f}.$

*Prove:* $\dfrac{a + c + e}{b + d + f} = \dfrac{a}{b}.$

*Hint:* Let $\dfrac{a}{b} = r.$

*Proof*:

| STATEMENTS | REASONS |
|---|---|
| 1. $\dfrac{a}{b} = \dfrac{c}{d} = \dfrac{e}{f} = r.$ | 1. Given. |
| 2. $a = br;\ c = dr;\ e = fr.$ | 2. Equals multiplied by equals are equal. |
| 3. $a + c + e = br + dr + fr.$ | 3. Equals added to equals are equal. |
| 4. $a + c + e = r(b + d + f).$ | 4. Use the distributive property to factor out $r$. |
| 5. $r = \dfrac{a + c + e}{b + d + f}.$ | 5. Equal quantities divided by equal nonzero quantities are equal. |
| 6. $\dfrac{a + c + e}{b + d + f} = \dfrac{a}{b}.$ | 6. Substitution.  □ |

## EXERCISE 6.1

In Exercises 1–20, express each phrase as a ratio in lowest terms.

**1.** 5 to 7 **2.** 3 to 5 **3.** 17 to 34 **4.** 19 to 38

**5.** 22 to 33 **6.** 14 to 21 **7.** 4 oz to 12 oz **8.** 3 in. to 15 in.

**9.** 12 min to 1 hr **10.** 8 oz to 1 lb **11.** 3 days to 1 week **12.** 2 qt to 1 gal

**13.** 4 in. to 2 yd **14.** 1 mile to 5280 ft **15.** 3 pints to 2 quarts **16.** 4 dimes to 8 pennies

**17.** 6 nickels to 1 quarter **18.** 3 people to 12 people **19.** 3 m to 12 cm **20.** 3 dollars to 3 quarters

In Exercises 21–28, tell whether each statement is a proportion.

**21.** $\dfrac{9}{7} = \dfrac{81}{70}$ **22.** $\dfrac{5}{2} = \dfrac{20}{8}$ **23.** $\dfrac{7}{3} = \dfrac{14}{6}$ **24.** $\dfrac{13}{19} = \dfrac{65}{95}$

**25.** $\dfrac{9}{19} = \dfrac{38}{80}$ **26.** $\dfrac{40}{29} = \dfrac{29}{22}$ **27.** $\dfrac{10.4}{3.6} = \dfrac{41.6}{14.4}$ **28.** $\dfrac{13.23}{3.45} = \dfrac{39.96}{11.35}$

In Exercises 29–42, solve for the variable in each proportion.

**29.** $\dfrac{2}{3} = \dfrac{x}{6}$ **30.** $\dfrac{3}{6} = \dfrac{x}{8}$ **31.** $\dfrac{5}{10} = \dfrac{3}{c}$ **32.** $\dfrac{7}{14} = \dfrac{2}{x}$

**33.** $\dfrac{6}{x} = \dfrac{8}{4}$ **34.** $\dfrac{4}{x} = \dfrac{2}{8}$ **35.** $\dfrac{x}{3} = \dfrac{9}{3}$ **36.** $\dfrac{x}{2} = \dfrac{18}{6}$

**37.** $\dfrac{x + 1}{5} = \dfrac{1}{5}$   **38.** $\dfrac{x - 1}{7} = \dfrac{2}{21}$   **39.** $\dfrac{n + 1}{n} = \dfrac{2}{3}$   **40.** $\dfrac{y + 1}{6} = \dfrac{y - 1}{4}$

**41.** $\dfrac{3(a + 5)}{2} = \dfrac{5(a - 2)}{3}$   **42.** $\dfrac{7(y + 6)}{6} = \dfrac{6(y + 3)}{5}$

**43.** Find the fourth proportional to 3, 9, and 17.

**44.** Find the fourth proportional to 6, 20, and 15.

**45.** The numbers 3, 7, $x$, and 21 form a proportion. Find $x$.

**46.** The numbers 2, $x$, 8, and 20 form a proportion. Find $x$.

In Exercises 47–58, set up and solve the required proportion.

**47.** Three pints of yogurt cost $1. How much will 51 pt cost?

**48.** Sport shirts are on sale at two for $25. How much will five shirts cost?

**49.** Garden seeds are on sale at three packets for 50 cents. How much will 39 packets cost?

**50.** A recipe for spaghetti sauce requires four 16-oz bottles of ketchup to make 2 gal of sauce. How many bottles of ketchup are needed to make 10 gal of sauce?

**51.** A car gets 42 miles per gallon of gas. How much gas is needed to drive 315 miles?

**52.** A truck gets 12 miles per gallon of gas. How far can the truck go on 17 gal of gas?

**53.** Bill earns $412 for a 40-hr week. Last week he missed 10 hr of work. How much did he get paid?

**54.** An HO-scale model railroad engine is 9 in. long. The HO scale is 87 ft to 1 ft. How long is a real engine?

**55.** An N-scale model railroad caboose is 3.5 in. long. The N scale is 169 ft to 1 ft. How long is a real caboose?

**56.** Standard doll house scale is 1 in. to 1 ft. Susan's doll house is 32 in. wide. How wide would it be if it were a real house?

**57.** A school board has determined that there should be 3 teachers for every 50 students. How many teachers are needed for an enrollment of 2700 students?

**58.** In a scale drawing, a 280-ft antenna tower is drawn 7 in. high. The building next to it is drawn 2 in. high. How tall is the actual building?

**59.** The instructions on a can of oil intended to be added to lawn mower gasoline read:

| Recommended | Gasoline | Oil |
|---|---|---|
| 50 to 1 | 6 gal | 16 oz |

Are these instructions correct? (*Hint*: There are 128 oz in 1 gal.)

**60.** The ratio of girls to boys in a kindergarten class is 4 to 5. In another class, it is 7 to 5. If there were 27 students in the first class and 36 students in the second class, how many girls and how many boys are there? What is the overall ratio of girls to boys?

**61.** *Prove Theorem 6.4*: In any proportion, the extremes can be interchanged to obtain another proportion.

**62.** *Prove Theorem 6.6*: A proportion can be written by subtraction to obtain another proportion.

**63.** *Prove Theorem 6.7*: If three terms of one proportion are equal, respectively, to three terms of a second proportion, then the fourth terms are equal.

## 6.2 SIMILAR TRIANGLES

Two triangles with the same shape are said to be **similar** triangles. Since congruent triangles have the same shape, they are similar triangles. However, similar triangles are not necessarily congruent. For two similar triangles to be congruent, they must also have the same area. Two similar triangles that are not congruent are shown in Figure 6.1. In symbols, we write

$\triangle ABC \sim \triangle DEF$      Read as "triangle $ABC$ is similar to triangle $DEF$."

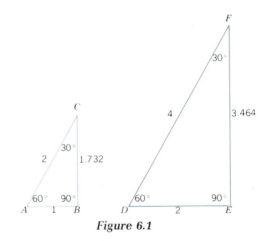

**Figure 6.1**

We shall define similar triangles in the following way:

---

**Definition 6.4** Two triangles are **similar triangles** if, and only if,

- three angles of one triangle are congruent to three angles of the second triangle, and
- the measures of all corresponding sides are in proportion.

---

In the similar triangles in Figure 6.1, the following relationships are true:

$$\angle A \cong \angle D \qquad \angle B \cong \angle E \qquad \angle C \cong \angle F$$

$$\frac{m(\overline{AB})}{m(\overline{DE})} = \frac{m(\overline{BC})}{m(\overline{EF})} = \frac{m(\overline{CA})}{m(\overline{FD})}$$

It is difficult to use Definition 6.4 to prove that two triangles are similar. The next postulate will be more useful.

> **Postulate 6.1** Two triangles are similar if, and only if, two angles of one triangle are congruent to two angles of the other triangle.

**EXAMPLE 1**

In Figure 6.2, $\overline{AB} \parallel \overline{DE}$. Show that $\triangle ABC \sim \triangle DEC$.

**Solution**

$\angle A$ and $\angle D$ form a pair of alternate interior angles, and because $\overline{AB} \parallel \overline{DE}$,

$$\angle A \cong \angle D$$

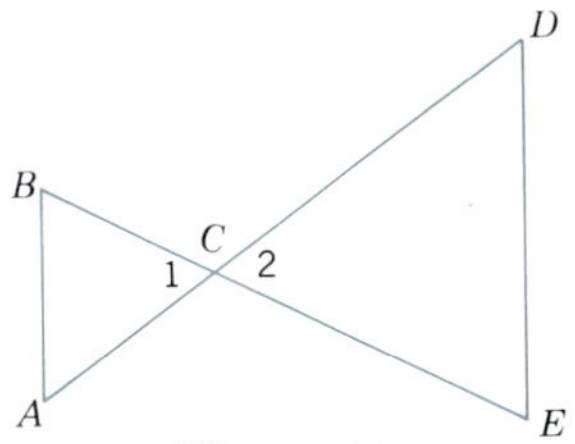

$\angle 1$ and $\angle 2$ form a pair of vertical angles, and because vertical angles are always congruent,

$$\angle 1 \cong \angle 2$$

**Figure 6.2**

Because two angles of $\triangle ABC$ are congruent to two angles of $\triangle DEC$, the triangles are similar. ∎

**EXAMPLE 2**

Refer to Figure 6.3 in which $\triangle ABC$ and $\triangle DEF$ have $\angle A \cong \angle D$ and $\angle C \cong \angle F$. Find (a) m($\overline{DE}$) and (b) m($\overline{EF}$).

**Solution**

Let m($\overline{DE}$) = $x$ and m($\overline{EF}$) = $y$. Because $\angle A \cong \angle D$ and $\angle C \cong \angle F$, two angles of $\triangle ABC$ are congruent to two angles of $\triangle DEF$ and the triangles are similar. Thus, the lengths of their corresponding sides are in proportion:

(1) $\dfrac{\text{m}(\overline{AC})}{\text{m}(\overline{DF})} = \dfrac{\text{m}(\overline{AB})}{\text{m}(\overline{DE})}$

and

(2) $\dfrac{\text{m}(\overline{AC})}{\text{m}(\overline{DF})} = \dfrac{\text{m}(\overline{BC})}{\text{m}(\overline{EF})}$

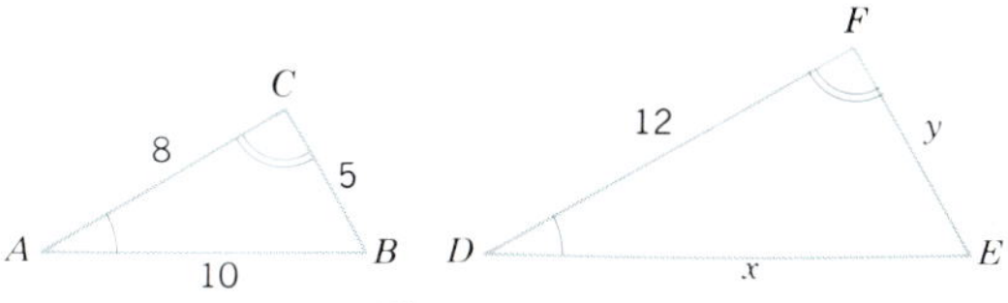

**Figure 6.3**

Substituting 8 for m($\overline{AC}$), 12 for m($\overline{DF}$), 10 for m($\overline{AB}$), and $x$ for m($\overline{DE}$) in Equation 1 and finding $x$ gives

$$\frac{8}{12} = \frac{10}{x}$$

$8x = 120$    In a proportion, the product of the extremes is equal to the product of the means.

$x = \dfrac{120}{8}$    Divide both sides by 8.

$= 15$

Substituting 8 for m($\overline{AC}$), 12 for m($\overline{DF}$), 5 for m($\overline{BC}$), and $y$ for m($\overline{EF}$)

in Equation 2 and finding $y$ gives

$$\frac{8}{12} = \frac{5}{y}$$

$8y = 60$     In a proportion, the product of the extremes is equal to the product of the means.

$y = \dfrac{60}{8}$     Divide both sides by 8.

$= \dfrac{15}{2}$     Simplify.

Thus, $m(\overline{DE}) = 15$ units and $m(\overline{EF}) = 7\dfrac{1}{2}$ units.     ■

**EXAMPLE 3**

In Figure 6.4, $\overline{DE} \parallel \overline{AB}$. Find $x$.

Solution

Since $\overline{DE} \parallel \overline{AB}$, the corresponding angles $\angle A$ and $\angle 1$ and the corresponding angles $\angle B$ and $\angle 2$ are congruent. Thus, two angles of $\triangle ABC$ are congruent to two angles of $\triangle DEC$ and the triangles are similar. Because the measures of corresponding sides of similar triangles are in proportion,

$$\frac{m(\overline{DC})}{m(\overline{AC})} = \frac{m(\overline{EC})}{m(\overline{BC})}$$

So,

$$\frac{8}{12} = \frac{x}{16}$$

$$128 = 12x$$

$$\frac{128}{12} = x$$

$$\frac{32}{3} = x$$

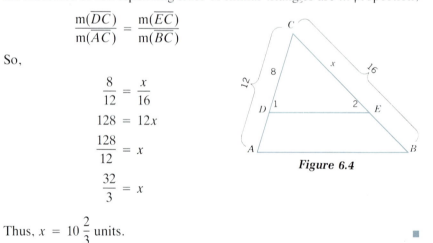

*Figure 6.4*

Thus, $x = 10\dfrac{2}{3}$ units.     ■

**EXAMPLE 4**

How can two Boy Scouts find the width of the river represented in Figure 6.5 without crossing to the other side?

Solution

The boys are able to measure sides $\overline{AB}$ and $\overline{BC}$ directly. They can estimate the measure of side $\overline{DE}$ indirectly by measuring $\overline{CF}$ ($F$ is on the bank straight across from $E$, and $C$ is on the bank straight across from $D$). The boys make the measurements and find that

$$m(\overline{AB}) = 3 \text{ m}$$
$$m(\overline{CB}) = 4 \text{ m}$$

and

$$m(\overline{CF}) = m(\overline{DE}) = 30 \text{ m}$$

Because $\angle ACB$ and $\angle ECD$ are vertical angles, they are congruent. Because $\angle B$ and $\angle D$ are right angles, they are congruent. Since two angles of $\triangle ABC$ are congruent to two angles of $\triangle EDC$,

$$\triangle ABC \sim \triangle EDC$$

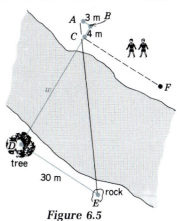

*Figure 6.5*

Thus, the boys can form the following proportion and solve for $w$:

$$\frac{3}{30} = \frac{4}{w}$$
$$120 = 3w$$
$$40 = w$$

The river is 40 m wide. ■

Postulate 6.1 can be used to prove many theorems involving similar triangles.

---

**Theorem 6.9** A line parallel to one side of a triangle divides the other two sides proportionally.

---

*Given*: $\triangle ABC$ with $\overline{AB} \parallel \overline{DE}$.

*Prove*: $\dfrac{m(\overline{AD})}{m(\overline{DC})} = \dfrac{m(\overline{BE})}{m(\overline{EC})}$.

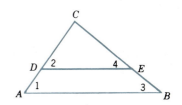

*Proof*:

| STATEMENTS | REASONS |
|---|---|
| **1.** $\triangle ABC$ with $\overline{AB} \parallel \overline{DE}$. | **1.** Given. |
| **2.** $\angle 1 \cong \angle 2$; $\angle 3 \cong \angle 4$. | **2.** If two parallel lines are cut by a transversal, then corresponding angles are congruent. |
| **3.** $\triangle ABC \sim \triangle DEC$. | **3.** If two angles of one triangle are congruent to two angles of |

| STATEMENTS | REASONS |
|---|---|
| | a second triangle, then the triangles are congruent. |
| **4.** $\dfrac{m(\overline{AC})}{m(\overline{DC})} = \dfrac{m(\overline{BC})}{m(\overline{EC})}$. | **4.** The lengths of the corresponding sides of similar triangles are in proportion. |
| **5.** $\dfrac{m(\overline{AC}) - m(\overline{DC})}{m(\overline{DC})}$ $= \dfrac{m(\overline{BC}) - m(\overline{EC})}{m(\overline{EC})}$. | **5.** A proportion can be written by subtraction. |
| **6.** $m(\overline{AD}) = m(\overline{AC}) - m(\overline{DC})$; $m(\overline{BE}) = m(\overline{BC}) - m(\overline{EC})$. | **6.** Postulate 1.10. |
| **7.** $\dfrac{m(\overline{AD})}{m(\overline{DC})} = \dfrac{m(\overline{BE})}{m(\overline{EC})}$. | **7.** Substitution.      □ |

**EXAMPLE 5**

In Figure 6.6, $\overline{DE} \parallel \overline{AB}$, $m(\overline{AC}) = 24$ units, $m(\overline{CE}) = 3$ units, and $m(\overline{EB}) = 5$ units. Find the length of $\overline{CD}$ and $\overline{DA}$.

**Solution**

We let $x$ represent $m(\overline{CD})$. Then $m(\overline{AD}) = 24 - x$. By Theorem 6.9, we know that $\overline{DE}$ divides sides $\overline{AC}$ and $\overline{BC}$ proportionally. Thus, we can set up the following proportion and solve it for $x$.

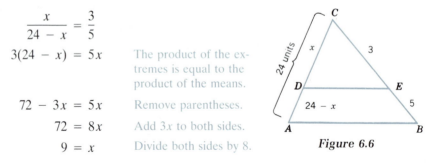

$$\frac{x}{24 - x} = \frac{3}{5}$$

$3(24 - x) = 5x$      The product of the extremes is equal to the product of the means.

$72 - 3x = 5x$      Remove parentheses.

$72 = 8x$      Add $3x$ to both sides.

$9 = x$      Divide both sides by 8.

*Figure 6.6*

Thus, $m(\overline{CD}) = 9$ units. Because $m(\overline{AC}) = 24$ units and $m(\overline{CD}) = 9$ units, it follows that $m(\overline{AD}) = (24 - 9)$ units, or 15 units.    ■

---

**Theorem 6.10** The bisector of one angle of a triangle divides the opposite side into the same ratio as the lengths of the other two sides.

*Given*: $\overline{CD}$ bisects $\angle C$.

*Prove*: $\dfrac{\mathrm{m}(\overline{AD})}{\mathrm{m}(\overline{DB})} = \dfrac{\mathrm{m}(\overline{AC})}{\mathrm{m}(\overline{CB})}$.

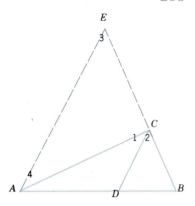

*Construction*: Through point $A$ construct a line parallel to line segment $\overline{DC}$. Extend side $\overline{BC}$ to intersect this line at point $E$.

*Proof*:

| STATEMENTS | REASONS |
| --- | --- |
| **1.** $\overline{CD}$ bisects $\angle C$. | **1.** Given. |
| **2.** $\angle 1 \cong \angle 2$. | **2.** An angle bisector divides an angle into two congruent angles. |
| **3.** $\overline{AE} \parallel \overline{DC}$. | **3.** A parallel to $\overline{CD}$ through point $A$ does exist. |
| **4.** $\angle 3 \cong \angle 2$. | **4.** If two parallel lines are cut by a transversal, then corresponding angles are congruent. |
| **5.** $\angle 1 \cong \angle 4$. | **5.** If two parallel lines are cut by a transversal, then alternate interior angles are congruent. |
| **6.** $\angle 3 \cong \angle 4$. | **6.** If two angles are congruent to two congruent angles, then they are congruent to each other. |
| **7.** $\overline{EC} \cong \overline{AC}$. | **7.** If two angles in a triangle are congruent, then the sides opposite those angles are congruent. |
| **8.** $\mathrm{m}(\overline{EC}) = \mathrm{m}(\overline{AC})$. | **8.** Congruent segments have the same measures. |
| **9.** $\dfrac{\mathrm{m}(\overline{AD})}{\mathrm{m}(\overline{DB})} = \dfrac{\mathrm{m}(\overline{EC})}{\mathrm{m}(\overline{CB})}$. | **9.** Theorem 6.9. |

| STATEMENTS | REASONS |
|---|---|
| **10.** $\dfrac{m(\overline{AD})}{m(\overline{DB})} = \dfrac{m(\overline{AC})}{m(\overline{CB})}.$ | **10.** Substitution. □ |

**EXAMPLE 6**

In Figure 6.7, right $\triangle ABC$ has legs that are 6 in. and 8 in. long, and $\overline{CD}$ bisects the right angle. Find $m(\overline{AD})$ and $m(\overline{DB})$.

*Solution*

By Theorem 6.10, we know that $\dfrac{6}{8} = \dfrac{m(\overline{AD})}{m(\overline{DB})}$. We write this proportion by addition to obtain

$$\frac{6 + 8}{8} = \frac{m(\overline{AD}) + m(\overline{DB})}{m(\overline{DB})}$$

However, $m(\overline{AD}) + m(\overline{DB}) = m(\overline{AB})$ and $\overline{AB}$ is the hypotenuse of a right triangle.

By the Pythagorean theorem, we find that $m(\overline{AB}) = 10$ inches. After substituting 10 for $m(\overline{AB})$, the proportion becomes

$$\frac{14}{8} = \frac{10}{m(\overline{DB})}$$

Because in a proportion the product of the extremes is equal to the product of the means, we have

$$14[m(\overline{DB})] = 8 \cdot 10$$
$$[m(\overline{DB})] = \frac{80}{14}$$
$$[m(\overline{DB})] = \frac{40}{7}$$

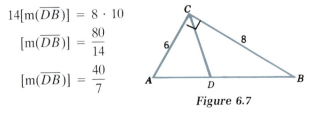

**Figure 6.7**

To find $m(\overline{AD})$, we note that

$$m(\overline{AD}) = m(\overline{AB}) - m(\overline{DB})$$

and substitute values for $m(\overline{AB})$ and $m(\overline{DB})$ and simplify.

$$m(\overline{AD}) = 10 - \frac{40}{7} = \frac{30}{7} \qquad \blacksquare$$

## EXERCISE 6.2

In Exercises 1–6, tell whether each statement is true. If a statement is false, explain why.

**1.** All equilateral triangles are similar.

**2.** All right triangles are similar.

**3.** All similar triangles are right triangles.

**4.** All similar triangles are isosceles triangles.

**5.** Some scalene triangles are similar.

**6.** Congruent triangles are similar.

In Exercises 7–10, the measures of certain parts of $\triangle ABC$ and $\triangle DEF$ are given. Draw the triangles and explain why they are similar.

**7.** $m(\angle A) = m(\angle D)$ and $m(\angle B) = m(\angle E)$.
**8.** $m(\angle A) = m(\angle D)$, $m(\angle A) = m(\angle B)$, and $m(\angle A) = m(\angle E)$.
**9.** $m(\angle A) = m(\angle D)$, $m(\angle A) = m(\angle B)$, and $m(\angle D) = m(\angle E)$.
**10.** $m(\angle A) = 60°$, $m(\angle D) = 60°$, $m(\angle B) = 50°$, and $m(\angle F) = 70°$.

In Exercises 11–14, refer to the similar triangles in Illustration 1, in which $\angle A \cong \angle D$ and $\angle B \cong \angle E$.

**11.** If $m(\overline{AC}) = 7$ cm, $m(\overline{DF}) = 4$ cm, and $m(\overline{AB}) = 15$ cm, find $m(\overline{DE})$.
**12.** If $m(\overline{CB}) = 12$ cm, $m(\overline{AB}) = 15$ cm, and $m(\overline{DE}) = 7$ cm, find $m(\overline{FE})$.
**13.** If $m(\overline{AC}) = 21$ cm, $m(\overline{BC}) = 25$ cm, and $m(\overline{FE}) = 8$ cm, find $m(\overline{DF})$.
**14.** If $m(\overline{DE}) = 6$ cm, $m(\overline{AC}) = 12$ cm, and $m(\overline{DF}) = 4$ cm, find $m(\overline{AB})$.

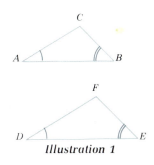

**Illustration 1**

In Exercises 15–20, refer to Illustration 2, in which $\overline{DE} \parallel \overline{AB}$.

**15.** If $m(\overline{DC}) = 9$, $m(\overline{AC}) = 12$, and $m(\overline{EC}) = 6$, find $m(\overline{BC})$.
**16.** If $m(\overline{DC}) = 7$, $m(\overline{EC}) = 6$, and $m(\overline{BC}) = 12$, find $m(\overline{AD})$.
**17.** If $m(\overline{AD}) = 5$, $m(\overline{DC}) = 12$, and $m(\overline{CB}) = 204$, find $m(\overline{CE})$.
**18.** If $m(\overline{AC}) = 15$, $m(\overline{CE}) = 16$, and $m(\overline{EB}) = 4$, find $m(\overline{DC})$.
**19.** If $m(\overline{BC}) = 65$, $m(\overline{EB}) = 10$, and $m(\overline{AC}) = 52$, find $m(\overline{DC})$.
**20.** If $m(\overline{AC}) = 42$, $m(\overline{AD}) = 6$, and $m(\overline{BC}) = 56$, find $m(\overline{EB})$.

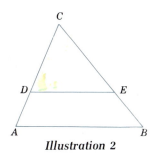

**Illustration 2**

In Exercises 21–26, refer to Illustration 3 in which $m(\angle ACB) = 90°$ and $m(\angle CDB) = 90°$.

**21.** If $m(\angle A) = 40°$, find $m(\angle 1)$.
**22.** If $m(\angle 2) = 60°$, find $m(\angle B)$.
**23.** Show that $\angle A \cong \angle 2$ and $\angle B \cong \angle 1$.
**24.** Show that $\triangle ADC \sim \triangle ACB$.
**25.** Show that $\triangle BDC \sim \triangle BCA$.
**26.** Show that $\triangle ADC \sim \triangle CDB$.

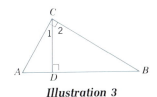

**Illustration 3**

In Exercises 27–28, refer to Illustration 4.

**27.** If $m(\overline{AE}) = 4$, $m(\overline{DE}) = 6$, $m(\overline{BE}) = 3$, and $\overline{AB} \parallel \overline{CD}$, find $m(\overline{CE})$.

**28.** If m($\overline{AE}$) = 4, m($\overline{DE}$) = 6, m($\overline{BC}$) = 8, and $\overline{AB} \parallel \overline{CD}$, find m($\overline{BE}$).

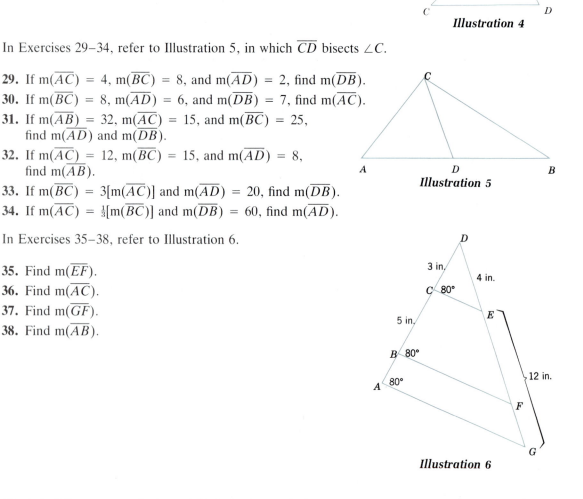

*Illustration 4*

In Exercises 29–34, refer to Illustration 5, in which $\overline{CD}$ bisects $\angle C$.

**29.** If m($\overline{AC}$) = 4, m($\overline{BC}$) = 8, and m($\overline{AD}$) = 2, find m($\overline{DB}$).
**30.** If m($\overline{BC}$) = 8, m($\overline{AD}$) = 6, and m($\overline{DB}$) = 7, find m($\overline{AC}$).
**31.** If m($\overline{AB}$) = 32, m($\overline{AC}$) = 15, and m($\overline{BC}$) = 25, find m($\overline{AD}$) and m($\overline{DB}$).
**32.** If m($\overline{AC}$) = 12, m($\overline{BC}$) = 15, and m($\overline{AD}$) = 8, find m($\overline{AB}$).
**33.** If m($\overline{BC}$) = 3[m($\overline{AC}$)] and m($\overline{AD}$) = 20, find m($\overline{DB}$).
**34.** If m($\overline{AC}$) = $\frac{1}{3}$[m($\overline{BC}$)] and m($\overline{DB}$) = 60, find m($\overline{AD}$).

*Illustration 5*

In Exercises 35–38, refer to Illustration 6.

**35.** Find m($\overline{EF}$).
**36.** Find m($\overline{AC}$).
**37.** Find m($\overline{GF}$).
**38.** Find m($\overline{AB}$).

*Illustration 6*

**39.** A building casts a shadow of 50 ft at the same time a person casts a shadow of 8 ft. If the person is 6 ft tall, how tall is the building?

**40.** A person $5\frac{1}{2}$ ft tall casts a 3-ft shadow. How long is the shadow of a 60-ft flagpole?

**41.** A straight road going up a constant grade rises 3 ft in the first 50 ft of roadway. If a car travels 1000 ft from the bottom of the hill to the top, how high is the hill?

**42.** A ski hill has a run $\frac{1}{2}$ mile long. If the hill drops 5 feet as the skier slides 9 ft, how high is the hill? (*Hint*: 5280 ft = 1 mile.)

**43.** Sally is 5 ft tall. To measure the height of a smokestack, she asks her 6-ft boyfriend to help her. They stand as in Illustration 7 with her line of sight to the top of the stack passing by his head. If the boyfriend stands 126 ft from the smokestack and the girl is 3 ft from her friend, how tall is the smokestack?

**44.** Find the width of the river shown in Illustration 8.

Illustration 7             *Illustration 8*

**45.** An airplane ascends 100 ft as it flies a horizontal distance of 1000 ft. How much altitude will it gain if it flies 1 mile?

**46.** In a landing approach, an airplane descends 45 ft as it flies a horizontal distance of 500 ft. How far does the plane fly as it drops 66 ft?

**47.** To estimate the height of the Space Needle in Seattle, Juan notes that the Space Needle casts a shadow of approximately 173 ft when a 7-ft pole casts a shadow of 2 ft. Approximately how tall is the Space Needle?

**48.** While walking in the woods in Wisconsin, you come upon a tall white-pine tree that has been estimated to be over 400 yr old. A small pine tree 3-ft tall casts a shadow of 2 ft when the large pine casts a shadow of 116 ft. How tall is the white pine?

**49.** *Prove:* A line that bisects one side of a triangle and is parallel to one of its other sides must bisect the third side.

**50.** *Prove:* The midpoint of the hypotenuse of a right triangle is equidistant from its vertices. (*Hint:* Construct a line from the midpoint of the hypotenuse that is parallel to one of the legs of the triangle.)

**51.** *Prove:* The ratio of the perimeters of two similar triangles is equal to the ratio of the lengths of a pair of corresponding sides.

**52.** Refer to Illustration 9, in which $\overline{AD}$ and $\overline{BE}$ intersect at point $C$ and $\overline{AB} \parallel \overline{DE}$. Prove that

$$\frac{m(\overline{AB})}{m(\overline{DE})} = \frac{m(\overline{AC})}{m(\overline{DC})}$$

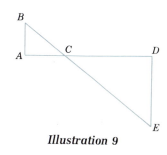

*Illustration 9*

**53.** Use the result of Exercise 52 to prove that $[m(\overline{DE})][m(\overline{AC})] = [(m\overline{AB})][m(\overline{DC})]$. (*Hint:* Consider Theorem 6.1.)

**54.** Refer to △$ABE$ in Illustration 10, in which ∠1 ≅ ∠$A$. Prove that

$$\frac{m(\overline{AE})}{m(\overline{AC})} = \frac{m(\overline{BE})}{m(\overline{BD})}$$

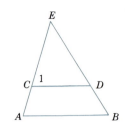

**Illustration 10**

**55.** Use the result of Exercise 54 to prove that $[m(\overline{AC})][m(\overline{BE})] = [m(\overline{AE})][m(\overline{BD})]$.

**56.** Refer to Illustration 11, in which △$ABC$ and △$ADE$ are right triangles. Prove that $[m(\overline{AD})][m(\overline{BC})] = [m(\overline{AB})][m(\overline{DE})]$.

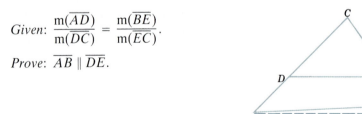

**Illustration 11**

# 6.3 MORE ON SIMILAR TRIANGLES

***Theorem 6.11*** If a line segment divides two sides of a triangle proportionally, then it is parallel to the third side.

*Given:* $\dfrac{m(\overline{AD})}{m(\overline{DC})} = \dfrac{m(\overline{BE})}{m(\overline{EC})}$.

*Prove:* $\overline{AB} \parallel \overline{DE}$.

*Construction:* Construct line segment $\overline{AF}$ parallel to line segment $\overline{DE}$, intersecting $\overleftrightarrow{BC}$ in some point, say $F$. This forms △$AFC$.

*Proof:*

| STATEMENTS | REASONS |
|---|---|
| **1.** $\overline{AF} \parallel \overline{DE}$. | **1.** By construction. |
| **2.** $\dfrac{m(\overline{AD})}{m(\overline{DC})} = \dfrac{m(\overline{FE})}{m(\overline{EC})}$. | **2.** Theorem 6.9. |

| STATEMENTS | REASONS |
|---|---|
| **3.** $\dfrac{m(\overline{AD})}{m(\overline{DC})} = \dfrac{m(\overline{BE})}{m(\overline{EC})}$. | **3.** Given. |
| **4.** $m(\overline{FB}) + m(\overline{BE}) = m(\overline{FE})$. | **4.** Postulate 1.10. |
| **5.** $m(\overline{BE}) = m(\overline{FE})$. | **5.** Theorem 6.7. |
| **6.** $m(\overline{FB}) = 0$. | **6.** Equals subtracted from equals are equal. |
| **7.** Points $F$ and $B$ must coincide. | **7.** If the length of a segment is 0, its endpoints must coincide. |
| **8.** $\overline{AB} \parallel \overline{DE}$. | **8.** $\overline{AF} \parallel \overline{DE}$ and $\overline{AF}$ and $\overline{AB}$ are the same line segment. |

The same proof applies if $F$ were to lie between $E$ and $B$. ☐

---

**Theorem 6.12** All congruent triangles are similar.

---

The proof of Theorem 6.12 is left as an exercise.

---

**Theorem 6.13** Two triangles similar to the same triangle are similar to each other.

---

The proof of Theorem 6.13 is left as an exercise.

---

**Theorem 6.14** If an angle in one triangle is congruent to an angle in a second triangle and the lengths of the respective sides including those angles are in proportion, then the triangles are similar.

---

*Given:* $\angle C \cong \angle F$.

$$\frac{m(\overline{DF})}{m(\overline{AC})} = \frac{m(\overline{EF})}{m(\overline{BC})}.$$

*Prove:* $\triangle ABC \sim \triangle DEF$.

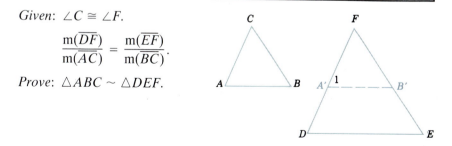

*Construction:* Construct point $A'$ on $\overline{FD}$ so that $\overline{FA'} \cong \overline{CA}$. Construct $B'$ on $\overline{FE}$ so that $\overline{FB'} \cong \overline{CB}$. Draw line segment $\overline{A'B'}$.

*Proof:*

| STATEMENTS | REASONS |
|---|---|
| **1.** $\angle C \cong \angle F$. | **1.** Given. |

| STATEMENTS | REASONS |
|---|---|
| **2.** $\overline{A'F} \cong \overline{AC}$; $\overline{B'F} \cong \overline{BC}$. | **2.** By construction. |
| **3.** $m(\overline{A'F}) = m(\overline{AC})$; $m(\overline{B'F}) = m(\overline{BC})$. | **3.** Congruent line segments have equal measures. |
| **4.** $\triangle ABC \cong \triangle A'B'F$. | **4.** SAS $\cong$ SAS. |
| **5.** $\triangle ABC \sim \triangle A'B'F$. | **5.** Congruent triangles are similar. |
| **6.** $\dfrac{m(\overline{DF})}{m(\overline{AC})} = \dfrac{m(\overline{EF})}{m(\overline{BC})}$. | **6.** Given. |
| **7.** $\dfrac{m(\overline{DF})}{m(\overline{A'F})} = \dfrac{m(\overline{EF})}{m(\overline{B'F})}$. | **7.** Substitution. |
| **8.** $\dfrac{m(\overline{DF}) - m(\overline{A'F})}{m(\overline{A'F})}$ $= \dfrac{m(\overline{EF}) - m(\overline{B'F})}{m(\overline{B'F})}$. | **8.** A proportion can be written by subtraction. |
| **9.** $m(\overline{DA'}) = m(\overline{DF}) - m(\overline{A'F})$. | **9.** Postulate 1.10. |
| **10.** $m(\overline{EB'}) = m(\overline{EF}) - m(\overline{B'F})$. | **10.** Postulate 1.10. |
| **11.** $\dfrac{m(\overline{DA'})}{m(\overline{A'F})} = \dfrac{m(\overline{EB'})}{m(\overline{B'F})}$. | **11.** Substitution. |
| **12.** $\overline{A'B'} \parallel \overline{DE}$. | **12.** Theorem 6.11. |
| **13.** $\angle D \cong \angle 1$. | **13.** If two parallel lines are cut by a transversal, then corresponding angles are congruent. |
| **14.** $\angle F \cong \angle F$. | **14.** An angle is congruent to itself. |
| **15.** $\triangle A'B'F \sim \triangle DEF$. | **15.** If two angles of one triangle are congruent to two angles of a second triangle, then the triangles are similar. |
| **16.** $\triangle ABC \sim \triangle DEF$. | **16.** If two triangles are similar to the same triangle, then they are similar to each other. □ |

---

**Theorem 6.15** If the lengths of three sides of one triangle are in proportion to the lengths of three corresponding sides of a second triangle, then the triangles are similar.

---

*Given:* $\dfrac{m(\overline{DF})}{m(\overline{AC})} = \dfrac{m(\overline{EF})}{m(\overline{BC})} = \dfrac{m(\overline{DE})}{m(\overline{AB})}$.

*Prove*: $\triangle DEF \sim \triangle ABC$.

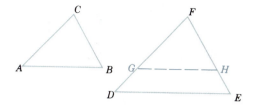

*Construction*: On sides $\overline{DF}$ and $\overline{EF}$ mark off points $G$ and $H$ so that $\overline{AC} \cong \overline{GF}$ and $\overline{BC} \cong \overline{HF}$.

*Proof*:

| STATEMENTS | REASONS |
| --- | --- |
| 1. $\angle F \cong \angle F$. | 1. Reflexive law. |
| 2. $\dfrac{m(\overline{DF})}{m(\overline{AC})} = \dfrac{m(\overline{EF})}{m(\overline{BC})}$. | 2. Given. |
| 3. $\overline{AC} \cong \overline{GF}$; $\overline{BC} \cong \overline{HF}$. | 3. By construction. |
| 4. $m(\overline{AC}) = m(\overline{GF})$; $m(\overline{BC}) = m(\overline{HF})$. | 4. Congruent segments have equal measures. |
| 5. $\dfrac{m(\overline{DF})}{m(\overline{GF})} = \dfrac{m(\overline{EF})}{m(\overline{HF})}$. | 5. Substitution. |
| 6. $\triangle DEF \sim \triangle GHF$. | 6. Theorem 6.14. |
| 7. $\dfrac{m(\overline{DF})}{m(\overline{GF})} = \dfrac{m(\overline{DE})}{m(\overline{GH})}$. | 7. The lengths of corresponding sides of similar triangles are in proportion. |
| 8. $\dfrac{m(\overline{DF})}{m(\overline{AC})} = \dfrac{m(\overline{DE})}{m(\overline{GH})}$. | 8. Substitution. |
| 9. $\dfrac{m(\overline{DF})}{m(\overline{AC})} = \dfrac{m(\overline{DE})}{m(\overline{AB})}$. | 9. Given. |
| 10. $m(\overline{AB}) = m(\overline{GH})$. | 10. Theorem 6.7 applied to Statements 8 and 9. |
| 11. $\overline{AB} \cong \overline{GH}$. | 11. If two segments have equal measures, then they are congruent. |
| 12. $\triangle ABC \cong \triangle GHF$. | 12. SSS $\cong$ SSS. |
| 13. $\triangle ABC \sim \triangle GHF$. | 13. Congruent triangles are similar. |
| 14. $\triangle ABC \sim \triangle DEF$. | 14. Theorem 6.13.    □ |

**Definition 6.5** If the second and third terms of a proportion are equal, then either the second or the third term is called the **mean proportional** between the first and fourth terms.

**EXAMPLE 1**    Find the positive number that is a mean proportional between 2 and 32.

Solution    We know that the first term of the proportion is 2 and that the fourth term is 32. Because we are finding a mean proportional, we let $x$ represent both the second and the third terms. We set up a proportion and solve for $x$.

$$\frac{2}{x} = \frac{x}{32}$$

$$x^2 = 64 \qquad \text{The product of the extremes equals the product of the means.}$$

$$x = 8 \text{ or } x = -8 \qquad \text{Take the square roots of both sides.}$$

Thus, the positive number that is a mean proportional between 2 and 32 is 8. Note that $-8$ is a negative number that is also a mean proportional between 2 and 32.    ■

---

**Theorem 6.16** If the altitude is drawn on the hypotenuse of a right triangle, then the length of either leg is the mean proportional between the length of the hypotenuse and the length of the segment of the hypotenuse adjacent to the leg.

---

*Given*: Right $\triangle ABC$ with $\angle C = 90°$.

$\overline{CD} \perp \overline{AB}$.

*Prove*: $\dfrac{m(\overline{AB})}{m(\overline{AC})} = \dfrac{m(\overline{AC})}{m(\overline{AD})}$.

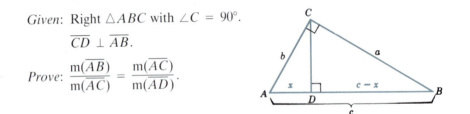

*Proof*:

| STATEMENTS | REASONS |
|---|---|
| **1.** Right $\triangle ABC$ with $\angle C = 90°$; $CD \perp \overline{AB}$. | **1.** Given. |
| **2.** $\angle A \cong \angle A$. | **2.** Reflexive law. |
| **3.** $\angle ADC \cong \angle ACB$. | **3.** All right angles are congruent. |
| **4.** $\triangle ABC \sim \triangle ACD$. | **4.** If two angles of one triangle are congruent to two angles of a second triangle, then the triangles are similar. |
| **5.** $\dfrac{m(\overline{AB})}{m(\overline{AC})} = \dfrac{m(\overline{AC})}{m(\overline{AD})}$. | **5.** The lengths of corresponding sides of similar triangles are in proportion. |

By similar reasoning we can prove that $\dfrac{m(\overline{AB})}{m(\overline{BC})} = \dfrac{m(\overline{BC})}{m(\overline{BD})}$.  ☐

In Chapter 5 we proved the Pythagorean theorem. Many people have enjoyed attempting to find new proofs of this theorem. Many such proofs exist. The one we present now is common to many textbooks. A third proof is given in Appendix I.

> ***Theorem 5.5 The Pythagorean Theorem.***  In a right triangle the square of the length of the hypotenuse is equal to the sum of the squares of the lengths of the other two sides.

*Given*: Right $\triangle ABC$.

      $\overline{CD} \perp \overline{AB}$.

*Prove*: $a^2 + b^2 = c^2$.

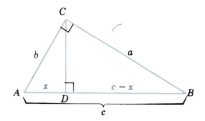

*Second Proof*:

| STATEMENTS | REASONS |
|---|---|
| **1.** Right $\triangle ABC$ with $\overline{CD} \perp \overline{AB}$. | **1.** Given. |
| **2.** $\dfrac{c}{b} = \dfrac{b}{x}; \dfrac{c}{a} = \dfrac{a}{c-x}$. | **2.** Theorem 6.16. |
| **3.** $b^2 = cx; a^2 = c^2 - cx$. | **3.** Theorem 6.1. |
| **4.** $a^2 + b^2 = c^2$. | **4.** Equals added to equals are equal.  ☐ |

> ***Theorem 6.17*** The length of the altitude on the hypotenuse of a right triangle is the mean proportional between the lengths of the segments of the hypotenuse.

*Given*: Right $\triangle ABC$.

*Prove*: $\dfrac{m(\overline{AD})}{m(\overline{CD})} = \dfrac{m(\overline{CD})}{m(\overline{BD})}$.

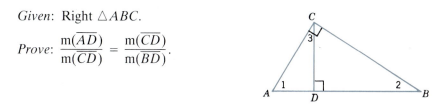

*Construction*: Construct $\overline{CD}$ perpendicular to $\overline{AB}$ to form right $\triangle ADC$.

*Proof*:

| STATEMENTS | REASONS |
|---|---|
| **1.** Right △$ABC$. | **1.** Given. |
| **2.** $\overline{CD} \perp \overline{AB}$; Right △$ADC$. | **2.** By construction. |
| **3.** ∠1 complementary ∠2. | **3.** The acute angles in a right triangle are complementary. |
| **4.** ∠1 complementary ∠3. | **4.** The acute angles in a right triangle are complementary. |
| **5.** m(∠2) = m(∠3). | **5.** Complements of the same angle are equal. |
| **6.** ∠2 ≅ ∠3. | **6.** If angles have equal measures, they are congruent. |
| **7.** ∠$ADC$ ≅ ∠$CDB$. | **7.** Perpendicular lines meet and form congruent adjacent angles. |
| **8.** △$ADC$ ~ △$CDB$. | **8.** Postulate 6.1. |
| **9.** $\dfrac{m(\overline{AD})}{m(\overline{CD})} = \dfrac{m(\overline{CD})}{m(\overline{BD})}$. | **9.** Corresponding sides of similar triangles are in proportion.    □ |

**EXAMPLE 2**

In Figure 6.8, right △$ABC$ has dimensions as shown. Find the length of (a) $\overline{AD}$ and (b) $\overline{CD}$.

Solution

(a) The length of either leg of a right triangle is the mean proportional between the length of the hypotenuse and the length of the segment of the hypotenuse adjacent to the leg. Thus, let $x$ represent m($\overline{AD}$), set up a proportion, and solve for $x$.

$$\frac{x}{5} = \frac{5}{13}$$

$$13x = 25 \qquad \text{The product of the extremes is equal to the product of the means.}$$

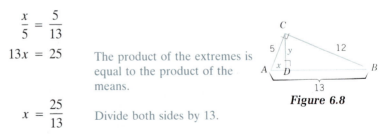

Figure 6.8

$$x = \frac{25}{13} \qquad \text{Divide both sides by 13.}$$

(b) The length of the altitude upon the base of a right triangle is the mean proportional between the lengths of the segments of the hypotenuse. Thus, let $y$ represent m($\overline{CD}$), set up a proportion, and solve for $y$.

$$\frac{\frac{25}{13}}{y} = \frac{y}{\frac{144}{13}} \qquad \text{Note that m(}\overline{DB}\text{)} = 13 - \frac{25}{13} \text{ or } \frac{144}{13}.$$

$$y^2 = \frac{3600}{169}$$ The product of the extremes is equal to the product of the means.

$$y = \frac{60}{13}$$ Take the square root of both sides.

Thus, m($\overline{AD}$) = $\frac{25}{13}$ and m($\overline{CD}$) = $\frac{60}{13}$. ∎

The concept of similarity of triangles may be generalized to include all polygons.

---

**Definition 6.6** Two **polygons are similar** if, and only if, all pairs of corresponding angles are congruent and the lengths of all pairs of corresponding sides are in proportion.

---

By definition of similar polygons, we know that the figures shown below are similar if $\angle A \cong \angle A'$, $\angle B \cong \angle B'$, $\angle C \cong \angle C'$, $\angle D \cong \angle D'$, $\angle E \cong \angle E'$ and that

$$\frac{m(\overline{AB})}{m(\overline{A'B'})} = \frac{m(\overline{BC})}{m(\overline{B'C'})} = \frac{m(\overline{CD})}{m(\overline{C'D'})} = \frac{m(\overline{DE})}{m(\overline{D'E'})} = \frac{m(\overline{EA})}{m(\overline{E'A'})}$$

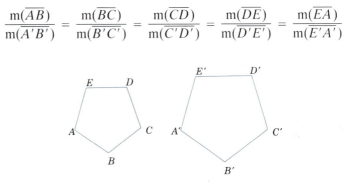

# EXERCISE 6.3

1. Is a triangle with sides measuring 3, 5, and 7 cm similar to a triangle with sides measuring 12, 20, and 28 cm? Explain.

2. Is a triangle with sides measuring 4, 6, and 9 m similar to a triangle with sides measuring 20, 35, and 50 m? Explain.

3. Is a triangle with sides measuring 3, 4, and 5 ft similar to a triangle with sides measuring 9, 12, and 15 cm? Explain.

4. Is a triangle with sides measuring 7, 8, and 10 in. similar to a triangle with sides measuring 28, 32, and 39 m? Explain.

5. Find the positive mean proportional between 4 and 16.

6. Find the positive mean proportional between 8 and 64.

7. Find two mean proportionals between 3 and 27.

8. Find two mean proportionals between 4 and 100.

In Exercises 9–24, refer to Illustration 1 in which triangles $ABC$, $ACD$, and $CBD$ are right triangles.

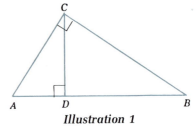

**9.** If m($\overline{AD}$) = 2 in. and m($\overline{AB}$) = 32 in., find m($\overline{AC}$).

**10.** If m($\overline{AD}$) = 24 m and m($\overline{DB}$) = 3 m, find m($\overline{CB}$).

**11.** If m($\overline{AD}$) = 3 ft and m($\overline{CD}$) = 6 ft, find m($\overline{DB}$).

**12.** If m($\overline{AD}$) = 4 ft and m($\overline{DB}$) = 16 ft, find m($\overline{CD}$).

**13.** If m($\overline{AC}$) = 6 yd and m($\overline{AD}$) = 1 yd, find m($\overline{DB}$).

**14.** If m($\overline{DB}$) = 2 in. and m($\overline{CB}$) = 10 in., find m($\overline{AD}$).

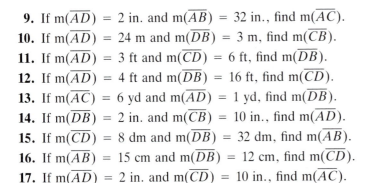

*Illustration 1*

**15.** If m($\overline{CD}$) = 8 dm and m($\overline{DB}$) = 32 dm, find m($\overline{AB}$).

**16.** If m($\overline{AB}$) = 15 cm and m($\overline{DB}$) = 12 cm, find m($\overline{CD}$).

**17.** If m($\overline{AD}$) = 2 in. and m($\overline{CD}$) = 10 in., find m($\overline{AC}$).

**18.** If m($\overline{AB}$) = 25 ft and m($\overline{AC}$) = 5 ft, find m($\overline{CD}$).

**19.** If m($\overline{AB}$) = 32 ft and m($\overline{CB}$) = 8 ft, find m($\overline{CD}$).

**20.** If m($\overline{CD}$) = 12 m and m($\overline{DB}$) = 36 m, find m($\overline{CB}$).

**21.** If m($\overline{AD}$) = 4 yd and m($\overline{DB}$) = 16 yd, find the area of $\triangle ABC$.

**22.** If m($\overline{AD}$) = 2 ft and m($\overline{AB}$) = 20 ft, find the area of $\triangle ABC$.

**23.** If m($\overline{AD}$) = 2 m and m($\overline{DB}$) = 72 m, find the area of $\triangle ADC$.

**24.** If m($\overline{CB}$) = 8 in. and m($\overline{DB}$) = 4 in., find the area of $\triangle ADC$.

**25.** A small tree 2-ft tall casts a shadow of 2.4 ft when a large tree casts a shadow of 45 ft. How tall is the larger tree?

**26.** A smokestack casts a shadow of 35 m while a meterstick casts a shadow of 1.5 m. How tall is the smokestack?

**27.** One leg of a right triangle is 5 cm long and its hypotenuse is 13 cm long. Find the area of the triangle.

**28.** An airplane ascends 950 ft as it flies a horizontal distance of 1 mile. How much altitude would the plane gain if it flew a horizontal distance of 10,000 ft? (*Hint*: 1 mile = 5280 ft.)

**29.** An isosceles triangle has sides of 41 cm and a base of 18 cm. Find its area.

**30.** Suppose that we wish to find the distance $d$ across a river but are unable to swim to the other side and measure the distance directly. By using similar triangles, we can find the distance easily. Find a marker on the opposite side of the river. A few feet from the river and on your

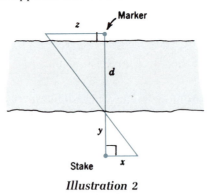

*Illustration 2*

side place a stake in the ground directly across from the marker. Measure a given distance on a line perpendicular to the line determined by the marker and the stake and form similar triangles as in Illustration 2. From your side of the river, you can measure sides $x$, $y$, and $z$. With these measures how would you determine the distance $d$?

**31.** Find the distance across the river in Exercise 30 if $x = 2$ ft, $y = 6$ ft, and $z = 50$ ft.

**32.** *Prove*: Two isosceles triangles are similar if the vertex angle of one triangle is equal to the vertex angle of the other triangle.

**33.** *Prove*: The line segments joining the midpoints of the sides of a triangle form a triangle similar to the given triangle.

**34.** *Prove*: In similar triangles the ratio of two corresponding medians is equal to the ratio formed by a pair of corresponding sides.

**35.** *Prove*: In similar triangles the ratio of the interior segments of two corresponding angle bisectors is equal to the ratio formed by a pair of corresponding sides.

**36.** *Prove*: In similar triangles the ratio of two corresponding altitudes is equal to the ratio formed by a pair of corresponding sides.

**37.** *Prove*: The perimeters of two similar triangles have the same ratio as the ratio formed by a pair of corresponding sides.

**38.** *Prove*: All squares are similar.

**39.** *Prove Theorem 6.12*: All congruent triangles are similar.

**40.** *Prove Theorem 6.13*: Two triangles similar to the same triangle are similar to each other.

# 6.4 INTRODUCTION TO TRIGONOMETRY; THE TANGENT RATIO

The properties of similar triangles provide the basis for a special area of mathematics called **trigonometry**. The word *trigonometry* comes from Greek words meaning *triangle measurement*. Trigonometry deals with relations among the sides and angles of triangles. These relationships enable us to measure angles and distances indirectly. Trigonometry is used in physics, carpentry, engineering, architecture, surveying, astronomy, and navigation.

Special ratios of the measures of pairs of sides of a right triangle are important in trigonometry. Before we discuss these ratios, we must consider two ideas. The first is a fact about similar triangles: If an acute angle of one right triangle is congruent to an acute angle of another right triangle, the triangles are similar.

In Figure 6.9, $\triangle ABC$ and $\triangle ADE$ are right triangles with $\angle A$ as a common acute angle. Thus, these triangles are similar. Because the measures of corresponding sides of similar triangles are in proportion, we have

$$\frac{m(\overline{BC})}{m(\overline{DE})} = \frac{m(\overline{AB})}{m(\overline{AD})}$$

By interchanging the means in this proportion, we obtain

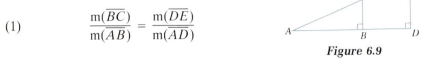

Figure 6.9

$$(1) \qquad \frac{m(\overline{BC})}{m(\overline{AB})} = \frac{m(\overline{DE})}{m(\overline{AD})}$$

The fraction on the left-hand side of Equation 1 is the ratio of two sides of $\triangle ABC$ shown in Figure 6.9. The fraction on the right-hand side is the ratio of the *corresponding two sides of* $\triangle ADE$. The values of these ratios depend on the size of the acute angle. They do not depend on the size of the triangles.

Likewise, it can be shown that measures of other pairs of sides in these triangles are also in proportion. In particular, the following proportions are true:

$$\frac{m(\overline{BC})}{m(\overline{AC})} = \frac{m(\overline{ED})}{m(\overline{AE})} \qquad \text{and} \qquad \frac{m(\overline{AB})}{m(\overline{AC})} = \frac{m(\overline{AD})}{m(\overline{AE})}$$

The second idea to be considered is the concept of **opposite** and **adjacent sides** in a right triangle. In right $\triangle ABC$ in Figure 6.10, the two acute angles are labeled $\angle A$ and $\angle B$. The right angle is designated as $\angle C$. The side that is across the triangle from $\angle A$ is named $a$, and we say that side $a$ is opposite $\angle A$. Side $b$ is one of the sides of $\angle A$ itself, and we say that side $b$ is adjacent to $\angle A$. The concepts of *opposite* and *adjacent* are relative. Side $a$ is opposite $\angle A$ but is adjacent to $\angle B$. Side $b$ is opposite $\angle B$ but is adjacent to $\angle A$. Side $c$ is the **hypotenuse**.

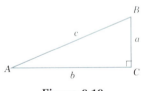

Figure 6.10

There are six trigonometric ratios. In this book we will discuss three. The first of these is the **tangent ratio**.

---

*Definition 6.7 The Tangent Ratio.*   The **tangent of an acute angle of a right triangle** is the ratio of the side opposite the angle to the side adjacent to the angle.

The tangent of $\angle A$ is abbreviated as **tan $A$**.

---

In Figure 6.11, the tangent of $\angle A$ is the ratio of the side opposite $\angle A$ to the side adjacent to $\angle A$.

$$\text{tangent of } \angle A = \frac{\text{side opposite } \angle A}{\text{side adjacent } \angle A}$$

or simply

$$\tan A = \frac{a}{b}$$

*Figure 6.11*

**EXAMPLE 1**    Refer to the right triangle in Figure 6.12. Find (a) tan $A$ and (b) tan $B$.

*Solution*    (a)    The side opposite $\angle A$ has a length of 4 cm. The side adjacent to $\angle A$ has a length of 3 cm. Thus,

$$\tan A = \frac{\text{side opposite } \angle A}{\text{side adjacent } \angle A}$$

$$= \frac{4}{3}$$

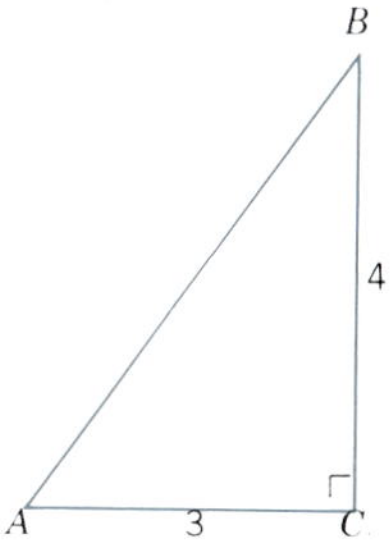

*Figure 6.12*

(b)    The side opposite $\angle B$ has a length of 3 cm. The side adjacent to $\angle B$ has a length of 4 cm. Thus,

$$\tan B = \frac{\text{side opposite } \angle B}{\text{side adjacent } \angle B}$$

$$= \frac{3}{4}$$

In this example, tan $A$ and tan $B$ are reciprocals of each other.    ■

**EXAMPLE 2**    Find tan 45°.

*Solution*    Draw a right triangle with an acute angle of 45° as in Figure 6.13. This triangle is an isosceles-right triangle. Because the values of the ratios of the sides of a right triangle depend on the angles of the triangle and not on the lengths of the sides, you can assume that the two legs of the triangle are each 1 unit long. Then, you have

$$\tan A = \frac{\text{side opposite } \angle A}{\text{side adjacent } \angle A}$$

$$\tan 45° = \frac{m(\overline{BC})}{m(\overline{AC})}$$

$$= \frac{1}{1}$$

$$= 1$$

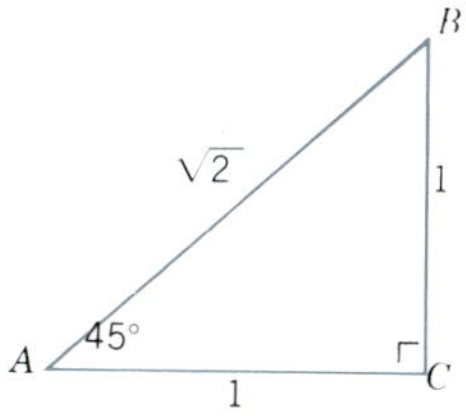

*Figure 6.13*

Thus, tan 45° = 1.    ■

It is possible to find the tangent of an angle without drawing a right triangle. To do so, we can use a scientific calculator set in degree mode. For example, to use a calculator to find tan 23°, we enter the number **23** and press the $\boxed{\text{TAN}}$ key. The number that appears on the display of the calculator is the tangent of 23°:

$$\tan 23° \approx 0.4244748162$$

**EXAMPLE 3**    A woman, sitting on the ground 47 ft from the base of a flagpole, looks up an angle of 38° to see its top. How tall is the flagpole?

*Solution*    The information of the example is summarized in Figure 6.14. The side opposite $\angle H$ is the height of the flagpole. Because that height is unknown, we will denote it with the variable $h$. The side adjacent to $\angle H$ is 47 ft long and m($\angle H$) is 38°. The tangent of 38° is the ratio of the measure of the side opposite $\angle H$ to the measure of the side adjacent to $\angle H$. Thus,

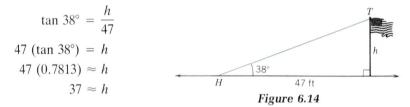

$$\tan 38° = \frac{h}{47}$$
$$47\,(\tan 38°) = h$$
$$47\,(0.7813) \approx h$$
$$37 \approx h$$

*Figure 6.14*

The flagpole is approximately 37 ft tall.    ■

EXERCISE 6.4

In Exercises 1–10, refer to Illustration 1 and use a calculator to find the value requested.

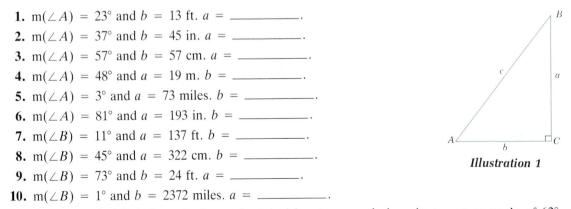

**1.** m($\angle A$) = 23° and $b$ = 13 ft. $a$ = _____.

**2.** m($\angle A$) = 37° and $b$ = 45 in. $a$ = _____.

**3.** m($\angle A$) = 57° and $b$ = 57 cm. $a$ = _____.

**4.** m($\angle A$) = 48° and $a$ = 19 m. $b$ = _____.

**5.** m($\angle A$) = 3° and $a$ = 73 miles. $b$ = _____.

**6.** m($\angle A$) = 81° and $a$ = 193 in. $b$ = _____.

**7.** m($\angle B$) = 11° and $a$ = 137 ft. $b$ = _____.

**8.** m($\angle B$) = 45° and $a$ = 322 cm. $b$ = _____.

**9.** m($\angle B$) = 73° and $b$ = 24 ft. $a$ = _____.

**10.** m($\angle B$) = 1° and $b$ = 2372 miles. $a$ = _____.

*Illustration 1*

**11.** Corey stands 123 ft from the base of a television tower and views its top at an angle of 62°. Find the height of the tower.

**12.** Martha stands 256 ft from the base of a tall building and views its top at an angle of 73°. Find the height of the building.

**13.** Jerry stands at a distance from the base of a 50-ft flagpole and views its top at an angle of 41°. How far from the flagpole does Jerry stand?

**14.** From the top of a tower 235 ft tall, a worker looks down toward a car at an angle of 77°. See Illustration 2. How far is the car from the base of the building?

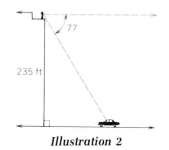

*Illustration 2*

**15.** When the sun is 37° above the horizon, a tree casts a 57-ft shadow. Find the height of the tree.

**16.** When the sun is 27° above the horizon, how long a shadow does a 63-ft tree cast?

**17.** Bill views the top of the 555-ft tall Washington Monument at an angle of 69°. How far is Bill from the base of the monument?

**18.** The measure of a base angle of an isosceles triangle is 37° and the length of its base is 14 ft. Find the height of the triangle. (*Hint:* The altitude bisects the base.)

**19.** One leg of a right triangle is 12 ft long and the angle opposite that leg measures 76°. Find the area of the triangle.

**20.** The length of the base of an isosceles triangle is 42 ft and each of the base angles measures 29°. Find the area of the triangle.

**21.** One leg of a right triangle is three times as long as the other leg. Find the tangent of the triangle's smallest angle.

**22.** An equilateral triangle has a perimeter of 24 cm. Find its area.

**23.** Use the properties of a 30°-60°-right triangle to find tan 30°.

**24.** Find tan 60°. Refer to Exercise 23.

# 6.5 THE SINE AND COSINE RATIOS

Another ratio of the measures of the sides of a right triangle is the **sine ratio**.

---

> *Definition 6.8* **The Sine Ratio.** The **sine of an acute angle of a right triangle** is the ratio of the length of the side opposite the angle to the length of the hypotenuse.
> The sine of $\angle A$ is abbreviated as sin $A$.

---

In Figure 6.15, the sine of $\angle A$ is the ratio of the length of the side opposite $\angle A$ to the length of the hypotenuse.

$$\text{Sine of } \angle A = \frac{\text{side opposite } \angle A}{\text{hypotenuse}}$$

or simply

$$\sin \angle A = \frac{a}{c}$$

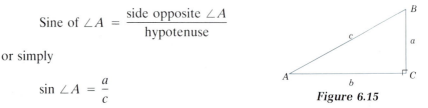

**Figure 6.15**

**EXAMPLE 1**

Refer to the right triangle in Figure 6.16 and find (a) sin $A$ and (b) sin $B$.

Solution

(a)  The length of the side opposite $\angle A$ is 4 units and the length of the hypotenuse is 5 units. Thus,

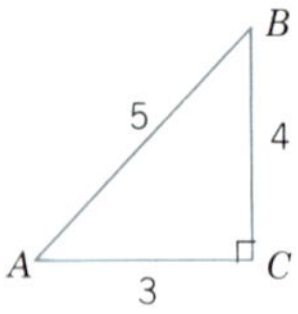

$$\sin A = \frac{\text{side opposite } \angle A}{\text{hypotenuse}}$$

$$= \frac{4}{5}$$

**Figure 6.16**

(b)  The length of the side opposite $\angle B$ is 3 units. The length of the hypotenuse is 5 units. Thus,

$$\sin B = \frac{\text{side opposite } \angle B}{\text{hypotenuse}}$$

$$= \frac{3}{5}$$

∎

**EXAMPLE 2**

Find sin 30°.

Solution

Draw a right triangle with an acute angle of 30° as in Figure 6.17. In this 30°-60°-right triangle, the side opposite the 30° angle is half as long as the hypotenuse.

Because the value of sin 30° depends on the size of the angle and not on the size of the triangle, we can assume that the length of the leg opposite the 30° angle is 1 unit long. Then, the hypotenuse will be 2 units long. The sine of 30° is the sine of $\angle A$. Thus,

$$\sin A = \frac{\text{side opposite } \angle A}{\text{hypotenuse}}$$

$$\sin 30° = \frac{m(\overline{BC})}{m(\overline{AB})}$$

$$= \frac{1}{2}$$

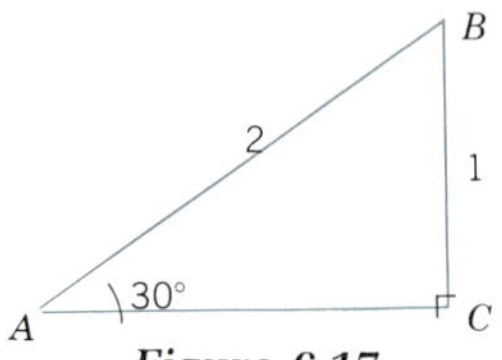

**Figure 6.17**

Thus, sin 30° $= \frac{1}{2}$.

∎

It is possible to use a scientific calculator to find the sine of an angle. For example, to find sin 57°, we **set the** calculator in degree mode, enter the number **57**, and press the $\boxed{\text{SIN}}$ key. The number that appears on the display of the calculator is the sine of 57°.

$$\sin 57° \approx 0.83867057$$

The sine ratio is also useful for measuring certain distances indirectly.

**EXAMPLE 3**

A circus tightrope walker ascends from the ground to a platform 75 ft above the arena by walking up a taut cable that makes an angle of 55° with the ground. How long is the cable?

*Solution*

The given information is summarized in Figure 6.18. To find the value of $x$, we can use the sine ratio as follows:

$$\sin 55° = \frac{\text{m}(\overline{BC})}{\text{m}(\overline{AC})}$$

$$\sin 55° = \frac{75}{x}$$

$$x(\sin 55°) = 75$$

$$x = \frac{75}{\sin 55°}$$

$$\approx \frac{75}{0.8191520}$$

$$\approx 91.56$$

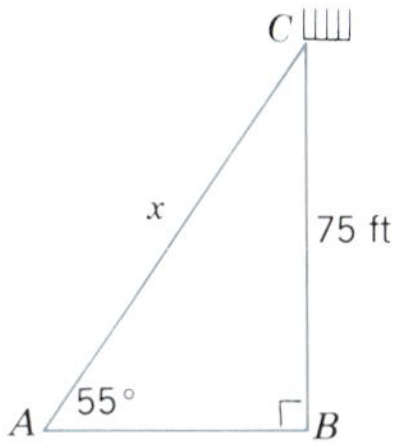

The cable is approximately 92 ft long. ∎

**Figure 6.18**

## THE COSINE RATIO

A third ratio of the measures of the sides of a right triangle is the **cosine ratio**.

> **Definition 6.9 The Cosine Ratio.** The **cosine of an acute angle of a right triangle** is the ratio of the length of the side adjacent the angle to the length of the hypotenuse.
> The cosine of $\angle A$ is abbreviated as cos $A$.

In Figure 6.19, the cosine of $\angle A$ is the ratio of the length of the side adjacent to $\angle A$ to the length of the hypotenuse.

$$\text{Cosine of } \angle A = \frac{\text{side adjacent to } \angle A}{\text{hypotenuse}}$$

or simply

$$\cos \angle A = \frac{b}{c}$$

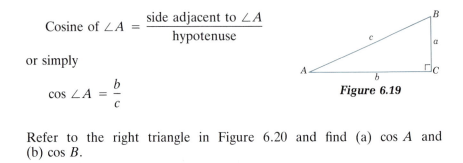

*Figure 6.19*

**EXAMPLE 1**   Refer to the right triangle in Figure 6.20 and find (a) cos *A* and (b) cos *B*.

SOLUTION   (a)  The length of the side adjacent to $\angle A$ is 3 units and the length of the hypotenuse is 5 units. Thus,

$$\cos A = \frac{\text{side adjacent } \angle A}{\text{hypotenuse}}$$

$$= \frac{3}{5}$$

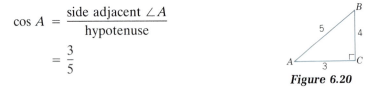

*Figure 6.20*

(b)  The length of the side adjacent to $\angle B$ is 4. The length of the hypotenuse is 5. Thus,

$$\cos B = \frac{\text{side adjacent } \angle B}{\text{hypotenuse}}$$

$$= \frac{4}{5}$$   ■

**EXAMPLE 2**   Find cos 30°.

Solution   Draw a right triangle with an acute angle of 30° as in Figure 6.21. In the 30°-60°-right triangle, the side adjacent to the 30° angle is $\sqrt{3}$ times as long as the side opposite the 30° angle, and the hypotenuse is twice as long as the side opposite the 30° angle.

Because cos 30° depends on the size of the angle and not the size of the triangle, we can assume that the length of the leg opposite the 30° angle is 1 unit long. Then, the hypotenuse will be 2 units long and the adjacent side $\sqrt{3}$ units long. The cosine of 30° is the cosine of $\angle A$. Thus,

$$\cos A = \frac{\text{side adjacent to } \angle A}{\text{hypotenuse}}$$

$$\cos 30° = \frac{m(\overline{AC})}{m(\overline{AB})}$$

$$= \frac{\sqrt{3}}{2}$$

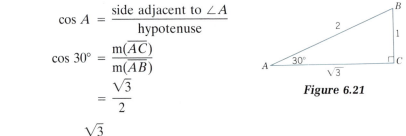

*Figure 6.21*

Thus, $\cos 30° = \dfrac{\sqrt{3}}{2}$.   ■

It is possible to use a scientific calculator to find the cosine of an angle. For example, to find cos 13°, we set the calculator in degree mode, enter the number **13**, and press the $\boxed{\text{COS}}$ key. The number that appears on the display of the calculator is the cosine of 13°.

$$\cos 13° \approx 0.97437006$$

The cosine ratio is also useful for measuring certain distances indirectly.

**EXAMPLE 3**      The side of a hill makes an angle of 8° with the horizontal. A dam is to be built to flood the valley and submerge 1500 ft of hillside. How far from the dam will the new shoreline be?

*Solution*      The given information is summarized in Figure 6.22. The distance between the dam and the shoreline is the length of $\overline{DS}$. Segments $\overline{DS}$ and $\overline{AB}$ are parallel making alternate interior angles $\angle DSA$ and $\angle SAB$ congruent angles. Use the cosine of $\angle DSA$ to calculate the length of $\overline{DS}$ as follows:

$$\cos \angle DSA = \frac{m(\overline{DS})}{m(\overline{AS})}$$

$$\cos 8° = \frac{m(\overline{DS})}{1500}$$

$$m(\overline{DS}) = 1500(\cos 8°)$$

$$m(\overline{DS}) \approx 1500(0.99026809)$$

$$m(\overline{DS}) \approx 1485$$

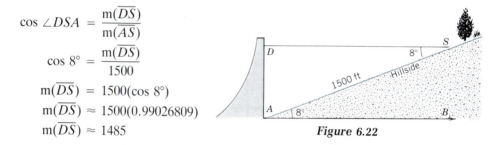

**Figure 6.22**

The new shoreline will be approximately 1485 ft from the dam.  ■

## EXERCISE 6.5

In Exercises 1–10, refer to Illustration 1. Use a calculator to find the required value.

**1.** $m(\angle A) = 47°$ and $a = 47$ ft. $c =$ _____.

**2.** $m(\angle A) = 32°$ and $a = 57$ in. $c =$ _____.

**3.** $m(\angle A) = 12°$ and $a = 86$ cm. $c =$ _____.

**4.** $m(\angle A) = 19°$ and $c = 90$ m. $a =$ _____.

**5.** $m(\angle A) = 53°$ and $c = 32$ miles. $a =$ _____.

**6.** $m(\angle A) = 87°$ and $c = 321$ in. $a =$ _____.

**7.** $m(\angle B) = 31°$ and $b = 457$ ft. $c =$ _____.

**8.** $m(\angle B) = 75°$ and $b = 533$ cm. $c =$ _____.

**9.** $m(\angle B) = 89°$ and $c = 12$ tt. $b =$ _____.

**10.** $m(\angle B) = 1°$ and $c = 5377$ m. $b =$ _____.

**Illustration 1**

In Exercises 11–20, refer to Illustration 2. Use a calculator to find the required value.

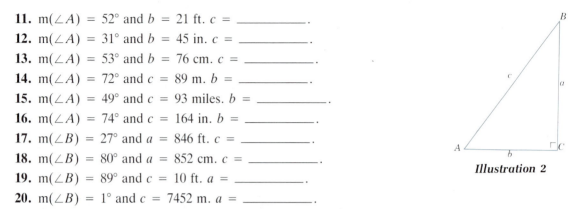

**11.** m($\angle A$) = 52° and $b$ = 21 ft. $c$ = _____.

**12.** m($\angle A$) = 31° and $b$ = 45 in. $c$ = _____.

**13.** m($\angle A$) = 53° and $b$ = 76 cm. $c$ = _____.

**14.** m($\angle A$) = 72° and $c$ = 89 m. $b$ = _____.

**15.** m($\angle A$) = 49° and $c$ = 93 miles. $b$ = _____.

**16.** m($\angle A$) = 74° and $c$ = 164 in. $b$ = _____.

**17.** m($\angle B$) = 27° and $a$ = 846 ft. $c$ = _____.

**18.** m($\angle B$) = 80° and $a$ = 852 cm. $c$ = _____.

**19.** m($\angle B$) = 89° and $c$ = 10 ft. $a$ = _____.

**20.** m($\angle B$) = 1° and $c$ = 7452 m. $a$ = _____.

*Illustration 2*

**21.** A high-wire artist ascends to a platform by walking 195 ft up an inclined wire that makes an angle of 33° with the horizontal. How far above the ground is the platform?

**22.** A 290-ft tall tower is supported by guy wires that run from the tower's top to the ground making angles of 75° with the horizontal. Find the length of each wire.

**23.** Carlos stands at a distance from the base of a 75-ft tall building and views its top at an angle of 57°. How far is Carlos from the *top* of the building?

**24.** From the top of a 235-ft tall building, a steeplejack looks down on a car at an angle of 77° from the horizontal. How far is the steeplejack from the car?

**25.** The vertex angle of an isosceles triangle is 64° and the length of its base is 24 in. Find the length of one of its congruent sides.

**26.** The vertex angle of an isosceles triangle is 72° and the length of one of its congruent sides is 27 ft. Find the length of the base of the triangle.

**27.** A stream flows down a mountain with an angle of descent of 6° from the horizontal. How far does the stream fall as it flows a distance of 2 miles?

**28.** The measure of a base angle of an isosceles triangle is 55°. The length of its base is 43 cm. Find the perimeter of the triangle.

**29.** The length of the hypotenuse of a right triangle is 37 in. and one acute angle measures 28°. Find the length of the altitude drawn to the hypotenuse.

**30.** Use the properties of a 30°-60°-right triangle to find sin 60°.

**31.** Use the properties of an isosceles-right triangle to find sin 45°.

**32.** A high-wire artist ascends to a platform by walking up a 174-ft inclined wire that makes an angle of 25° with the horizontal. How far from the base of the tower is the wire anchored?

**33.** A transmitter tower is supported by guy wires that run from the tower's top to the ground. Each wire makes an angle of 67° with the horizontal and is 423 ft long. How far from the base of the tower is each wire anchored?

**34.** A hillside has a slope of 12°. A new dam at the base of the hill will cause 925 ft of the hillside to be flooded by water. How far will the dam be from the shoreline?

**35.** Sighted from a lighthouse, a boat is 23° east of due north. Another boat is 21 nautical miles due east of the lighthouse. If the first boat is due north of the second boat, find the distance between the first boat and the lighthouse.

**36.** The vertex angle of an isosceles triangle is 48°. If the altitude drawn upon the base is 14 in. long, find the length of one of its congruent sides.

**37.** The vertex angle of an isosceles triangle is 54° and one of its congruent sides is 27 ft long. Find the length of the triangle's altitude drawn upon its base.

**38.** If a base angle of an isosceles triangle is 42° and the base is 8 ft long, find its perimeter.

**39.** A base angle of an isosceles triangle is 35°. If the altitude drawn upon its base is 32 cm long, find the perimeter of the triangle.

**40.** A regular pentagon has a perimeter of 40 in. Find the radius of the circle that passes through its five vertices.

**41.** Use the properties of a 30°-60°-right triangle to find the cosine of 60°.

**42.** Use the properties of an isosceles-right triangle to find the cosine of 45°.

**43.** Is $\cos 30° = \sin 60°$?

**44.** Is $\cos 45° = \sin 45°$?

## CHAPTER SIX REVIEW EXERCISES

In Review Exercises 1–2, express each phrase as a ratio in lowest terms.

**1.** 5 liters to 12 liters    **2.** 7 oz to 2 lb

In Review Exercises 3–4, tell whether each expression is a proportion.

**3.** $\dfrac{5}{6} = \dfrac{30}{38}$    **4.** $\dfrac{30}{48} = \dfrac{22.4}{27.2}$

In Review Exercises 5–8, solve each proportion.

**5.** $\dfrac{x}{4} = \dfrac{6}{8}$    **6.** $\dfrac{5}{x} = \dfrac{12}{17}$    **7.** $\dfrac{15}{8} = \dfrac{y-2}{4}$    **8.** $\dfrac{9}{y+9} = \dfrac{18}{y}$

In Review Exercises 9–10, refer to Illustration 1.

**9.** Prove that $\triangle ABC \sim \triangle DEC$.

**10.** $m(\overline{DE}) = $ _____ .

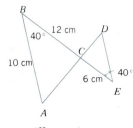

*Illustration 1*

In Review Exercises 11–14, refer to Illustration 2 in which $\overline{DE} \parallel \overline{AB}$.

**11.** $m(\overline{BC}) = $ _____.

**12.** $m(\overline{EB}) = $ _____.

**13.** $m(\overline{DE}) = $ _____.

**14.** Find the perimeter of $\triangle ABC$.

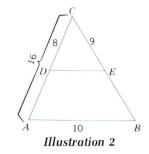

**Illustration 2**

In Review Exercises 15–18, refer to Illustration 3.

**15.** $m(\angle 2) = $ _____.

**16.** $m(\angle CDB) = $ _____.

**17.** $m(\angle 1) = $ _____.

**18.** $m(\angle B) = $ _____.

**Illustration 3**

In Review Exercises 19–20, refer to Illustration 4 in which $\overline{CD}$ bisects $\angle C$.

**19.** If $m(\overline{AD}) = 8$, $m(\overline{DB}) = 4$, and $m(\overline{AC}) = 12$, find $m(\overline{BC})$.

**20.** If $m(\overline{AB}) = 14$, $m(\overline{AC}) = 20$, and $m(\overline{BC}) = 15$, find $m(\overline{AD})$.

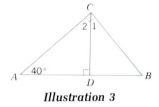

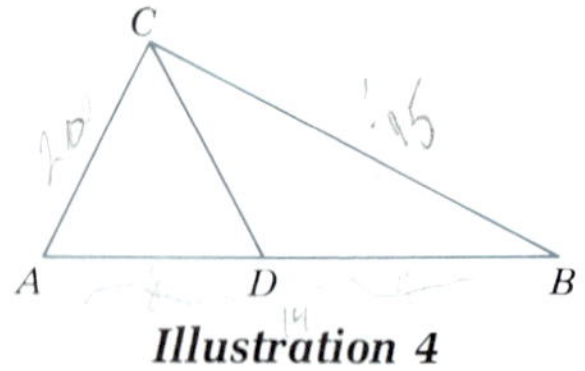

**Illustration 4**

**21.** A building casts a shadow of 75 ft at the same time a 6-ft person casts a shadow of 5 ft. How tall is the building?

**22.** If a 22-ft ladder safely reaches a window 16 ft above the ground, how high will a 30-ft ladder reach? Assume the same angle with the horizontal.

**23.** A 15-ft ladder rests against the side of a building. The base of the ladder is 9 ft from the wall. If a painter climbs 10 ft up the ladder, how far is he from the wall?

**24.** The top of a 20-ft slide is 12 ft above the ground. If a girl slides 5 ft down the slide and stops, how far is she above the ground?

**25.** Find the positive mean proportional between 6 and 48.

**26.** Find the negative mean proportional between 5 and 125.

In Review Exercises 27–30, refer to Illustration 5 in which triangles $ABC$, $ADC$, and $BDC$ are right triangles.

**27.** If $m(\overline{AD}) = 2$ and $m(\overline{AB}) = 8$, find $m(\overline{AC})$.

**28.** If $m(\overline{AB}) = 12$ and $m(\overline{AC}) = 3$, find $m(\overline{AD})$.

**29.** If $m(\overline{AD}) = 5$ and $m(\overline{CD}) = 10$, find $m(\overline{DB})$.

**30.** If $m(\overline{AD}) = 3$ and $m(\overline{AB}) = 15$, find $m(\overline{CD})$.

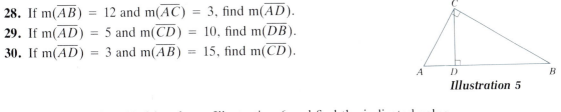

*Illustration 5*

In Review Exercises 31–36, refer to Illustration 6 and find the indicated value.

**31.** $\tan A = $ _____.

**32.** $\sin A = $ _____.

**33.** $\cos A = $ _____.

**34.** $\cos B = $ _____.

**35.** $\sin B = $ _____.

**36.** $\tan B = $ _____.

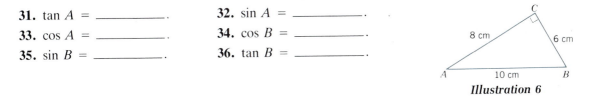

*Illustration 6*

In Review Exercises 37–44, do not use a calculator.

**37.** Find $\sin 30°$.

**38.** Find $\tan 30°$.

**39.** Find $\cos 30°$.

**40.** Find $\tan 60°$.

**41.** Find $\cos 45°$.

**42.** Find $\sin 45°$.

**43.** Find $\tan 45°$.

**44.** Find $\cos 60°$.

In Review Exercises 45–52, use a calculator to find each value. Round to the nearest ten thousandth.

**45.** Find $\cos 20°$.

**46.** Find $\sin 70°$.

**47.** Find $\tan 80°$.

**48.** Find $\sin 50°$.

**49.** Find $\sin 27°$.

**50.** Find $\cos 53°$.

**51.** Find $\tan 65°$.

**52.** Find $\tan 18°$.

In Review Exercises 53–58, use a calculator to help find each value.

**53.** On an approach to an airport a plane is descending at an angle of 3°. How much altitude is lost as the plane travels a horizontal distance of 17.5 miles?

**54.** An observer looks up at an angle of 32° and sees a plane directly over a landmark that is 1500 m from the observer. How far is the observer from the plane?

**55.** A 30-ft ladder leans against a building. If the ladder makes an angle of 70° with the horizontal, how far will it reach up the building?

**56.** A child slides down a 90-ft chute that makes an angle of 35° with the ground. How far above the ground is the high end of the chute?

**57.** A 30-ft ladder leans against a building. How far is the base of the ladder from the building when the ladder makes an angle of 65° with the ground?

**58.** A person sits and watches the launch of a hot-air balloon from a position 500 ft from the balloon. What is the altitude of the balloon when the observer looks up at an angle of 37° to see it?

## CHAPTER SIX TEST

In Exercises 1–30, classify each statement as true or false.

**1.** If $\dfrac{a}{b} = \dfrac{c}{d}$, then $ad = bc$.

**2.** If $\dfrac{a}{b} = \dfrac{c}{d}$ and $a = c$, then $b = d$.

**3.** If $\dfrac{a}{b} = \dfrac{c}{d}$, then $\dfrac{d}{b} = \dfrac{c}{a}$.

**4.** If $\dfrac{a}{b} = \dfrac{c}{d}$, then $\dfrac{a+c}{b+d} = \dfrac{a}{c}$.

**5.** If two triangles are similar, then each pair of corresponding parts must be congruent.

**6.** In $\triangle ABC$ shown in Illustration 1, $m(\overline{BD}) = \dfrac{20}{3}$.

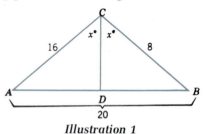

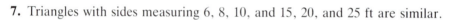

*Illustration 1*

**7.** Triangles with sides measuring 6, 8, 10, and 15, 20, and 25 ft are similar.

**8.** The mean proportional to 4 and 6 is 5.

**9.** If an altitude is drawn upon the hypotenuse of a right triangle, the lengths of the legs of the triangle are in proportion to the lengths of the segments of the hypotenuse.

**10.** A triangle with sides measuring 5, 12, and 13 units is a right triangle.

**11.** If two legs of a right triangle are 15 and 20 cm, the length of the hypotenuse is 25 cm.

**12.** All right triangles are similar.

**13.** If two isosceles triangles have one pair of corresponding base angles that are congruent, then the triangles are similar.

**14.** If two triangles are congruent, then they are similar.

**15.** A proportion is a statement that two ratios are equal.

**16.** In two similar triangles, the ratio of the measures of two corresponding angles is equal to the ratio of the lengths of two corresponding sides.

**17.** In two similar triangles, the ratio of the areas is equal to the ratio of the lengths of two corresponding sides.

**18.** Adding a constant to both the numerator and the denominator of a fraction does not change the value of the fraction.

**19.** If an acute angle of one right triangle is complementary to an acute angle of another right triangle, then the triangles are similar.

**20.** If two angles of a triangle are complementary, then the triangle is a right triangle.

**21.** If $\angle A$ is an acute angle in a right triangle, then $\sin A$ is the ratio of the length of the side opposite $\angle A$ to the length of the side adjacent to $\angle A$.

**22.** If $\angle A$ is an acute angle in a right triangle, then tan $A$ is the ratio of the length of the side opposite $\angle A$ to the length of the side adjacent to $\angle A$.

**23.** If $\angle A$ is an acute angle of an isosceles right triangle, then cos $A$ = sin $A$.

In Statements 24–29, refer to Illustration 2.

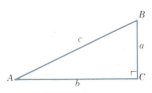

**24.** If m($\angle A$) = 30° and $a$ = 3 units, then $b$ = $3\sqrt{3}$ units.

**25.** If m($\angle A$) = 45° and $a$ = 3 units, then $c$ = $3\sqrt{2}$ units.

**26.** If m($\angle A$) = 60° and $b$ = $2\sqrt{3}$ units, then $c$ = 4 units.

**27.** If m($\angle A$) = 25° and $a$ = 7 units, then $b$ = 7(tan 25°) units.

**28.** If m($\angle A$) = 72° and $c$ = 7 units, then $b$ = 7(cos 72°) units.

**29.** If m($\angle A$) = 31° and $b$ = 7 units, then $a$ = 7(tan 31°) units.

***Illustration 2***

**30.** If $\angle A$ and $\angle B$ are the two acute angles of a right triangle, then sin $A$ = cos $B$.

# 7

Archimedes (287–212 B.C.). Archimedes is famous for his work in geometry and applied mathematics. Many of his inventions were machines of war that were so successful that besieging armies were held off from Syracuse for three years. When the city was captured, the enemy soldiers of Claudius Marcellus were under orders not to kill Archimedes because of his scientific genius. However, a soldier found Archimedes drawing geometric figures in the sand and ordered him to move. When Archimedes ignored the order, he was killed (Alinari—Art Reference Bureau).

# Circles and More on Similarity

*I*n this chapter we will prove many theorems involving circles, their arcs, and related angles. In the development of this material, we will also review some algebraic techniques.

## 7.1 CIRCLES

Recall that a **circle** is a set of all points in the same plane that are a fixed distance from a point called its **center**. A line segment drawn from the center of a circle to one of the points on the circle is called a **radius** of the circle. The plural of *radius* is *radii*. From the definition of a circle, it follows that all radii of the same circle are congruent.

The center of a circle is usually denoted by the letter $O$. Thus, whenever we see a circle with an interior point named $O$, we will assume that point $O$ is the center of the circle.

**Definition 7.1** A line segment connecting two points on a circle is called a **chord** of the circle.

**Definition 7.2** A chord that passes through the center of a circle is called a **diameter** of the circle.

**Theorem 7.1** The length of a diameter of a circle is twice the length of any of its radii.

The proof of Theorem 7.1 is left as an exercise.

**Definition 7.3** A line that has exactly one point of intersection with a circle is called a **tangent** to the circle. The point of intersection is called a **point of tangency**.

A circle, a radius, a chord, a diameter, and a tangent are shown in Figure 7.1.

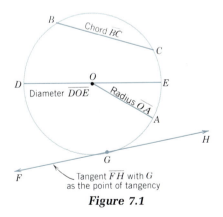

**Figure 7.1**

**Postulate 7.1** A line drawn from the center of a circle to a point of tangency is perpendicular to the tangent that passes through the point of tangency.

> **Postulate 7.2** If a line is perpendicular to a radius at the point where the radius intersects a circle, then the line is a tangent to the circle.

Postulate 7.2 provides a method for constructing a tangent to a circle (with a known center) that passes through a point on the circle. For example, suppose we wish to construct a tangent to a circle with center at $O$ that passes through a given point $A$ as in Figure 7.2. To do so, we first draw a radius to point $A$, and then construct a line that is perpendicular to that radius at point $A$. This line will be the desired tangent.

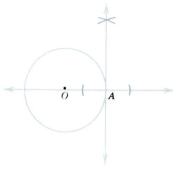

*Figure 7.2*

> **Definition 7.4** A continuous portion of a circle is called an **arc** of the circle.

> **Definition 7.5** A **semicircle** is an arc of a circle whose endpoints are endpoints of a diameter of the circle.

> **Definition 7.6** An arc that is greater than a semicircle is called a **major arc**. An arc that is less than a semicircle is called a **minor arc**.

Arcs are often named by using three points that lie on the arc. For example, in Figure 7.3 arc $ABC$ (usually written as $\overset{\frown}{ABC}$) has endpoints $A$ and $C$. Because points $A$ and $C$ are also endpoints of the diameter $\overline{AOC}$, arc $\overset{\frown}{ABC}$ is a semicircle. Diameter $\overline{AC}$ divides the circle into two different semicircles–one above $\overline{AC}$ and one below $\overline{AC}$. When naming an arc, the purpose of the middle letter is to identify which of the two arcs is being considered. In Figure 7.3, $\overset{\frown}{ABC}$ is the semicircle lying above the diameter $\overline{AC}$, $\overset{\frown}{BDC}$ is a minor arc, and $\overset{\frown}{BAC}$ is a major arc.

   If only two letters are used to name an arc, we will assume that the arc is a minor arc.

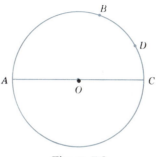

**Figure 7.3**

   Circles with congruent radii have the same area, and they are called **congruent circles**.

---

**Definition 7.7 Two circles are congruent** if, and only if, their radii are congruent.

---

**Theorem 7.2** Two circles are congruent if, and only if, their diameters are congruent.

---

The proof of Theorem 7.2 is left as an exercise.
   There are many types of angles that are associated with a circle. Perhaps the most important of these is a **central angle**.

---

**Definition 7.8** An angle whose vertex is at the center of a circle and whose sides are radii of the circle is called a **central angle**.

---

   In Figure 7.4, $\angle 1$, $\angle 2$, $\angle 3$, and $\angle 4$ are central angles. Because points $A$ and $D$ are the intersection points of the sides of $\angle 1$ and the arc $\overset{\frown}{AD}$, we say that $\angle 1$ intercepts arc $\overset{\frown}{AD}$.

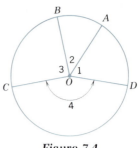

*Figure 7.4*

---

**Definition 7.9** The **measure of an arc** is equal to the number of degrees in the central angle that intercepts the arc.

---

In Figure 7.5, the measure of $\overset{\frown}{AB}$, denoted as m($\overset{\frown}{AB}$), is 30°, the same as the measure of the 30°-degree central angle that intercepts the arc. The measure of $\overset{\frown}{CD}$, denoted as m($\overset{\frown}{CD}$), is 100°.

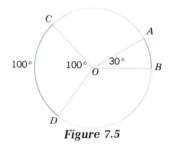

*Figure 7.5*

It is important to remember that the measure of an arc is not defined to be its length. Instead, the measure of an arc is defined to be the number of degrees in the central angle that intercepts the arc. In Figure 7.5, arc $\overset{\frown}{AB}$ is the set of all points on the arc between the points $A$ and $B$, including points $A$ and $B$. However, the measure of $\overset{\frown}{AB}$ is the number of degrees in the arc's measure: m($\overset{\frown}{AB}$) = 30°.

---

**Definition 7.10** If two arcs in the same circle or congruent circles have the same measure, they are called **congruent arcs**.

---

From Definitions 7.9 and 7.10, it follows that congruent central angles in the same circle or in congruent circles have congruent arcs.

Recall that if a circle is divided into 360 equal parts and lines are drawn from the center of a circle to consecutive points of division, the angle formed by those lines has a measure of 1°. Thus, there are 360° in the arc that forms an entire circle, and there are 180° in the measure of a semicircle. A minor arc has a measure of less than 180°, and a major arc has a measure of greater than 180°.

As with line segments and angles, we will make some postulates governing the measures of an arc and the measures of parts of the arc. When reading Postulate 7.3, refer to Figure 7.6.

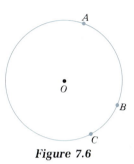

**Figure 7.6**

**Postulate 7.3** If $A$, $B$, and $C$ are three points on the same arc, and $B$ is between $A$ and $C$, then

$$m(\widehat{AC}) = m(\widehat{AB}) + m(\widehat{BC})$$
$$m(\widehat{BC}) = m(\widehat{AC}) - m(\widehat{AB})$$
$$m(\widehat{AB}) = m(\widehat{AC}) - m(\widehat{BC})$$

Another important angle is the inscribed angle.

**Definition 7.11** An angle whose vertex is on a circle and whose sides are chords of the circle is called an **inscribed angle**.

In Figure 7.7, $\angle 1$ and $\angle 2$ are inscribed angles.

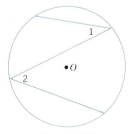

**Figure 7.7**

We are now ready to prove an important theorem about inscribed angles. Theorem 7.3 will be the key to many proofs of other theorems in this chapter.

---

**Theorem 7.3** The measure of an inscribed angle is equal to one-half the measure of its intercepted arc.

---

In the proof of this theorem, it will be necessary to consider three separate cases. Case 1 considers the possibility of the center of the circle being on one side of the inscribed angle. Case 2 considers the possibility of the center of the circle being on the interior of the angle, and Case 3 considers the possibility of the center of the circle being on the exterior of the angle.

CASE 1

*Given*: $\angle ABC$ is an inscribed angle.
       Point $O$ is the center of the circle.

*Prove*: $m(\angle B) = \frac{1}{2}[m(\widehat{AC})]$.

*Construction*: Draw radius $\overline{OC}$.

*Proof*:

| STATEMENTS | REASONS |
|---|---|
| **1.** $\angle ABC$ is an inscribed angle. Point $O$ is the center of the circle. | **1.** Given. |
| **2.** Draw radius $\overline{OC}$. | **2.** Possible construction. |
| **3.** $m(\angle 1) = m(\widehat{AC})$. | **3.** A central angle has the same measure as its intercepted arc. |
| **4.** $\overline{OB} \cong \overline{OC}$. | **4.** All radii of the same circle are congruent. |
| **5.** $\angle B \cong \angle C$. | **5.** In a triangle, angles opposite congruent sides are congruent. |
| **6.** $m(\angle B) = m(\angle C)$. | **6.** Congruent angles have equal measures. |

| STATEMENTS | REASONS |
|---|---|
| **7.** m($\angle B$) + m($C$) = m($\angle 1$). | **7.** The measure of an exterior angle of a triangle is equal to the sum of the measures of the nonadjacent interior angles. |
| **8.** m($\angle B$) + m($\angle B$) = m($\angle 1$). | **8.** Substitution. |
| **9.** $2[m(\angle B)]$ = m($\angle 1$). | **9.** Combine terms. |
| **10.** m($\angle B$) = $\frac{1}{2}$[m($\angle 1$)]. | **10.** Equals divided by equals are equal. |
| **11.** m($\angle B$) = $\frac{1}{2}$[m($\overset{\frown}{AC}$)]. | **11.** Substitution. |

CASE 2

*Given*: $\angle ABC$ is an inscribed angle.
   Point $O$ is the center of the circle.

*Prove*: m($\angle ABC$) = $\frac{1}{2}$[m($\overset{\frown}{AC}$)].

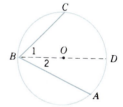

*Construction*: Draw line segment $\overline{OB}$ and extend it until it intersects the circle at point $D$.

*Proof*:

| STATEMENTS | REASONS |
|---|---|
| **1.** $\angle ABC$ is an inscribed angle. Point $O$ is the center of the circle. | **1.** Given. |
| **2.** Draw line segment $\overline{BOD}$. | **2.** Possible construction. |
| **3.** m($\angle 1$) = $\frac{1}{2}$[m($\overset{\frown}{DC}$)]; m($\angle 2$) = $\frac{1}{2}$[m($\overset{\frown}{AD}$)]. | **3.** Result from Case 1. |
| **4.** m($\angle 2$) + m($\angle 1$) = $\frac{1}{2}$[m($\overset{\frown}{AD}$)] + $\frac{1}{2}$[m($\overset{\frown}{DC}$)]. | **4.** Equals added to equals are equal. |

| STATEMENTS | REASONS |
|---|---|

**5.** $m(\angle 2) + m(\angle 1)$
$= \frac{1}{2}[m(\widehat{AD}) + m(\widehat{DC})].$

**5.** Distributive law.

**6.** $m(\widehat{AD}) + m(\widehat{DC}) = m(\widehat{AC}).$

**6.** Postulate 7.3.

**7.** $m(\angle 1) + m(\angle 2)$
$= m(\angle ABC).$

**7.** Postulate 1.11.

**8.** $m(\angle ABC) = \frac{1}{2}[m(\widehat{AC})].$

**8.** Substitution.

CASE 3

*Given*: $\angle ABC$ is an inscribed angle.
Point $O$ is the center of the circle.

*Prove*: $m(\angle ABC) = \frac{1}{2}[m(AC)].$

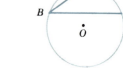

The proof of Case 3 is left as an exercise.    □

**EXAMPLE 1**

In Figure 7.8, $\angle AOB$ is a central angle with a measure of 100° and $\angle C$ is an inscribed angle. Find $m(\angle C)$.

*Solution*

We are given that $\angle AOB$ is a central angle with a measure of 100°. Because an arc has the same measure as its central angle, it follows that $m(\widehat{AB}) = 100°$. Because $\angle C$ is an inscribed angle that intercepts $AB$, its measure must be one-half of 100°. Thus, $m(\angle C) = 50°$.

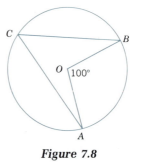

**Figure 7.8**    ■

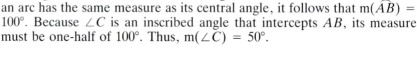

**Theorem 7.4** The measure of an angle formed by a tangent and a chord is equal to one-half the measure of its intercepted arc.

*Given*: ∠*ABC* is formed by a tangent
and a chord.

*Prove*: $m(\angle ABC) = \frac{1}{2}[m(\widehat{AB})]$.

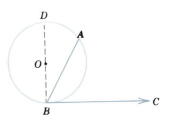

*Construction*: Draw diameter $\overline{DB}$ to form semicircle $\widehat{DAB}$.

*Proof*:

| STATEMENTS | REASONS |
|---|---|
| **1.** ∠*ABC* is formed by a tangent and a chord. | **1.** Given. |
| **2.** Draw diameter $\overline{DB}$ to form semicircle $\widehat{DAB}$. | **2.** Possible construction. |
| **3.** $m(\angle ABC) = m(\angle DBC) - m(\angle DBA)$. | **3.** Postulate 1.11. |
| **4.** $m(\angle DBC) = 90°$. | **4.** A line passing through the center of a circle is perpendicular to the tangent at the point of tangency. |
| **5.** $m(\angle DBA) = \frac{1}{2}[m(\widehat{DA})]$. | **5.** An inscribed angle has a measure of one-half its intercepted arc. |
| **6.** $m(\angle ABC) = 90° - \frac{1}{2}[m(\widehat{DA})]$. | **6.** Substitution. |
| **7.** $2[m(\angle ABC)] = 180° - m(\widehat{DA})$. | **7.** Equals multiplied by equals are equal. |
| **8.** $m(\widehat{DAB}) = 180°$. | **8.** The measure of a semicircle is 180°. |
| **9.** $2[m(\angle ABC)] = m(\widehat{DAB}) - m(\widehat{DA})$. | **9.** Substitution. |
| **10.** $m(\widehat{AB}) = m(\widehat{DAB}) - m(\widehat{DA})$. | **10.** Postulate 7.3. |
| **11.** $2[m(\angle ABC)] = m(\widehat{AB})$. | **11.** Substitution. |
| **12.** $m(\angle ABC) = \frac{1}{2}[m(\widehat{AB})]$. | **12.** Equals divided by equals are equal. □ |

> **Theorem 7.5** If two chords intersect within a circle, the measure of each angle formed is equal to one-half the sum of the measures of its intercepted arc and the intercepted arc of its vertical angle.

*Given*: $\overline{AB}$ and $\overline{DC}$ are chords intersecting at point $E$.

*Prove*: $m(\angle 1) = \dfrac{1}{2}[m(\widehat{BC}) + m(\widehat{AD})]$.

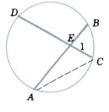

*Construction*: Draw segment $\overline{AC}$.

*Proof*:

| STATEMENTS | REASONS |
|---|---|
| 1. $\overline{AB}$ and $\overline{DC}$ are chords intersecting at point $E$. | 1. Given. |
| 2. $m(\angle A) = \dfrac{1}{2}[m(\widehat{BC})]$; $\quad m(\angle C) = \dfrac{1}{2}[m(\widehat{AD})]$. | 2. Theorem 7.3. |
| 3. $m(\angle 1) = m(\angle A) + m(\angle C)$. | 3. The measure of an exterior angle of a triangle is equal to the sum of the nonadjacent interior angles. |
| 4. $m(\angle 1) = \dfrac{1}{2}[m(\widehat{BC})] + \dfrac{1}{2}[m(\widehat{AD})]$. | 4. Substitution. |
| 5. $m(\angle 1) = \dfrac{1}{2}[m(\widehat{BC}) + m(\widehat{AD})]$. | 5. Use the distributive law to factor out $\frac{1}{2}$. |

**EXAMPLE 2**

In Figure 7.9, $m(\widehat{DB}) = 80°$ and $m(\widehat{AC}) = 110°$. Find $m(\angle 1)$.

*Solution*

Because of Theorem 7.5, $m(\angle 1)$ is equal to one-half the sum of its intercepted arc and the intercepted arc of its vertical angle. Thus, we have

$$m(\angle 1) = \frac{1}{2}[m(\widehat{AC}) + m(\widehat{BD})]$$

$$= \frac{1}{2}(110° + 80°)$$

$$= \frac{1}{2}(190°)$$

$$= 95°$$

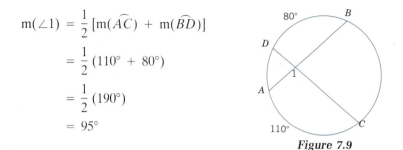

*Figure 7.9*

Thus, $m(\angle 1) = 95°$. ■

---

**Definition 7.12** A line that intersects a circle in two points is called a **secant**.

---

**Theorem 7.6** The measure of an angle formed by the intersection of two secants outside a circle is equal to one-half the difference of the measures of the intercepted arcs.

*Given*: Segments $\overline{CE}$ and $\overline{CA}$ lie on secants intersecting at point $C$.

*Prove*: $m(\angle C) = \frac{1}{2}[m(\widehat{AE}) - m(\widehat{BD})]$.

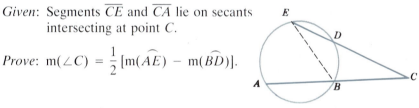

The proof of Theorem 7.6 is left as an exercise.

---

**Theorem 7.7** The measure of an angle formed by the intersection of a tangent and a secant outside a circle is equal to one-half the difference of the measures of the intercepted arcs.

*Given*: Secant $\overline{AC}$ and tangent $\overline{CD}$ intersect at point $C$.

*Prove*: $m(\angle C) = \frac{1}{2}[m(\widehat{AD}) - m(\widehat{BD})]$.

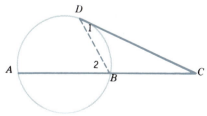

The proof of Theorem 7.7 is left as an exercise.

**EXAMPLE 3**    Refer to Figure 7.10, in which $m(\overset{\frown}{AE}) = 80°$ and $m(\angle C) = 20°$. Find (a) $m(\angle 2)$, (b) $m(\overset{\frown}{BD})$, and (c) $m(\angle 1)$.

Solution

(a)  Since $\angle 2$ is an inscribed angle that intercepts an arc of 80°, $m(\angle 2)$ must be one-half the measure of its intercepted arc. Therefore, $m(\angle 2) = 40°$.

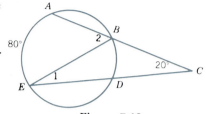

**Figure 7.10**

(b)  By Theorem 7.6, we know that $m(\angle C) = \dfrac{1}{2}[m(\overset{\frown}{AE}) - m(\overset{\frown}{BD})]$.

Substituting the given values, we have

$$20° = \frac{1}{2}[(80° - m(\overset{\frown}{BD})]$$

$$40° = 80° - m(\overset{\frown}{BD}) \qquad \text{Multiply both sides by 2.}$$
$$m(\overset{\frown}{BD}) = 40° \qquad \text{Add } m(\overset{\frown}{BD}) - 40° \text{ to both sides.}$$

(c)  Since $\angle 1$ is an inscribed angle that intercepts a 40° arc, $m(\angle 1)$ must be one-half of 40°, or 20°. ∎

## EXERCISE 7.1

In Exercises 1–4, refer to Illustration 1.

**1.** Find $m(\angle BOC)$.
**2.** Find $m(\angle BAC)$.
**3.** If $m(\angle DAB) = 40°$, find $m(\overset{\frown}{DC})$.
**4.** If $m(\angle DAB) = 60°$, find $m(\overset{\frown}{AB})$.

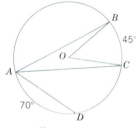

**Illustration 1**

In Exercises 5–8, refer to Illustration 2 in which $\overline{CD}$ and $\overline{BD}$ are tangents to the circle.

**5.** Find $m(\angle 1)$.
**6.** Find $m(\overset{\frown}{CB})$.
**7.** Find $m(\angle 2)$.
**8.** Find $m(\angle D)$.

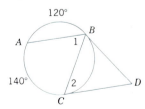

**Illustration 2**

In Exercises 9–12, refer to Illustration 3.

**9.** Find m($\angle 2$).
**10.** Find m($\angle 1$).
**11.** Find m($\angle 4$).
**12.** Find m($\angle BAD$).

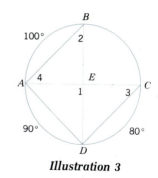

*Illustration 3*

In Exercises 13–16, refer to Illustration 4.

**13.** If m($\overset{\frown}{BD}$) = 60° and m($\overset{\frown}{AE}$) = 100°, find m($\angle C$).
**14.** If m($\angle C$) = 30° and m($\overset{\frown}{BD}$) = 80°, find m($\overset{\frown}{AE}$).
**15.** If m($\angle C$) = 42° and m($\overset{\frown}{AE}$) = 120°, find m($\overset{\frown}{BD}$).
**16.** If m($\overset{\frown}{BD}$) = 15° and m($\overset{\frown}{AE}$) = 3[m($\overset{\frown}{BD}$)], find m($\angle C$).

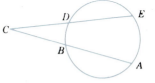

*Illustration 4*

In Exercises 17–20, refer to Illustration 5 in which $\overline{AC}$ is tangent to circle $O$ and $\overline{AB}$ is a secant.

**17.** Find m($\overset{\frown}{CD}$).
**18.** Find m($\angle ACD$).
**19.** Find m($\angle BDC$).
**20.** Find m($\overset{\frown}{BD}$).

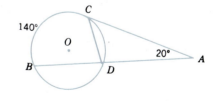

*Illustration 5*

In Exercises 21–24, refer to Illustration 6 in which $\overline{BC}$ is tangent to circle $O$ and $\overline{BD}$ is a diameter.

**21.** Find m($\overset{\frown}{AB}$).
**22.** Find m($\overset{\frown}{AD}$).
**23.** Find m($\angle A$).
**24.** Find m($\angle BDA$).

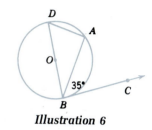

*Illustration 6*

**25.** In Illustration 7, $\overline{AB} \parallel \overline{CD}$, $C$ is a point of tangency, and m($\overset{\frown}{AC}$) = 130°. Find m($\overset{\frown}{BC}$).
**26.** In Illustration 8, m($\angle 1$) = 40° and m($\angle 2$) = 10°. Find m($\overset{\frown}{AD}$) and m($\overset{\frown}{BC}$).

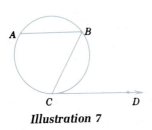

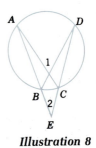

Illustration 7

Illustration 8

**27.** In Illustration 9, m($\angle 1$) = 55° and m($\angle D$) = 40°. Find m($\overset{\frown}{CD}$) and m($\angle B$).

**28.** In Illustration 10, $\overline{AB}$ and $\overline{BC}$ are tangents to the circle and m($\angle 1$) = 70°. Find m($\angle B$).

Illustration 9

Illustration 10

**29.** *Prove Theorem 7.1*: The length of a diameter of a circle is twice the length of a radius of the circle.

**30.** *Prove Theorem 7.2*: Two circles are congruent if, and only if, their diameters are congruent. (*Hint*: Prove both the statements **1**. *If two circles are congruent, then their diameters are congruent* and **2**. *If the diameters of two circles are congruent, then the circles are congruent*.)

**31.** *Prove Case 3 of Theorem 7.3*: The measure of an inscribed angle is equal to one-half the measure of its intercepted arc. (Assume point *O* is on the exterior of the inscribed angle.)

**32.** *Prove Theorem 7.6*: The measure of an angle formed by the intersection of two secants outside a circle is equal to one-half the difference of the measures of the intercepted arcs.

**33.** *Prove Theorem 7.7*: The measure of an angle formed by the intersection of a tangent and a secant outside a circle is equal to one-half the difference of the measures of the intercepted arcs.

**34.** Explain why congruent central angles in the same circle have congruent arcs.

**35.** Explain why congruent inscribed angles in the same circle have congruent arcs.

## 7.2 THEOREMS INVOLVING CHORDS, TANGENTS, AND SECANTS OF CIRCLES

In this section, we will prove several more theorems about circles and their chords, tangents, and secants.

---

**Definition 7.13** A line **bisects an arc** if the line divides the arc into two congruent arcs.

---

**Theorem 7.8** If a line is drawn from the center of a circle perpendicular to a chord, then it bisects the chord and its arc.

---

*Given*: Point $O$ is the center of the circle.
$\overline{OD} \perp \overline{AB}$.

*Prove*: $\overline{OD}$ bisects chord $\overline{AB}$.
$\overline{OD}$ bisects arc $\widehat{AB}$.

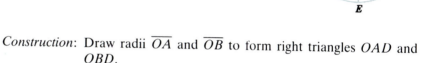

*Construction*: Draw radii $\overline{OA}$ and $\overline{OB}$ to form right triangles $OAD$ and $OBD$.

*Proof*:

STATEMENTS

**1.** $O$ is the center of the circle; $\overline{OD} \perp \overline{AB}$.

**2.** $\overline{OA}$ and $\overline{OB}$ are radii forming right triangles $OAD$ and $OBD$.

**3.** $\overline{OA} \cong \overline{OB}$.

**4.** $\overline{OD} \cong \overline{OD}$.

**5.** $\triangle OAD \cong \triangle OBD$.

**6.** $\overline{AD} \cong \overline{BD}$; $\angle 1 \cong \angle 2$.

**7.** $\widehat{AE} \cong \widehat{EB}$.

**8.** $\overline{OD}$ bisects $\widehat{AB}$.

**9.** $\overline{OD}$ bisects $\overline{AB}$.

REASONS

**1.** Given.

**2.** By construction.

**3.** Radii in the same circle are congruent.

**4.** Reflexive law.

**5.** $hl = hl$.

**6.** cpctc.

**7.** Equal central angles have congruent arcs.

**8.** If a line divides an arc into two congruent arcs, then it bisects the arc.

**9.** If a line divides a segment into two congruent segments, then it bisects the segment.    □

> **Theorem 7.9** A line drawn from the center of a circle to the midpoint of a chord that is not a diameter or to the midpoint of its arc is perpendicular to the chord.

*Given*: M is the midpoint of chord $\overline{AB}$.

*Prove*: $\overline{OM} \perp \overline{AB}$.

*Construction*: Draw radii $\overline{OA}$ and $\overline{OB}$.

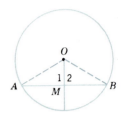

*Proof*:

| STATEMENTS | REASONS |
|---|---|
| 1. M is the midpoint of $\overline{AB}$. | 1. Given. |
| 2. $\overline{AM} \cong \overline{MB}$. | 2. A midpoint divides a segment into two congruent segments. |
| 3. $\overline{OB} \cong \overline{OA}$. | 3. Radii in the same circle are congruent. |
| 4. $\overline{OM} \cong \overline{OM}$. | 4. Reflexive law. |
| 5. $\triangle AOM \cong \triangle BOM$. | 5. SSS $\cong$ SSS. |
| 6. $\angle 1 \cong \angle 2$. | 6. cpctc. |
| 7. $\overline{OM} \perp \overline{AB}$. | 7. If two lines meet and form congruent adjacent angles, then the lines are perpendicular. ☐ |

The case where $\overline{OM}$ is drawn to the midpoint of $\overset{\frown}{AB}$ is left as an exercise.

> **Theorem 7.10** The perpendicular bisector of a chord passes through the center of the circle.

*Given*: $\overline{DE}$ is the perpendicular bisector of $\overline{AB}$.

*Prove*: $\overline{DE}$ passes through the center of the circle.

*Construction*: Draw radii $\overline{OA}$ and $\overline{OB}$.

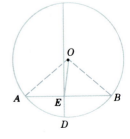

*Proof*:

| STATEMENTS | REASONS |
|---|---|
| 1. $\overline{OA} \cong \overline{OB}$. | 1. Radii in the same circle are congruent. |
| 2. $\triangle AOB$ is isosceles. | 2. If a triangle has two congruent sides, then it is isosceles. |
| 3. $\overline{DE}$ is the perpendicular bisector of $\overline{AB}$. | 3. Given. |
| 4. $\overline{AB}$ is the base of the isosceles $\triangle AOB$. | 4. The third side of an isosceles triangle is its base. |
| 5. $\overline{DE}$ passes through the center of the circle. | 5. Theorem 2.4. |

Theorem 7.10 provides a method of locating the center of an arc or a circle. For example, to find the center of the arc $\overset{\frown}{ACBD}$ shown in Figure 7.11 we first draw a chord $\overline{AB}$ in the arc. Then we construct the perpendicular bisector of $\overline{AB}$ and call it $\overleftrightarrow{NM}$. By Theorem 7.10, line $\overleftrightarrow{NM}$ passes through the center of the arc. We then draw chord $\overline{CD}$ and construct its perpendicular bisector $\overleftrightarrow{RS}$. By Theorem 7.10, line $\overleftrightarrow{RS}$ will pass through the center of the arc. The intersection of $\overleftrightarrow{RS}$ and $\overleftrightarrow{NM}$ will be the center of the arc. Hence, point $O$ is the center of the arc. Point $O$ is also the center of the circle that makes the arc.

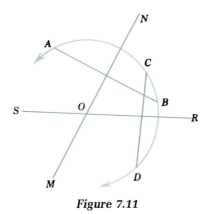

**Figure 7.11**

---

**Theorem 7.11** In the same circle or congruent circles, congruent chords have congruent arcs.

*Given*: $\overline{AB} \cong \overline{CD}$.

*Prove*: $\overparen{AB} \cong \overparen{CD}$.

*Construction*: Draw radii $\overline{OA}, \overline{OB}, \overline{OC},$ and $\overline{OD}$.

*Proof*:

| STATEMENTS | REASONS |
|---|---|
| **1.** $\overline{AB} \cong \overline{CD}$. | **1.** Given. |
| **2.** $\overline{OB} \cong \overline{OC}$; $\overline{OA} \cong \overline{OD}$. | **2.** Radii in the same circle are congruent. |
| **3.** $\triangle AOB \cong \triangle DOC$. | **3.** SSS $\cong$ SSS. |
| **4.** $\angle 1 \cong \angle 2$. | **4.** cpctc. |
| **5.** $AB \cong DC$. | **5.** Congruent central angles have congruent arcs. |

The case for congruent circles is left as an exercise.

---

**Theorem 7.12** In the same circle or congruent circles, congruent arcs have congruent chords.

---

The proof of Theorem 7.12 is left as an exercise.

---

**Theorem 7.13** In the same circle or congruent circles, congruent chords are equidistant from the center.

---

*Given*: $\overline{AB} \cong \overline{CD}$.

*Prove*: $\overline{AB}$ and $\overline{CD}$ are equidistant from $O$.

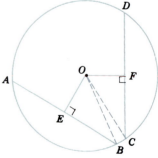

*Construction*: Construct $\overline{OE} \perp \overline{AB}$ and $\overline{OF} \perp \overline{CD}$ to form right triangles $OEB$ and $OFC$. $\overline{OE}$ and $\overline{OF}$ represent the distances the chords are from the center of the circle.

*Proof*:

| STATEMENTS | REASONS |
|---|---|
| 1. $\overline{AB} \cong \overline{CD}$. | 1. Given. |
| 2. $m(\overline{AB}) = m(\overline{CD})$. | 2. If two segments are congruent, they have equal measures. |
| 3. $\overline{OE} \perp \overline{AB}$; $\overline{OF} \perp \overline{CD}$; right triangles $OEB$ and $OFC$. | 3. By construction. |
| 4. $\overline{AE} \cong \overline{EB}$; $\overline{DF} \cong \overline{FC}$. | 4. A line from the center of a circle that is perpendicular to a chord bisects the chord. |
| 5. $m(\overline{AE}) = m(\overline{EB})$; $m(\overline{DF}) = m(\overline{FC})$. | 5. If two segments are congruent, they have equal measures. |
| 6. $m(\overline{AB}) = m(\overline{AE}) + m(\overline{EB})$; $m(\overline{CD}) = m(\overline{DF}) + m(\overline{FC})$; | 6. Postulate 1.10. |
| 7. $m(\overline{AB}) = m(\overline{EB}) + m(\overline{EB})$; $m(\overline{CD}) = m(\overline{FC}) + m(\overline{FC})$; | 7. Substitution. |
| 8. $m(\overline{AB}) = 2[m(\overline{EB})]$; $m(\overline{CD}) = 2[m(\overline{FC})]$. | 8. Combine terms. |
| 9. $2[m(\overline{EB})] = 2[m(\overline{FC})]$. | 9. Substitution. |
| 10. $m(\overline{EB}) = m(\overline{FC})$. | 10. Equals divided by equals are equal. |
| 11. $\overline{EB} \cong \overline{FC}$. | 11. If two segments have equal measures, they are congruent. |
| 12. $\overline{OB} \cong \overline{OC}$. | 12. Radii in the same circle are congruent. |
| 13. $\triangle OEB \cong \triangle OFC$. | 13. $hl \cong hl$. |
| 14. $\overline{OE} \cong \overline{OF}$. | 14. cpctc. |
| 15. $\overline{AB}$ and $\overline{CD}$ are equidistant from point $O$. | 15. Their distances from point $O$ are equal. |

The case for congruent circles is left as an exercise.

**Theorem 7.14** In the same circle or congruent circles, chords equidistant from the center are congruent.

The proof of Theorem 7.14 is left as an exercise.

EXERCISE 7.2

In Exercises 1–6, refer to Illustration 1 in which $\overline{OC}$ is perpendicular to $\overline{AB}$.

1. If m($\overline{AE}$) = 3 in., find m($\overline{AB}$).
2. If m($\angle A$) = 30°, find m($\overset{\frown}{AC}$).
3. If m($\angle AOB$) = 110°, find m($\overset{\frown}{DB}$).
4. If m($\overset{\frown}{AD}$) = 140°, find m($\overset{\frown}{AB}$).

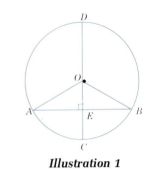

*Illustration 1*

5. If m($\overline{OD}$) = 5 in. and chord $\overline{AB}$ is 3 in. from the center of the circle, find the length of chord $\overline{AB}$.
6. If m($\overline{OC}$) = 13 cm and m($\overline{AE}$) = 5 cm, find the length of segment $\overline{EC}$.

In Exercises 7–10, refer to Illustration 2 in which $\overline{AB} \cong \overline{CD}$.

7. If m($\overset{\frown}{AB}$) = 120° and m($\overset{\frown}{CB}$) = 40°, find m($\overset{\frown}{BD}$).
8. If m($\overset{\frown}{AB}$) = 140°, and m($\overset{\frown}{DB}$) = 80°, find m($\overset{\frown}{AC}$).
9. If m($\angle A$) = 30° and m($\overset{\frown}{BC}$) = 80°, find m($\overset{\frown}{AB}$).
10. If m($\overset{\frown}{BD}$) = 60° and m($\overset{\frown}{CB}$) = 20°, find m($\angle 1$).

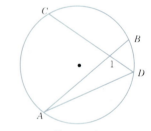

*Illustration 2*

In Exercises 11–14, refer to Illustration 3 in which circles $O$ and $O'$ are congruent and $\overline{AC} \cong \overline{BC}$.

11. If m($\angle BAC$) = 60°, find m($\overset{\frown}{BGC}$).
12. If m($\overset{\frown}{BDC}$) = 200°, find m($\overset{\frown}{AEC}$).
13. If m($\angle F$) = 60°, find m($\overset{\frown}{BDC}$).
14. If m($BDC$) = 210°, find m($\angle F$).

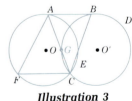

*Illustration 3*

15. *Complete the proof of Theorem 7.9*: A line drawn from the center of a circle to the midpoint of a chord that is not a diameter or to the midpoint of its arc is perpendicular to the chord.
16. *Complete the proof of Theorem 7.11*: In the same circle or congruent circles, equal chords have equal arcs.

**17.** *Prove Theorem 7.12*: In the same circle or congruent circles, congruent arcs have congruent chords.

**18.** *Complete the proof of Theorem 7.13*: In the same circle or congruent circles, congruent chords are equidistant from the center.

**19.** *Prove Theorem 7.14*: In the same circle or congruent circles, chords equidistant from the center are congruent.

**20.** Draw a circle of a convenient size. Use only a compass and a straightedge to locate the center of the circle.

**21.** Draw a rectangle with sides approximately 2 in. and 3 in. Construct a square that has the same area as the rectangle.

# 7.3 MORE THEOREMS INVOLVING TANGENTS TO CIRCLES

There is a way to construct a tangent to a circle from a point outside the circle. See Figure 7.12. To construct a tangent to circle $O$ from a given point $A$ that is outside the circle, we must find a point $C$ on the circle such that radius $\overline{OC}$ is perpendicular to the desired tangent $\overleftrightarrow{CA}$. To find this point we join point $O$ and point $A$, and bisect this segment to obtain point $P$. We then construct a circle using point $P$ as center and segment $\overline{PA}$ as radius. The points of intersection of the two circles, points $C$ and $D$, are the required points of tangency. Lines $\overleftrightarrow{CA}$ and $\overleftrightarrow{DA}$ are tangents to the original circle. The proof of this construction is left as an exercise.

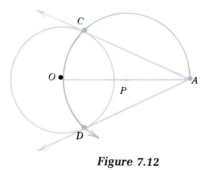

*Figure 7.12*

---

**Theorem 7.15** If two tangent segments are drawn to a circle from a point outside the circle, then the segments are congruent.

---

*Given*: $\overline{AB}$ and $\overline{AC}$ are tangents to circle $O$.

*Prove*: $\overline{AB} \cong \overline{AC}$.

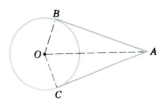

*Construction*: Draw radii $\overline{OB}$, $\overline{OC}$, and segment $\overline{OA}$.

*Proof*:

| STATEMENTS | REASONS |
|---|---|
| 1. $\overline{AB}$ and $\overline{AC}$ are tangents to circle $O$. | 1. Given. |
| 2. $\overline{OB} \perp \overline{AB}$; $\overline{OC} \perp \overline{CA}$. | 2. A line from the center of a circle to a point of tangency is perpendicular to the tangent passing through the point of tangency. |
| 3. $\angle ABO$ and $\angle ACO$ are right angles. | 3. Perpendicular lines meet and form right angles. |
| 4. Triangles $OAB$ and $OAC$ are right triangles. | 4. If a triangle has a right angle, then it is a right triangle. |
| 5. $\overline{OC} \cong \overline{OB}$. | 5. Radii in the same circle are congruent. |
| 6. $\overline{OA} \cong \overline{OA}$. | 6. Reflexive law. |
| 7. $\triangle OAB \cong \triangle OAC$. | 7. $hl = hl$. |
| 8. $\overline{AB} \cong \overline{AC}$. | 8. cpctc.     □ |

Two circles may intersect in two, one, or no points. If two circles intersect in a single point, they are said to be *tangent* to each other. If one circle is on the interior of the other, except for the point of tangency, the circles are said to be **tangent internally**. See Figure 7.13*a*. Otherwise they are said to be **tangent externally**. See Figure 7.13*b*.

Two circles that do not intersect may have a common tangent. If the two circles lie on different sides of a common tangent, the tangent is called a **common internal tangent**. See Figure 7.13*c*. If the circles lie on the same side of a common tangent, the tangent is called a **common external tangent**. See Figure 7.13*d*.

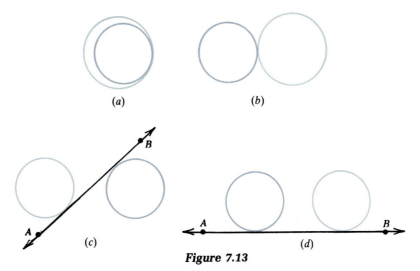

*(a)*          *(b)*

*(c)*          *(d)*

**Figure 7.13**

---

**Definition 7.14** The line passing through the centers of two circles is called a **line of centers**.

---

**Definition 7.15** If three or more points lie on the same line, then the points are called **collinear**.

---

**Theorem 7.16** If two circles are tangent, either internally or externally, then their point of tangency and the center of each circle are collinear.

---

*Given*: Circles $O$ and $O'$ are tangent externally with point $E$ as point of tangency. $\overleftrightarrow{EF}$ is the common internal tangent.

*Prove*: $\overleftrightarrow{OO'}$ passes through point $E$.

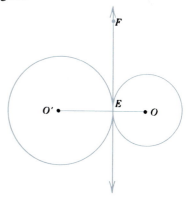

*Proof*:

| STATEMENTS | REASONS |
|---|---|
| **1.** Circles $O$ and $O'$ are tangent externally with point $E$ as point of tangency. $\overleftrightarrow{EF}$ is the common internal tangent. | **1.** Given. |
| **2.** Draw $\overleftrightarrow{O'E}$. | **2.** Two points determine a line. |
| **3.** $\overleftrightarrow{O'E} \perp \overleftrightarrow{EF}$ at point $E$. | **3.** A line from the center of a circle is perpendicular to a tangent at the point of tangency. |
| **4.** Draw $\overleftrightarrow{OE}$. | **4.** Same as reason 2. |
| **5.** $\overleftrightarrow{OE} \perp \overleftrightarrow{EF}$ at point $E$. | **5.** Same as reason 3. |
| **6.** $\overleftrightarrow{O'E}$ and $\overleftrightarrow{OE}$ coincide. | **6.** There is only one perpendicular to $\overleftrightarrow{EF}$ at point $E$. |
| **7.** $\overleftrightarrow{OO'}$ passes through point $E$. | **7.** $\overleftrightarrow{OO'}$ is another name for line $\overleftrightarrow{OE}$, or for $\overleftrightarrow{O'E}$. |

The case where the circles are tangent internally is left as an exercise.

## EXERCISE 7.3

**1.** From a given point outside a circle, how many tangents can be drawn to the circle?

**2.** From a given point outside a circle, how many secants can be drawn to the circle?

**3.** How many internal tangents can two circles have? Discuss the possibilities.

**4.** How many external tangents can two circles have? Discuss the possibilities.

In Exercises 5–8, refer to Illustration 1 in which $\overleftrightarrow{AB}$ and $\overleftrightarrow{CB}$ are tangents to circle $O$.

**5.** If m($\overline{AB}$) = 12 cm, find m($\overline{CB}$).

**6.** If m($\angle BAC$) = 70°, find m($\angle B$).

**7.** If m($\angle B$) = 30°, find m($\widehat{AC}$).

**8.** If m($\widehat{ADC}$) = 200°, find m($\angle B$).

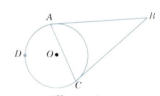

***Illustration 1***

In Exercises 9–10, refer to Illustration 2 in which line $l$ is tangent to circle $O$ at point $E$.

**9.** Find m($\angle 1$).

**10.** Find m($\angle 2$).

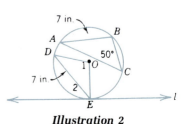

***Illustration 2***

In Exercises 11–12, refer to Illustration 3 in which $\overline{AB}$ is a tangent to circle $O$ and $\overline{AD}$ is a secant.

**11.** Find m($\overset{\frown}{DB}$).

**12.** Find m($\angle ABD$).

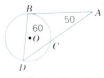

***Illustration 3***

In Exercises 13–16, refer to Illustration 4 in which $\overline{AB}$ and $\overline{CB}$ are tangent to circle $O$.

**13.** If m($\overline{OC}$) = 6 cm and m($\overline{AB}$) = 8 cm, find m($\overline{OB}$).

**14.** If m($\angle OCA$) = 30°, find m($\angle ABC$).

**15.** If m($\angle ABC$) = 50°, find m($\overset{\frown}{AC}$).

**16.** If m($\overset{\frown}{ADC}$) = 200°, find m($\angle OCA$).

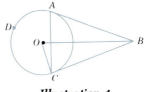

***Illustration 4***

**17.** *Complete the proof of Theorem 7.16*: If two circles are tangent, either internally or externally, then their point of tangency and the center of each circle are collinear.

**18.** In Illustration 5, segments $\overline{AB}$ and $\overline{DB}$ are tangent to circle $O'$, and $\overline{DB}$ and $\overline{BC}$ are tangent to circle $O$. Prove that $\overline{AB} \cong \overline{BC}$.

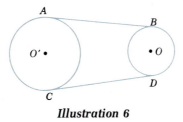

***Illustration 5***

**19.** In Illustration 6, segments $\overline{AB}$ and $\overline{CD}$ are common external tangents to circles $O'$ and $O$. Prove that $\overline{AB} \cong \overline{CD}$. (*Hint*: Consider extending $\overline{AB}$ and $\overline{CD}$.)

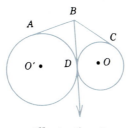

***Illustration 6***

**20.** *Prove*: An angle inscribed in a semicircle is a right angle.

**21.** Prove the construction given on page 255.

## 7.4 MORE THEOREMS INVOLVING CIRCLES

> **Theorem 7.17** If two circles intersect in two points, then their line of centers is the perpendicular bisector of their common chord.

*Given*: Circle $O'$ intersects circle $O$ at points $A$ and $B$.

*Prove*: $\overline{OO'}$ is the perpendicular bisector of $\overline{AB}$.

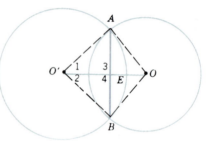

*Construction*: Draw radii $\overline{O'A}$, $\overline{O'B}$, $\overline{OA}$, and $\overline{OB}$.

*Proof*:

| STATEMENTS | REASONS |
|---|---|
| **1.** Circle $O'$ intersects circle $O$ at points $A$ and $B$. | **1.** Given. |
| **2.** $\overline{O'B} \cong \overline{O'A}$; $\overline{OA} \cong \overline{OB}$. | **2.** Radii in the same circle are congruent. |
| **3.** $\overline{OO'} \cong \overline{OO'}$. | **3.** Reflexive law. |
| **4.** $\triangle O'AO \cong \triangle O'BO$. | **4.** SSS $\cong$ SSS. |
| **5.** $\angle 1 \cong \angle 2$. | **5.** cpctc. |
| **6.** $\overline{O'E} \cong \overline{O'E}$. | **6.** Reflexive law. |
| **7.** $\triangle O'AE \cong \triangle O'BE$. | **7.** SAS $\cong$ SAS. |
| **8.** $\overline{BE} \cong \overline{AE}$. | **8.** cpctc. |
| **9.** $\overline{O'O}$ bisects $\overline{AB}$. | **9.** If a segment divides a segment into two congruent segments, then it bisects the segment. |
| **10.** $\angle 3 \cong \angle 4$. | **10.** cpctc. |

| STATEMENTS | REASONS |
|---|---|
| **11.** $\overline{O'O}$ is perpendicular to $\overline{AB}$. | **11.** If two segments intersect and form congruent adjacent angles, then they are perpendicular. |
| **12.** $\overline{OO'}$ is the perpendicular bisector of $\overline{AB}$. | **12.** If a line is both perpendicular to and a bisector of another segment, then it is the perpendicular bisector of the segment.      □ |

---

**Theorem 7.18** In a circle, parallel lines intercept congruent arcs.

---

The proof of Theorem 7.18 is left as an exercise.

---

**Theorem 7.19** If two chords intersect within a circle, then the product of the lengths of the segments of one chord is equal to the product of the lengths of the segments of the other chord.

---

*Given*: Chords $\overline{AB}$ and $\overline{CD}$ intersect at $E$.

*Prove*: $[m(\overline{AE})][m(\overline{BE})] = [m(\overline{EC})][m(\overline{DE})]$.

*Construction*: Draw segments $\overline{AD}$ and $\overline{BC}$.

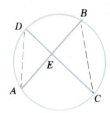

*Proof*:

| STATEMENTS | REASONS |
|---|---|
| **1.** Chords $\overline{AB}$ and $\overline{CD}$ intersect at $E$. | **1.** Given. |
| **2.** $m(\angle D) = \frac{1}{2}[m(\overline{AC})]$;  $m(\angle B) = \frac{1}{2}[m(\overline{AC})]$. | **2.** An inscribed angle has a measure of half its intercepted arc. |
| **3.** $m(\angle D) = m(\angle B)$. | **3.** Substitution. |
| **4.** $\angle D \cong \angle B$. | **4.** If two angles have equal measures, they are congruent. |

| STATEMENTS | REASONS |
|---|---|
| **5.** $m(\angle A) = \frac{1}{2}[m(\overline{DB})];$ $m(\angle C) = \frac{1}{2}[m(\overline{DB})].$ | **5.** An inscribed angle has a measure of half its intercepted arc. |
| **6.** $m(\angle A) = m(\angle C).$ | **6.** Substitution. |
| **7.** $\angle A \cong \angle C.$ | **7.** If two angles have equal measures, they are congruent. |
| **8.** $\triangle ADE \sim \triangle CBE.$ | **8.** If two angles of one triangle are congruent to two angles of a second triangle, then the triangles are similar. |
| **9.** $\dfrac{m(\overline{AE})}{m(\overline{CE})} = \dfrac{m(\overline{DE})}{m(\overline{BE})}.$ | **9.** The lengths of the corresponding sides of similar triangles are in proportion. |
| **10.** $[m(\overline{AE})][m(\overline{BE})] = [m(\overline{CE})][m(\overline{DE})].$ | **10.** In a proportion, the product of the means equals the product of the extremes.    □ |

**EXAMPLE 1**

In Figure 7.14, chords $\overline{AB}$ and $\overline{DC}$ intersect at point $E$. Find $m(\overline{BE})$.

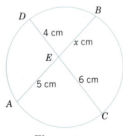

**Figure 7.14**

*Solution*

Because of Theorem 7.19, we have

$$[m(\overline{AE})][m(\overline{BE})] = [m(\overline{EC})][m(\overline{DE})]$$

Substitute values for the lengths of these segments and solve for $x$.

$$5x = 6(4)$$
$$5x = 24$$
$$x = \frac{24}{5}$$

The length of $\overline{BE}$ is 4.8 cm.    ∎

**Theorem 7.20** If a secant and a tangent are drawn to a circle, then the length of the tangent is the mean proportional between the lengths of the secant and its external segment.

*Given*: $\overline{AB}$ is tangent to circle $O$ and $\overline{AD}$ is a secant.

*Prove*: $\dfrac{m(\overline{AD})}{m(\overline{AB})} = \dfrac{m(\overline{AB})}{m(\overline{AC})}$.

*Construction*: Draw segments $\overline{DB}$ and $\overline{BC}$.

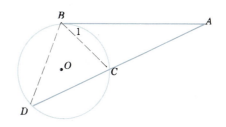

*Proof*:

| STATEMENTS | REASONS |
|---|---|
| **1.** $\overline{AB}$ is a tangent and $\overline{AD}$ is a secant. | **1.** Given. |
| **2.** $\angle A \cong \angle A$. | **2.** Reflexive law. |
| **3.** $m(\angle D) = \dfrac{1}{2}[m(\overline{BC})]$. | **3.** An inscribed angle measures half of its intercepted arc. |
| **4.** $m(\angle 1) = \dfrac{1}{2}[m(\overline{BC})]$. | **4.** An angle formed by a tangent and a chord measures half of its intercepted arc. |
| **5.** $m(\angle D) = m(\angle 1)$. | **5.** Substitution. |
| **6.** $\angle D \cong \angle 1$. | **6.** If two angles have equal measures, they are congruent. |
| **7.** $\triangle ABD \sim \triangle ACB$. | **7.** If two angles of one triangle are congruent to two angles of a second triangle, then the triangles are similar. |
| **8.** $\dfrac{m(\overline{AD})}{m(\overline{AB})} = \dfrac{m(\overline{AB})}{m(\overline{AC})}$. | **8.** The lengths of corresponding sides of similar triangles are in proportion. |

**EXAMPLE 2**    In Figure 7.15, $\overline{AB}$ is a tangent to the circle and $\overline{AD}$ is a secant. Find $m(\overline{DC})$.

Solution    Because of Theorem 7.20, we have

$$\frac{m(\overline{AD})}{m(\overline{AB})} = \frac{m(\overline{AB})}{m(\overline{AC})}$$

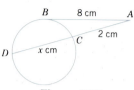

**Figure 7.15**

Substitute values for the lengths of these segments and solve for $x$.

$$\frac{x + 2}{8} = \frac{8}{2}$$

$2x + 4 = 64$    The product of the means equals the product of the extremes.

$2x = 60$    Add $-4$ to both sides.

$x = 30$    Divide both sides by 2.

The length of $\overline{DC}$ is 30 cm.  ∎

---

**Theorem 7.21** If two secants are drawn to a circle from a point outside the circle, then the products of the lengths of the secants and their external segments are equal.

---

*Given:* $\overline{AC}$ and $\overline{AE}$ are secants from point $A$.

*Prove:* $[m(\overline{AC})][m(\overline{AB})]$
$= [m(\overline{AE})][m(\overline{AD})]$.

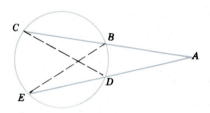

*Construction:* Draw segments $\overline{EB}$ and $\overline{CD}$.

*Proof:*

| STATEMENTS | REASONS |
|---|---|
| **1.** $\overline{AC}$ and $\overline{AE}$ are secants from point $A$. | **1.** Given. |
| **2.** $\angle A \cong \angle A$. | **2.** Reflexive law. |

| STATEMENTS | REASONS |
|---|---|
| 3. $m(\angle C) = \frac{1}{2}[m(\overline{BD})]$;<br><br>$m(\angle E) = \frac{1}{2}[m(\overline{BD})]$. | 3. An inscribed angle measures half of its intercepted arc. |
| 4. $m(\angle C) \cong m(\angle E)$. | 4. Substitution. |
| 5. $\angle C \cong \angle E$. | 5. If two angles have equal measures, they are congruent. |
| 6. $\triangle ACD \sim \triangle AEB$. | 6. If two angles of one triangle are congruent to two angles of a second triangle, then the triangles are similar. |
| 7. $\frac{m(\overline{AC})}{m(\overline{AE})} = \frac{m(\overline{AD})}{m(\overline{AB})}$. | 7. The lengths of the corresponding sides of similar triangles are in proportion. |
| 8. $[m(\overline{AC})][m(\overline{AB})]$ $= [m(\overline{AE})][m(\overline{AD})]$. | 8. The product of the means equals the product of the extremes.    □ |

*EXAMPLE 3*    In Figure 7.16, $\overline{AC}$ and $\overline{AE}$ are secants to circle $O$. Find the length of $\overline{AC}$.

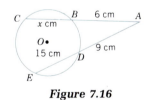

*Figure 7.16*

*Solution*    Because of Theorem 7.21, we have

$$[m(\overline{AC})][m(\overline{AB})] = [m(\overline{AE})][m(\overline{AD})]$$

Substitute values for the lengths of these segments and solve for $x$.

$(6 + x)(6) = 24(9)$

$36 + 6x = 216$    Remove parentheses.

$6x = 180$    Add $-36$ to both sides.

$x = 30$    Divide both sides by 6.

The length of $\overline{CB}$ is 30 cm. Hence, the length of $A\overline{C}$ is 30 cm + 6 cm, or 36 cm.    ∎

If a polygon has its vertices on a circle, the polygon is said to be **inscribed** within the circle. If all sides of the polygon are tangent to a circle, the polygon is said to be **circumscribed** about the circle. In Figure 7.17*a*, $\triangle ABC$ is inscribed within a circle, and the circle is said to be circumscribed about the triangle. In Figure 7.17*b*, $\triangle DEF$ is circumscribed about a circle, and the circle is said to be inscribed within the triangle.

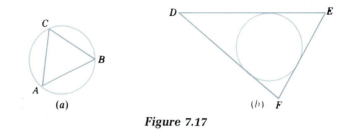

(*a*)                          (*b*)

**Figure 7.17**

---

**Theorem 7.22**  The opposite angles of an inscribed quadrilateral are supplementary.

---

*Given*: $ABCD$ is an inscribed quadrilateral.

*Prove*: $\angle A$ is supplementary to $\angle C$.

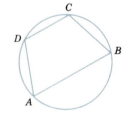

*Proof*:

| STATEMENTS | REASONS |
|---|---|
| **1.** $ABCD$ is an inscribed quadrilateral. | **1.** Given. |
| **2.** $m(\angle A) = \frac{1}{2}[m(\overset{\frown}{DCB})]$; $m(\angle C) = \frac{1}{2}[m(\overset{\frown}{BAD})]$. | **2.** The measure of an inscribed angle is half the measure of its intercepted arc. |
| **3.** $m(\overset{\frown}{DCB}) + m(\overset{\frown}{BAD}) = m(\overset{\frown}{DCBAD})$. | **3.** Postulate 7.3. |
| **4.** $m(\overset{\frown}{DCBAD}) = 360°$. | **4.** A circle contains 360°. |

| STATEMENTS | REASONS |
|---|---|
| **5.** m($\angle A$) + m($\angle C$) $= \frac{1}{2}$ [m($\overset{\frown}{DCB}$)] + $\frac{1}{2}$ [m($\overset{\frown}{BAD}$)]. | **5.** Equal quantities added to equal quantities are equal. |
| **6.** m($\angle A$) + m($\angle C$) = $\frac{1}{2}$ [m($\overset{\frown}{DCB}$) + m($\overset{\frown}{BAD}$)]. | **6.** Use the distributive law to factor out $\frac{1}{2}$. |
| **7.** m($\angle A$) + m($\angle C$) = $\frac{1}{2}$ (360°) = 180°. | **7.** Substitution. |
| **8.** $\angle A$ is supplementary to $\angle C$. | **8.** If the sum of the measures of two angles is 180°, then the angles are supplementary. □ |

By a similar argument we can show that $\angle D$ is supplementary to $\angle B$.

---

**Theorem 7.23** If a parallelogram is inscribed within a circle, then it is a rectangle.

---

*Given*: *ABCD* is an inscribed parallelogram.

*Prove*: *ABCD* is a rectangle.

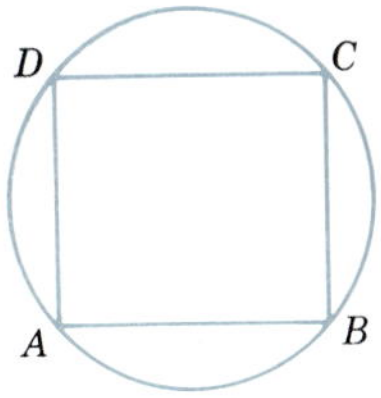

*Proof*:

| STATEMENTS | REASONS |
|---|---|
| **1.** *ABCD* is a parallelogram. | **1.** Given. |
| **2.** $\angle A \cong \angle C$. | **2.** Opposite angles in a parallelogram are congruent. |
| **3.** $\angle A$ is supplementary to $\angle C$. | **3.** Opposite angles in an inscribed quadrilateral are supplementary. |
| **4.** m($\angle A$) = m($\angle C$) = 90°. | **4.** If two angles are congruent and supplementary, then each has a measure of 90°. |

| STATEMENTS | REASONS |
|---|---|
| **5.** $\angle A$ and $\angle C$ are right angles. | **5.** If an angle measures 90°, then it is a right angle. |
| **6.** $ABCD$ is a rectangle. | **6.** If a parallelogram has a right angle, then it is a rectangle. □ |

## EXERCISE 7.4

In Exercises 1–5, refer to Illustration 1 in which $\overline{AB}$ and $\overline{CD}$ are two chords intersecting at $E$.

**1.** Assume that m($\overline{AE}$) = 8 cm, m($\overline{CE}$) = 6 cm, and m($\overline{ED}$) = 4 cm. Find m($\overline{EB}$).
**2.** Assume that m($\overline{AE}$) = 9 cm, m($\overline{CE}$) = 6 cm, and m($\overline{ED}$) = 6 cm. Find m($\overline{AB}$).
**3.** Assume that m($\overline{CD}$) = 11 in., m($\overline{ED}$) = 5 in., and m($\overline{EB}$) = 3 in. Find m($\overline{AB}$).
**4.** Assume that m($\overline{AB}$) = m($\overline{CD}$), m($\overline{EB}$) = 2 in., and m($\overline{CE}$) = 6 in. Find m($\overline{ED}$).
**5.** Assume that m($\overline{AE}$) bisects m($\overline{CD}$), m($\overline{AE}$) = 9 cm, and m($\overline{BE}$) = 4 cm. Find m($\overline{CD}$).

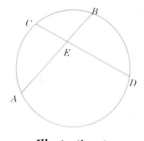

*Illustration 1*

In Exercises 6–12, refer to Illustration 2 which $\overline{AB}$ is a tangent and $\overline{DB}$ is a secant intersecting at $B$.

**6.** Assume that m($\overline{DB}$) = 12 m and m($\overline{CB}$) = 3 m. Find m($\overline{AB}$).
**7.** Assume that m($\overline{AB}$) = 12 yd and m($\overline{BC}$) = 4 yd. Find m($\overline{DC}$).
**8.** Assume that m($\overline{AB}$) = 25 ft and m($\overline{CB}$) = 5 ft. Find m($\overline{DB}$).

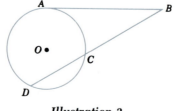

*Illustration 2*

9. Assume that m($\overline{BD}$) = 20 cm and m($\overline{DC}$) = 12 cm. Find m($\overline{AB}$).
10. Assume that m($\overline{AB}$) = 16 cm and m($\overline{CB}$) = 4 cm. Find m($\overline{DC}$).
11. Assume that m($\overline{AB}$) = 10 in. and m($\overline{BD}$) = 80 in. Find m($\overline{DC}$).
12. Assume that m($\overline{DC}$) = 95 yd and m($\overline{DB}$) = 100 yd. Find m($\overline{AB}$).

In Exercises 13–18, refer to Illustration 3 in which $\overline{AC}$ and $\overline{EC}$ are secants intersecting at C.

13. Assume that m($\overline{AC}$) = 15 cm, m($\overline{BC}$) = 5 cm, and m($\overline{DC}$) = 3 cm. Find m($\overline{EC}$).
14. Assume that m($\overline{AC}$) = 12 ft, m($\overline{BA}$) = 4 ft, and m($\overline{CE}$) = 16 ft. Find m($\overline{ED}$).
15. Assume that m($\overline{AB}$) = 38 cm, m($\overline{BC}$) = 2 cm, and m($\overline{DC}$) = 4 cm. Find m($\overline{ED}$).
16. Assume that m($\overline{AC}$) = 20 mi, m($\overline{BC}$) = 4 mi, and m($\overline{EC}$) = 16 mi. Find m($\overline{ED}$).
17. Assume that m($\overline{AB}$) and m($\overline{DC}$) are both 4 m. Find m($\overline{ED}$) if m($\overline{BC}$) = 6 m.
18. Assume that m($\overline{AB}$) = 12 cm, m($\overline{BC}$) = 4 cm, and m($\overline{EC}$) is twice m($\overline{AC}$). Find m($\overline{ED}$).

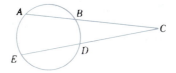

**Illustration 3**

In Exercises 19–23, refer to Illustration 4.

19. If m($\angle D$) = 80°, find m($\angle B$).
20. If $\angle A$ = 100°, find m($\overset{\frown}{DAB}$).
21. If $\overset{\frown}{ABC}$ = 200°, find m($\angle B$).

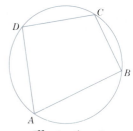

**Illustration 4**

22. If $\overline{AD}$ and $\overline{BC}$ are both 6 cm long and $\overline{DC}$ and $\overline{AB}$ are both 10 cm long, find the area of quadrilateral $ABCD$.
23. If m($\angle A$) = 60°, m($\overset{\frown}{DC}$) = 100°, and m($\overset{\frown}{AB}$) = 80°, find m($\angle B$).
24. *Prove Theorem 7.18*: In a circle, parallel lines intercept congruent arcs.
25. *Prove*: If two tangents are drawn to a circle from a point outside the circle, then the line segment joining the center of the circle and the point of intersection of the tangents bisects the angle formed by the tangents.
26. In Illustration 5, circle $O$ is congruent to circle $O'$, and $\overline{AB}$ and $\overline{CD}$ are common external tangents. Prove that $\overline{AB} \cong \overline{CD}$.

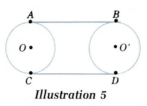

**Illustration 5**

**27.** A rather surprising result first discovered in 1821 is illustrated by this construction:

(a) Draw any triangle in the center of a sheet of paper. Make the sides about 2 or 3 in. long.
(b) Find the midpoints of the three sides and label them $A$, $B$, and $C$.
(c) Construct the three altitudes of the triangle. Use letters $D$, $E$, and $F$ to label the feet of the altitudes.
(d) If you have been careful, the three altitudes meet in a single point. Bisect each of the three line segments joining the point of intersection of the altitudes to a vertex. Label these midpoints $G$, $H$, and $I$.

If your drawing is cluttered, go back and mark more heavily the nine labeled points. What do you notice about these nine points? If you have done the construction accurately, all nine points will lie on a circle, called the **nine-point circle**.

## CHAPTER SEVEN REVIEW EXERCISES

In Review Exercises 1–8, refer to Illustration 1 in which $\overline{AOB}$ is a diameter, $\overline{AE}$ is a tangent, $\overline{AD} \parallel \overline{OC}$, and $m(\widehat{BD}) = 80°$.

**1.** Find $m(\angle 1)$.

**2.** Find $m(\angle 2)$.

**3.** Find $m(\widehat{BC})$.

**4.** Find $m(\widehat{DC})$.

**5.** Find $m(\widehat{AD})$.

**6.** Find $m(\angle 5)$.

**7.** Find $m(\angle 3)$.

**8.** Find $m(\angle 4)$.

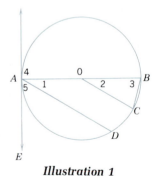

**Illustration 1**

In Review Exercises 9–14, refer to Illustration 2.

**9.** If $m(\widehat{AD}) = 140°$ and $m(\widehat{BC}) = 30°$, find $m(\angle 1)$.
**10.** If $m(\widehat{AD}) = 140°$ and $m(\widehat{BC}) = 30°$, find $m(\angle 2)$.
**11.** If $m(\widehat{AD}) = 140°$ and $m(\widehat{BC}) = 30°$, find $m(\angle C)$.
**12.** If $m(\widehat{AD}) = 140°$ and $m(\angle 1) = 80°$, find $m(\widehat{BC})$.

**13.** If m($\angle 2$) = 110° and m($\widehat{DB}$) = 100°, find m($\widehat{AC}$).

**14.** If m($\angle 2$) = 120° and m($\widehat{BC}$) = 40°, find m($\widehat{AD}$).

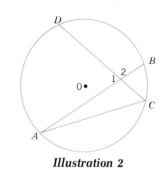

*Illustration 2*

In Reveiw Exercises 15–18, refer to Illustration 3 in which $\overline{AC}$ and $\overline{EC}$ are secants and $\overleftrightarrow{CF}$ is a tangent.

**15.** If m($\widehat{AE}$) = 80° and m($\widehat{BD}$) = 10°, find m($\angle 1$).

**16.** If m($\widehat{AE}$) = 100° and m($\angle 1$) = 20°, find m($\widehat{BD}$).

**17.** If m($\widehat{EF}$) = 85° and m($\widehat{DF}$) = 30°, find m($\angle 2$).

**18.** If m($\angle 2$) = 15° and m($\widehat{DF}$) = 40°, find m($\widehat{EF}$).

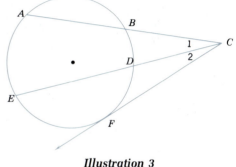

*Illustration 3*

In Review Exercises 19–22, refer to Illustration 4 in which $\overline{DOC}$ is a diameter and $\overline{AB}$ is a chord.

**19.** If m($\overline{AE}$) = 8 cm, find m($\overline{AB}$).

**20.** If m($\widehat{ACB}$) = 120°, find m($\angle AOE$).

**21.** If m($\widehat{BD}$) = 110°, find m($\angle OAE$).

**22.** If m($\angle B$) = 50°, find m($\widehat{BDA}$).

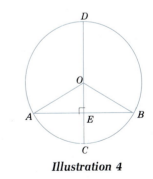

*Illustration 4*

In Review Exercises 23–26, refer to Illustration 5 in which $\overline{AB}$ and $\overline{BC}$ are tangents to circle $O$.

**23.** If m($\overline{AB}$) = 8 m, find m($\overline{CB}$).

**24.** If m($\angle BAC$) = 65°, find m($\angle B$).

**25.** If m($\angle B$) = 40°, find m($\overparen{AC}$).

**26.** If m($\overparen{ADC}$) = 200°, find m($\angle B$).

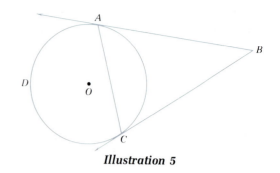

**Illustration 5**

In Review Exercises 27–30, refer to Illustration 6 in which $\overline{AB}$ and $\overline{CD}$ are chords of the circle.

**27.** If m($\overline{DE}$) = 5, m($\overline{AE}$) = 6, m($\overline{EB}$) = 2, find m($\overline{EC}$).

**28.** If m($\overline{AE}$) = 8, m($\overline{EB}$) = 4, m($\overline{DE}$) = 3, find m($\overline{DC}$).

**29.** If m($\overline{AB}$) = 12, m($\overline{AE}$) = 9, m($\overline{DE}$) = 4, find m($\overline{DC}$).

**30.** If m($\overline{AE}$) = m($\overline{EC}$) and m($\overline{DE}$) = 4, find m($\overline{EB}$).

**Illustration 6**

In Review Exercises 31–34, refer to Illustration 7 in which $\overline{AB}$ is a tangent to the circle and $\overline{AC}$ and $\overline{AD}$ are secants.

**31.** If m($\overline{AB}$) = 6 and m($\overline{AC}$) = 10, find m($\overline{AF}$).

**32.** If m($\overline{AC}$) = 12 and m($\overline{AF}$) = 3, find m($\overline{AB}$).

**33.** If m($\overline{AC}$) = 10, m($\overline{AF}$) = 6 and m($\overline{AD}$) = 12, find m($\overline{AE}$).

**34.** If m($\overline{AC}$) = 6, m($\overline{CF}$) = 4 and m($\overline{AE}$) = 3, find m($\overline{DE}$).

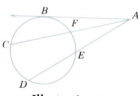

**Illustration 7**

## CHAPTER SEVEN TEST

In Exercises 1–20, classify each statement as true or false.

**1.** A diameter is a chord of a circle.

**2.** A minor arc is an arc that is less than 90°.

**3.** An inscribed angle is equal to one-half the measure of its intercepted arc.

**4.** An inscribed angle is twice as large as a central angle.

**5.** If two secants are drawn to a circle from a point outside the circle, the product of the lengths of the segments of one secant equals the product of the lengths of the segments of the other secant.

**6.** Equal chords are equidistant from the center of their circle.

**7.** If a chord measures 7 in., its intercepted arc measures 7 in.

**8.** Secants drawn to a circle from a point outside the circle are equal.

**9.** If two circles are tangent internally, they have no internal tangents.

**10.** Each bisector of a chord passes through the center of the circle.

**11.** Opposite angles of a quadrilateral inscribed in a circle are equal.

**12.** An inscribed angle can never be an obtuse angle.

**13.** If two chords intersect within a circle, four central angles will be formed.

**14.** If a regular pentagon is inscribed within a circle, the circle will be divided into five equal arcs.

**15.** A regular hexagon is inscribed within a circle. The measure of each arc of the circle is 60°.

**16.** If two circles intersect, their line of centers bisects their common chord.

**17.** The opposite angles of a quadrilateral inscribed within a circle are supplementary.

**18.** The measure of an inscribed angle depends on the radius of the circle.

**19.** A secant intersects a circle in two points.

**20.** If two circles are tangent externally, their line of centers is parallel to their common external tangents.

# 8

*Leonhard Euler (1707–1783). Euler, a Swiss mathematician, was a major founder of advanced mathematics. His work involved the topics of trigonometry, analytic geometry, calculus, and topology. Euler was the first to give a formula that described the relationship between the vertices, edges, and faces of a regular polyhedron. If V represents the number of vertices, E represents the number of edges, and F represents the number of faces, then*

$$V - E + F = 2$$

# Inequalities

$T$he topic of inequalities is introduced in a beginning algebra course. In this chapter we shall review much of this material and investigate some new ideas. Throughout this discussion we shall regard the term **positive number** as an undefined term.

## 8.1 FUNDAMENTAL PROPERTIES OF INEQUALITIES

The symbol "<" is read as "is less than," and the symbol ">" is read as "is greater than." These inequalities are called **strict inequalities**. In this section, we will limit the discussion to these inequalities.

> **Definition 8.1** If $a$ and $b$ are two real numbers, then $a < b$ if, and only if, there is a positive number $p$ such that $b = a + p$. If $a$ and $b$ are two numbers, then $a > b$ if, and only if, $b < a$.

Numbers that are less than zero are called **negative numbers**.

*EXAMPLE 1*     Explain why $2 < 7$.

*Solution*     The statement $2 < 7$ is true if, and only if, there is a positive number $p$ such that $7 = 2 + p$. The number 5 is such a number because $7 = 2 + 5$. Thus, by Definition 8.1, we have $2 < 7$.  ■

Recall from algebra the commutative and associative laws of addition and multiplication, and the distributive law. These laws state that for any numbers $a$, $b$, and $c$,

| | |
|---|---|
| $a + b = b + a$ | Commutative law of addition |
| $ab = ba$ | Commutative law of multiplication |
| $(a + b) + c = a + (b + c)$ | Associative law of addition |
| $(ab)c = a(bc)$ | Associative law of multiplication |
| $a(b + c) = ab + ac$ | Distributive law |

Also recall these rules governing the arithmetic of signed numbers.

1.  The product of a positive number and a negative number is a negative number.
2.  The negative of a negative number is a positive number.
3.  The sum of two negative numbers is a negative number.

We are now ready to prove many theorems involving inequalities.

> **Theorem 8.1** If the same quantity is added to both sides of an inequality, then the sums are unequal and in the same order.
>
> If $a < b$, then $a + c < b + c$.

*Given:* $a < b$.

*Prove:* $a + c < b + c$.

*Proof*:

| STATEMENTS | REASONS |
|---|---|
| **1.** $a < b$. | **1.** Given. |
| **2.** $b = a + p$ where $p$ is a positive number. | **2.** Definition 8.1. |
| **3.** $b + c = a + p + c$. | **3.** If equal quantities are added to equal quantities, then the results are equal quantities. |
| **4.** $b + c = a + c + p$. | **4.** Commutative law of addition. |
| **5.** $a + c < b + c$. | **5.** Definition 8.1. $\square$ |

---

**Theorem 8.2** If equal quantities are added to unequal quantities, then the sums are unequal and in the same order.

If $a < b$ and $c = d$, then $a + c < b + d$.

---

*Given*: $a < b$.
    $c = d$.

*Prove*: $a + c < b + d$.

*Proof*:

| STATEMENTS | REASONS |
|---|---|
| **1.** $a < b$. | **1.** Given. |
| **2.** $a + c < b + c$. | **2.** Theorem 8.1 (Add $c$ to both sides.) |
| **3.** $c = d$. | **3.** Given. |
| **4.** $a + c < b + d$. | **4.** Substitution. $\square$ |

Note that Theorems 8.1 and 8.2 allow for the subtraction of the same or equal quantities from both sides of an inequality. This is because the subtraction of a quantity $c$ is simply the addition of the quantity $-c$.

---

**Theorem 8.3** If unequal quantities are subtracted from equal quantities, then the differences are unequal but in the *opposite* order.

If $a < b$ and $c = d$, then $c - a > d - b$.

---

*Given*: $a < b$.
   $c = d$.

*Prove*: $c - a > d - b$.

*Proof*:

| STATEMENTS | REASONS |
|---|---|
| **1.** $a < b$. | **1.** Given. |
| **2.** $c = d$. | **2.** Given. |
| **3.** $a + d < b + c$. | **3.** Theorem 8.2. |
| **4.** $d - b < c - a$. | **4.** Theorem 8.1 (Add $-b - a$ to both sides.) |
| **5.** $c - a > d - b$. | **5.** Definition 8.1. □ |

---

**Theorem 8.4** If both sides of an inequality are multiplied by a positive number, then the products are unequal and in the same order.

If $a < b$ and $c$ is a positive number, then $ac < bc$.

---

*Given*: $a < b$.
   $c$ is a positive number.

*Prove*: $ac < bc$.

*Proof*:

| STATEMENTS | REASONS |
|---|---|
| **1.** $a < b$. | **1.** Given. |
| **2.** $a - b < 0$. | **2.** Theorem 8.1. (Add $-b$ to both sides.) |
| **3.** $c$ is a positive number. | **3.** Given. |
| **4.** $c(a - b) < 0$. | **4.** The product of a positive number and a negative number is a negative number. |
| **5.** $c(a - b) = ac - bc$. | **5.** Distributive law and commutative law of multiplication. |
| **6.** $ac - bc < 0$. | **6.** Substitution. |
| **7.** $ac < bc$. | **7.** Theorem 8.1. (Add $bc$ to both sides.) □ |

> **Theorem 8.5** If both sides of an inequality are multiplied by a negative number, then the products are unequal but in the *opposite* order.
>
> If $a < b$ and $c$ is a negative number, then $ac > bc$.

*Given*: $a < b$.
  $c$ is a negative number.

*Prove*: $ac > bc$.

*Proof*:

| STATEMENTS | REASONS |
|---|---|
| 1. $a < b$. | 1. Given. |
| 2. $c$ is a negative number. | 2. Given. |
| 3. $-c$ is a positive number. | 3. The negative of a negative number is a positive number. |
| 4. $-c(a) < -c(b)$. | 4. Theorem 8.4. (Multiply both sides of Statement 1 by $-c$.) |
| 5. $-ac < -bc$. | 5. Associative and commutative laws of multiplication. |
| 6. $bc < ac$. | 6. Theorem 8.1. (Add $bc + ac$ to both sides.) |
| 7. $ac > bc$. | 7. Definition 8.1. □ |

Note that Theorems 8.4 and 8.5 allow for the division of both sides of an inequality by the same nonzero number. This is because dividing by a number is equivalent to multiplying by its reciprocal. However, we must remember to reverse the order of the inequality when dividing both sides by a negative number.

> **Theorem 8.6** The relation "$<$" is transitive.
>
> If $a < b$ and $b < c$, then $a < c$.

*Given*: $a < b$.
  $b < c$.

*Prove*: $a < c$.

*Proof*:

| STATEMENTS | REASONS |
|---|---|
| **1.** $a < b$. | **1.** Given. |
| **2.** $a - b < 0$. | **2.** Theorem 8.1. (Add $-b$ to both sides.) |
| **3.** $b < c$. | **3.** Given. |
| **4.** $b - c < 0$. | **4.** Theorem 8.1. (Add $-c$ to both sides.) |
| **5.** $a - c < 0$. | **5.** The sum of two negative numbers is a negative number. (Add $a - b$ and $b - c$.) |
| **6.** $a < c$. | **6.** Theorem 8.1. (Add $c$ to both sides.)  □ |

In the exercises you will be asked to prove that the relation ">" is transitive.

---

**Theorem 8.7** If unequal quantities are added to unequal quantities in the same order, then the sums are unequal quantities in the same order.

If $a < b$ and $c < d$, then $a + c < b + d$.

---

*Given*: $a < b$.
      $c < d$.

*Prove*: $a + c < b + d$.

*Proof*:

| STATEMENTS | REASONS |
|---|---|
| **1.** $a < b$. | **1.** Given. |
| **2.** $a + c < b + c$. | **2.** Theorem 8.1. (Add $c$ to both sides.) |
| **3.** $c < d$. | **3.** Given. |
| **4.** $b + c < b + d$. | **4.** Theorem 8.1. (Add $b$ to both sides.) |
| **5.** $a + c < b + d$. | **5.** Theorem 8.6. (Transitive law of "<.")  □ |

> **Theorem 8.8** In a triangle an exterior angle is greater than either nonadjacent interior angle.

*Given*: $\triangle ABC$ with $m(\angle C) > 0°$.

*Prove*: $m(\angle 1) > m(\angle 2)$.

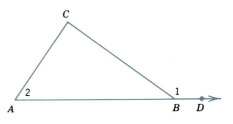

*Proof*:

| STATEMENTS | REASONS |
|---|---|
| **1.** $\triangle ABC$ with $m(\angle C) > 0°$. | **1.** Given. |
| **2.** $m(\angle 1) = m(\angle 2) + m(\angle C)$. | **2.** In a triangle the measure of an exterior angle equals the sum of the measures of the nonadjacent interior angles. |
| **3.** $m(\angle 1) > m(\angle 2)$. | **3.** Definition 8.1. (If $m(\angle 1) = m(\angle 2) + m(\angle C)$ and $m(\angle C)$ is positive, then $m(\angle 1) > m(\angle 2)$.) |

By a similar argument $m(\angle 1)$ can be shown to be greater than $m(\angle C)$.

□

*EXAMPLE 2*

Solve the inequality $2x + 2 < x + 7$.

*Solution*

By Theorem 8.1, we can add $-2$ to both sides of the inequality and simplify to get another inequality with the same order.

$$2x + 2 < x + 7$$
$$2x + 2 + (-2) < x + 7 + (-2)$$
$$2x < x + 5$$

Similarly, we can add $-x$ to both sides of the inequality and simplify to get yet another inequality with the same order.

$$2x + (-x) < x + 5 + (-x)$$
$$x < 5$$

Thus, the original inequality will be satisfied if $x$ is any number less than 5.    ■

**EXAMPLE 3**      Solve the inequality $5(3x - 2) > 2(5x - 10)$.

*Solution*       Solve for $x$ as follows.

$$5(3x - 2) > 2(5x - 10)$$
$$15x - 10 > 10x - 20 \qquad \text{Remove parentheses.}$$
$$15x > 10x - 10 \qquad \text{Add 10 to both sides.}$$
$$5x > -10 \qquad \text{Add } -10x \text{ to both sides.}$$
$$x > -2 \qquad \text{Divide both sides by 5.}$$

Thus, the original inequality will be satisfied if $x$ is any number greater than $-2$. ∎

**EXAMPLE 4**      Solve the inequality $x(x - 3) < x^2 + 9$.

Solution       Solve for $x$ as follows.

$$x(x - 3) < x^2 + 9$$
$$x^2 - 3x < x^2 + 9 \qquad \text{Remove parentheses.}$$
$$-3x < 9 \qquad \text{Add } -x^2 \text{ to both sides.}$$

Theorem 8.5 allows division of both sides by $-3$ provided that we reverse the direction of the inequality. Hence, we have

$$x > -3 \qquad \text{Divide both sides by } -3.$$

The original inequality will be satisfied if $x$ is any number greater than $-3$. ∎

**EXAMPLE 5**      If possible, solve the inequality $x(x + 1) > x^2 + x + 2$.

Solution       Solve for $x$ as follows.

$$x(x + 1) > x^2 + x + 2$$
$$x^2 + x > x^2 + x + 2 \qquad \text{Remove parentheses.}$$
$$x > x + 2 \qquad \text{Add } -x^2 \text{ to both sides.}$$
$$0 > 2 \qquad \text{Add } -x \text{ to both sides.}$$

Of course, the result $0 > 2$ is a false statement. This means that there are no numbers $x$ that will satisfy the original inequality. The given inequality has no solutions. ∎

### EXERCISE 8.1

In Exercises 1–14, solve each inequality, if possible. Be able to give a reason for each step.

**1.** $3x < 2x + 4$                            **2.** $5x > -3x - 16$

**3.** $6x - 2 < 3(x - 4)$                   **4.** $4(x + 2) > 2(x - 6)$

**5.** $-5(2x - 3) - 3 > -2(3x - 4)$

**6.** $-3(2x + 1) < 3(2x - 7) - 6$

**7.** $x(x - 1) < x^2 - x - 3$

**8.** $\dfrac{3x}{2} < 5 + x$

**9.** $x(x^2 - 7) > 3x + x^3 - 4$

**10.** $x(2x + 3) + x > 2x(x + 2) - 1$

**11.** $7x - 3 > \dfrac{2}{3}x + 7$

**12.** $2x - \dfrac{1}{3} > x + 3$

**13.** $2x^2 > x(x - 7) + x(x + 12)$

**14.** $(x + 7) < 2(x + 2) - x + \dfrac{6}{x}$ ($x$ is a positive number)

**15.** *Prove*: The measure of an exterior angle associated with a base angle of an isosceles triangle is unequal to the measure of the vertex angle.

**16.** *Prove*: The relation ">" is transitive.

# 8.2 INEQUALITIES INVOLVING TRIANGLES AND CIRCLES

We now consider several theorems involving sides and angles of triangles.

---

**Theorem 8.9** If the lengths of two sides of a triangle are unequal, then the measures of the angles opposite those sides are unequal in the same order.

---

*Given*: $m(\overline{BC}) > m(\overline{AC})$.

*Prove*: $m(\angle BAC) > m(\angle B)$.

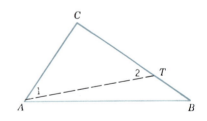

*Construction*: On side $\overline{CB}$ construct a segment $\overline{CT}$ so that $\overline{CT} \cong \overline{CA}$.

*Proof*:

| STATEMENTS | REASONS |
|---|---|
| **1.** $\overline{CT} \cong \overline{CA}$. | **1.** By construction. |
| **2.** $\angle 1 \cong \angle 2$. | **2.** In a triangle angles opposite congruent sides are congruent. |
| **3.** $m(\angle 1) = m(\angle 2)$. | **3.** If two angles are congruent, they have equal measures. |

| STATEMENTS | REASONS |
|---|---|
| **4.** m($\angle BAC$) = m($\angle 1$) + m($\angle TAB$). | **4.** Postulate 1.11. |
| **5.** m($\angle BAC$) > m($\angle 1$). | **5.** Definition 8.1. |
| **6.** m($\angle BAC$) > ($\angle 2$). | **6.** Substitution. |
| **7.** m($\angle 2$) > m($\angle B$). | **7.** Theorem 8.8. |
| **8.** m($\angle BAC$) > m($\angle B$). | **8.** Transitive law of ">." □ |

> ***Postulate 8.1  The Trichotomy Principle***   If $a$ and $b$ are two numbers, then exactly one of these relationships is true.
>
> $$a < b, \qquad a = b, \qquad \text{or} \qquad a > b$$

> **Theorem 8.10**  If the measures of two angles of a triangle are unequal, then the lengths of the sides opposite those angles are unequal in the same order.

*Given*: m($\angle A$) > m($\angle B$).

*Prove*: m($\overline{BC}$) > m($\overline{AC}$).

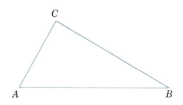

*Plan*: The proof is indirect. We first assume that m($\overline{BC}$) = m($\overline{AC}$) and then assume that m($\overline{BC}$) < m($\overline{AC}$) and show that each of these assumptions leads to a contradiction. By the trichotomy principle, m($\overline{BC}$) > m($\overline{AC}$) is the only remaining possibility.

*Proof*:

| STATEMENTS | REASONS |
|---|---|
| **1.** m($\overline{BC}$) = m($\overline{AC}$). | **1.** By assumption. |
| **2.** m($\angle A$) = m($\angle B$). | **2.** If the lengths of two sides of a triangle are equal, then the measures of the angles opposite those sides are equal. |

| STATEMENTS | REASONS |
|---|---|
| **3.** $m(\angle A) > m(\angle B)$. | **3.** Given. |
| **4.** Statement 1 is false. | **4.** It leads to a contradiction. See Statements 2 and 3. |
| **5.** $m(\overline{BC}) < m(\overline{AC})$. | **5.** By assumption. |
| **6.** $m(\angle A) < m(\angle B)$. | **6.** Theorem 8.9. |
| **7.** Statement 5 is false. | **7.** It leads to a contradiction. See Statements 3 and 6. |
| **8.** $m(\overline{BC}) > m(\overline{AC})$ or $m(\overline{BC}) = m(\overline{AC})$ or $m(\overline{BC}) < m(\overline{AC})$. | **8.** The trichotomy principle. |
| **9.** $m(\overline{BC}) > m(\overline{AC})$. | **9.** It is the only remaining possibility.    □ |

---

**Postulate 8.2  The Triangle Inequality**   The sum of the lengths of any two sides of a triangle is greater than the length of the third side.

---

The triangle inequality implies that in Figure 8.1 we have

$$m(\overline{AB}) + m(\overline{BC}) > m(\overline{AC})$$
$$m(\overline{BC}) + m(\overline{CA}) > m(\overline{AB})$$

and

$$m(\overline{CA}) + m(\overline{AB}) > m(\overline{BC})$$

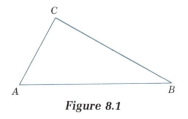

**Figure 8.1**

---

**Theorem 8.11**  If two sides of one triangle are congruent to two sides of a second triangle, and the included angle of the first is greater than the included angle of the second, then the third side of the first triangle is longer than the third side of the second triangle.

---

*Plan*: The proof of this theorem must be considered in three cases depending on whether the endpoint of constructed line segment $\overline{CG}$ falls outside, on, or inside a given triangle.

CASE 1   (Point *G* falls outside $\triangle ABC$.)

*Given*: $\overline{AC} \cong \overline{DF}$.
$\qquad\quad \overline{BC} \cong \overline{EF}$.
$\qquad\quad \text{m}(\angle C) > \text{m}(\angle F)$.

*Prove*: $\text{m}(\overline{AB}) > \text{m}(\overline{DE})$.

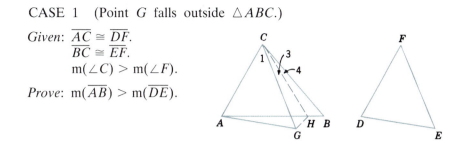

*Construction*: Construct $\angle 1 \cong \angle F$, $\overline{GC} \cong \overline{EF}$ and draw segment $\overline{AG}$. Construct $\overline{CH}$, the bisector of $\angle GCB$, and draw segment $\overline{GH}$.

*Proof*:

| STATEMENTS | REASONS |
|---|---|
| 1. $\overline{AC} \cong \overline{DF}$; $\text{m}(\angle C) > \text{m}(\angle F)$. | 1. Given. |
| 2. $\angle 1 \cong \angle F$; $\overline{GC} \cong \overline{EF}$; draw segment $\overline{AG}$; $\overline{GC}$ is on the interior of $\angle C$. | 2. Possible construction. |
| 3. $\triangle ACG \cong \triangle DFE$. | 3. SAS $\cong$ SAS. |
| 4. $\overline{AG} \cong \overline{DE}$. | 4. cpctc. |
| 5. $\text{m}(\overline{AG}) = \text{m}(\overline{DE})$. | 5. If two segments are congruent, they have equal measures. |
| 6. $\overline{CH}$ bisects $\angle GCB$. | 6. Possible construction. |
| 7. $\angle 3 \cong \angle 4$. | 7. An angle bisector divides an angle into two congruent angles. |
| 8. $\overline{CH} \cong \overline{CH}$. | 8. Reflexive law. |
| 9. $\overline{BC} \cong \overline{EF}$. | 9. Given. |
| 10. $\overline{BC} \cong \overline{GC}$. | 10. If two segments are congruent to the same segment, they are congruent to each other. |
| 11. $\triangle GCH \cong \triangle BCH$. | 11. SAS $\cong$ SAS. |
| 12. $\overline{HG} \cong \overline{HB}$. | 12. cpctc. |
| 13. $\text{m}(\overline{HG}) = \text{m}(\overline{HB})$. | 13. If two semgents are congruent, they have equal measures. |
| 14. $\text{m}(\overline{AH}) + \text{m}(\overline{HG}) > \text{m}(\overline{AG})$. | 14. Postulate 8.2. |
| 15. $\text{m}(\overline{AH}) + \text{m}(\overline{HB}) > \text{m}(\overline{AG})$. | 15. Substitution. |

| STATEMENTS | REASONS |
|---|---|
| **16.** m($\overline{AH}$) + m($\overline{HB}$) = m($\overline{AB}$). | **16.** Postulate 1.10. |
| **17.** m($\overline{AB}$) > m($\overline{DE}$). | **17.** Substitution. |

CASE 2    (Point *G* falls on △*ABC*.)

*Given*: $\overline{AC} \cong \overline{DF}$.
$\qquad\overline{BC} \cong \overline{EF}$.
$\qquad$ m($\angle C$) > m($\angle F$).

*Prove*: m($\overline{AB}$) > m($\overline{DE}$).

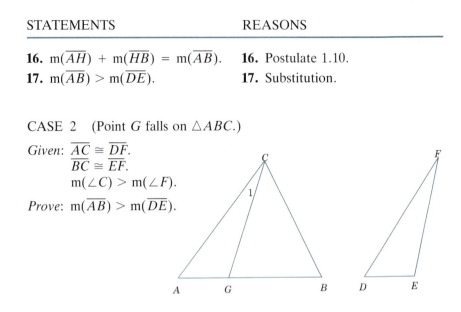

*Construction*: Construct $\angle 1 \cong \angle F$ and $\overline{GC} \cong \overline{EF}$.

The proof of Case 2 is left as an exercise.

CASE 3    (Point *G* falls inside △*ABC*.)

*Given*: $\overline{AC} \cong \overline{DF}$.
$\qquad\overline{BC} \cong \overline{EF}$.
$\qquad$ m($\angle C$) > m($\angle F$).

*Prove*: m($\overline{AB}$) > m($\overline{DE}$).

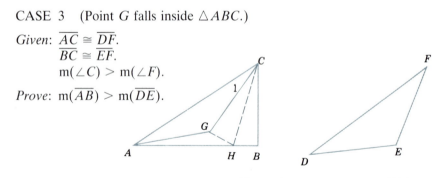

*Construction*: Construct $\angle 1 \cong \angle F$ and $\overline{GC} \cong \overline{EF}$. Also construct $\overline{CH}$, the bisector of $\angle GCB$, and draw segment $\overline{GH}$.

The proof of Case 3 is left as an exercise.                    □

---

**Theorem 8.12**  If two sides of one triangle are congruent to two sides of a second triangle, and if the third side of the first is longer than the third side of the second, then the angle opposite the third side of the first triangle is greater than the angle opposite the third side of the second triangle.

*Given:* $\overline{AC} \cong \overline{DF}.$
$\overline{CB} \cong \overline{FE}.$
$m(\overline{AB}) > m(\overline{DE}).$

*Prove:* $m(\angle C) > m(\angle F).$

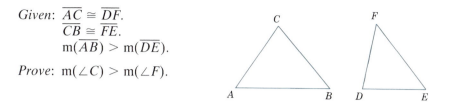

*Plan:* The proof is indirect. We will assume $\angle C \cong \angle F$ and then assume $m(\angle C) < m(\angle F)$ and show that each assumption leads to a contradiction.

*Proof:*

| STATEMENTS | REASONS |
| --- | --- |
| **1.** $\angle C \cong \angle F.$ | **1.** By asumption. |
| **2.** $\overline{AC} \cong \overline{DF}; \overline{CB} \cong \overline{FE}.$ | **2.** Given. |
| **3.** $\triangle ABC \cong \triangle DEF.$ | **3.** SAS $\cong$ SAS. |
| **4.** $\overline{AB} \cong \overline{DE}.$ | **4.** cpctc. |
| **5.** $m(\overline{AB}) = m(\overline{DE}).$ | **5.** If two segments are congruent, they have equal measures. |
| **6.** $m(\overline{AB}) > m(\overline{DE}).$ | **6.** Given. |
| **7.** Statement 1 is false. | **7.** It leads to a contradiction. See Statements 4 and 5. |
| **8.** $m(\angle C) < m(\angle F).$ | **8.** By assumption. |
| **9.** $m(\overline{AB}) < m(\overline{DE}).$ | **9.** Theorem 8.11. |
| **10.** Statement 8 is false. | **10.** It leads to a contradiction. See Statements 6 and 9. |
| **11.** $m(\angle C) = m(\angle F), m(\angle C) < (\angle F),$ or $m(\angle C) > m(\angle F).$ | **11.** The trichotomy principle. |
| **12.** $m(\angle C) > m(\angle F).$ | **12.** It is the only remaining possibility. □ |

---

**Theorem 8.13** In the same circle or in congruent circles, the greater of two central angles will intercept the greater arc.

*Given:* ∠1 and ∠2 are central angles.
  m(∠1) > m(∠2).

*Prove:* m($\overset{\frown}{AB}$) > m($\overset{\frown}{CD}$).

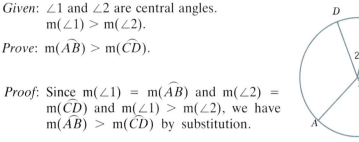

*Proof:* Since m(∠1) = m($\overset{\frown}{AB}$) and m(∠2) = m($\overset{\frown}{CD}$) and m(∠1) > m(∠2), we have m($\overset{\frown}{AB}$) > m($\overset{\frown}{CD}$) by substitution.

The proof of this theorem for the case involving congruent circles is left as an exercise.

---

**Theorem 8.14**  In the same circle or in congruent circles, the greater of two arcs will be intercepted by the greater of the two central angles.

---

The proof of Theorem 8.14 is left as an exercise.

## EXERCISE 8.2

1. In △ABC, m($\overline{AC}$) = 5 cm, m($\overline{AB}$) = 7 cm, and m($\overline{BC}$) = 10 cm. Which angle in the triangle is the largest?

2. In right △ABC, legs $\overline{AC}$ an $\overline{BC}$ measure 10 in. and 15 in., respectively. Which angle in the triangle is the smallest? Explain.

3. In △ABC, $\overline{AC}$ ≅ $\overline{BC}$ and m(∠C) = 70°. Which side of the triangle is the longest? Explain.

4. In △ABC, $\overline{AC}$ ≅ $\overline{BC}$ and m(∠B) = 65°. Which side of the triangle is the shortest? Explain.

5. Can a triangle have sides of 8 m, 9 m, and 20 m? Explain.

6. Can a triangle have sides of 9 in., 11 in., and 13 in.? Explain.

In Exercises 7–10, refer to Illustration 1 in which $\overline{AC}$ ≅ $\overline{DF}$ and $\overline{BC}$ ≅ $\overline{EF}$.

7. Which of ∠C or ∠F is the larger angle if m($\overline{AB}$) > m($\overline{DE}$)?

8. Which of $\overline{AB}$ or $\overline{DE}$ is the longer side if m(∠C) < m(∠F)?

9. Is m($\overline{AB}$) − m($\overline{DE}$) a positive or a negative number if m(∠C) > m(∠F)?

10. Which of $\overline{AC}$ or $\overline{AB}$ is shorter if $\overline{AC}$ ≅ $\overline{EF}$, ∠C ≅ ∠F, and m(∠E) = 30°?

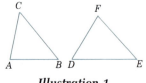

***Illustration 1***

In Exercises 11–14, refer to Illustration 2.

**11.** If $m(\widehat{AD}) = 80°$ and $m(\widehat{BC}) = 60°$, which of $\angle 1$ and $\angle 2$ is the greater?

**12.** If $m(\widehat{AD}) = 80°$ and $m(\widehat{BC}) = 60°$, which of $\angle DOB$ and $\angle AOC$ is the greater?

**13.** If $m(\angle 1) = 55°$ and $m(\angle 2) = 50°$, which of $\widehat{DAB}$ an $\widehat{ABC}$ is the greater?

**14.** If $m(\widehat{DCB}) = 210°$ and $m(\widehat{ADC}) = 190°$, which of $\angle 1$ and $\angle 2$ is the greater?

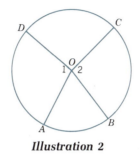

*Illustration 2*

**15.** *Prove*: The diagonal of a rectangle is longer than any side of the rectangle.

**16.** Given the information in Illustration 3, prove that $m(\overline{AB}) > m(\overline{AE})$.

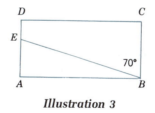

*Illustration 3*

**17.** Given the information in Illustration 4, prove that $m(\overline{OA}) > m(\overline{PA})$. Assume $\overline{PA}$ is a tangent to the circle.

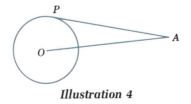

*Illustration 4*

**18.** *Prove*: The greater of two inscribed angles will intercept the greater arc.

**19.** Refer to Illustration 5 in which $\overline{AD} \cong \overline{DB}$, $\angle A \cong \angle 1$, and $m(\angle 1) < 90°$. Prove that $m(\overline{CD}) < m(\overline{CB})$.

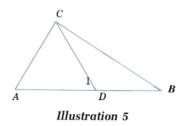

***Illustration 5***

**20.** *Prove*: The shortest distance from a point to a line is the length of a perpendicular from the point to the line.

**21.** Prove Case 2 of Theorem 8.11.

**22.** Prove Case 3 of Theorem 8.11.

**23.** *Complete the proof of Theorem 8.13*: In the same circle or in congruent circles, the greater of two central angles will intercept the greater arc.

**24.** *Prove Theorem 8.14*: In the same circle or in congruent circles, the greater of two arcs will be intercepted by the greater of the two central angles.

## 8.3 INEQUALITIES INVOLVING CHORDS OF CIRCLES

> ***Theorem 8.15*** In the same circle or in congruent circles, the greater of two chords intercepts the greater minor arc.

*Given*: $m(\overline{AB}) > m(\overline{CD})$.

*Prove*: $m(\overset{\frown}{AB}) > m(\overset{\frown}{CD})$.

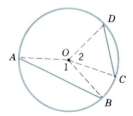

*Construction*: Draw radii $\overline{OA}$, $\overline{OB}$, $\overline{OC}$, and $\overline{OD}$.

*Proof*:

| STATEMENTS | REASONS |
|---|---|
| **1.** $m(\overline{AB}) > m(\overline{CD})$. | **1.** Given. |
| **2.** $\overline{OA} \cong \overline{OB} \cong \overline{OC} \cong \overline{OD}$. | **2.** All radii in the same circle are congruent. |

| STATEMENTS | REASONS |
|---|---|
| **3.** $m(\angle 1) > m(\angle 2)$. | **3.** Theorem 8.12. |
| **4.** $m(\widehat{AB}) > m(\widehat{CD})$. | **4.** Theorem 8.13. |

The case for congruent circles is left as an exercise.    □

> **Theorem 8.16**  In the same circle or in congruent circles, the greater of two minor arcs has the greater chord.

The proof of Theorem 8.16 is left as an exercise.

> **Theorem 8.17**  If two unequal chords form an inscribed angle within a circle, then the shorter chord is the farther from the center of the circle.

To prove Theorem 8.17 we must consider three cases: when the center of the circle is on the interior of the inscribed angle, when the center of the circle is on the inscribed angle, and when the center of the circle is on the exterior of the inscribed angle. We give the proof for the first case.

CASE 1

*Given*: Point $O$ is on the interior of $\angle BAC$.
  $m(\overline{AB}) > m(\overline{AC})$.
  $\overline{OD} \perp \overline{AB}$.
  $\overline{OE} \perp \overline{AC}$.

*Prove*: $m(\overline{OE}) > m(\overline{OD})$.

*Construction*: Draw segment $\overline{DE}$.

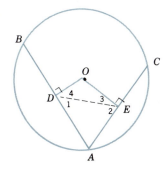

*Proof*:

| STATEMENTS | REASONS |
|---|---|
| **1.** Point $O$ is on the interior of $\angle BAC$; $m(\overline{AB}) > m(\overline{AC})$; $\overline{OD} \perp \overline{AB}$; and $\overline{OE} \perp \overline{AC}$. | **1.** Given. |

| STATEMENTS | REASONS |
|---|---|
| 2. $\overline{OD}$ bisects $\overline{AB}$;<br>   $\overline{OE}$ bisects $\overline{AC}$. | 2. A line drawn from the center of a circle and perpendicular to a chord bisects the chord. |
| 3. $m(\overline{AD}) = \dfrac{1}{2}[m(\overline{AB})]$;<br><br>   $m(\overline{AE}) = \dfrac{1}{2}[m(\overline{AC})]$. | 3. Theorem 1.5: The midpoint of a given line segment divides the segment into two equal segments, each half as long as the given segment. |
| 4. $2[m(\overline{AD})] = m(\overline{AB})$;<br>   $2[m(\overline{AE})] = m(\overline{AC})$. | 4. Equals multiplied by equals are equal. |
| 5. $2[m(\overline{AD})] > 2[m(\overline{AE})]$. | 5. Substitution. |
| 6. $m(\overline{AD}) > m(\overline{AE})$. | 6. Theorem 8.4. |
| 7. $m(\angle OEA) = m(\angle ODA) = 90°$. | 7. Perpendicular lines meet and form 90° angles. |
| 8. $m(\angle 2) > m(\angle 1)$. | 8. Theorem 8.9. |
| 9. $m(\angle 3) < m(\angle 4)$ or<br>   $m(\angle 4) > m(\angle 3)$. | 9. Theorem 8.3. |
| 10. $m(\overline{OE}) > m(\overline{OD})$. | 10. Theorem 8.10. |

The proofs for the other two cases are left as exercises.    □

---

**Theorem 8.18**  In the same circle or in congruent circles, the shorter of two chords is the greater distance from the center of the circle.

---

The proof of Theorem 8.18 is left as an exercise.

## EXERCISE 8.3

In Exercises 1–6, refer to Illustration 1.

1. Is $\overset{\frown}{AB}$ or $\overset{\frown}{DC}$ the greater if $m(\overline{AB}) = 10$ cm and $m(\overline{DC}) = 8$ cm?
2. Is $\overline{EC}$ or $\overline{BF}$ the longer if $m(\overset{\frown}{DC}) > m(\overset{\frown}{AB})$?
3. Is $\overset{\frown}{AB}$ or $\overset{\frown}{DC}$ the greater if $m(\overline{OF}) > m(\overline{OE})$?
4. Is $\overline{AB}$ or $\overline{CD}$ nearer the center if $m(\overset{\frown}{DC}) > m(\overset{\frown}{AB})$?
5. If $m(\overset{\frown}{ABD}) = 240°$ and $m(\overset{\frown}{BDC}) = 200°$, which of $\overline{AB}$ or $\overline{CD}$ is the longer?
6. If $m(\overline{OF}) = 5$ in. and $m(\overline{OE}) = 6$ in., which of $\overset{\frown}{CAB}$ or $\overset{\frown}{DCA}$ is the greater?
7. *Prove*: If two unequal chords form an inscribed angle within a circle, then the chord that is farther from the center is the shorter.

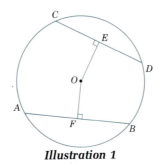

*Illustration 1*

8. *Complete the proof of Theorem 8.15*: In the same circle or in congruent circles, the greater of two chords intercepts the greater minor arc.

9. *Prove Theorem 8.16*: In the same circle or in congruent circles, the greater of two minor arcs has the greater chord.

10. *Prove Theorem 8.18*: In the same circle or in congruent circles, the shorter of two chords is the greater distance from the center of the circle.

11. Prove Theorem 8.17 for the case where the center of the circle lies on the inscribed angle.

12. Prove Theorem 8.17 for the case where the center of the circle lies on the exterior of the inscribed angle.

## CHAPTER EIGHT REVIEW EXERCISES

In Review Exercises 1–12, solve each inequality.

1. $x + 2 < 6$

2. $x - 3 > 12$

3. $3y - 5 > 10$

4. $5y + 2 < 22$

5. $6t + 2 < t + 17$

6. $4t - 3 > -t + 12$

7. $9(a - 11) > 13 + 7a$

8. $3(b - 8) < 5b + 6$

9. $8(5 - y) < 10(8 - y)$

10. $17(3 + c) > 3 - 13c$

11. $\dfrac{x - 3}{2} > x - 2$

12. $\dfrac{2(z + 5)}{3} < 3z - 6$

In Review Exercises 13–16, refer to Illustration 1, in which the triangles are *not drawn to scale*. If the triangles were drawn to scale, answer the following questions.

13. Which angle in $\triangle DBC$ is the largest?

14. Which angle in $\triangle DBC$ is the smallest?

15. Which angle in $\triangle ABD$ is the smallest?

16. Which angle in $\triangle ABD$ is the largest?

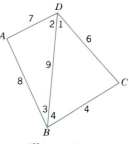

*Illustration 1*

**17.** Does a triangle exist with sides measuring 13, 16, and 30 cm?

**18.** Does a triangle exist with sides measuring 5, 8, and 12 m?

In Review Exercises 19–24, refer to Illustration 2.

**19.** If $m(\widehat{AB}) = 80°$ and $m(\widehat{CD}) = 100°$, which of $\angle 1$ and $\angle 3$ is the larger?

**20.** If $m(\widehat{AD}) = 120°$ and $m(\widehat{BC}) = 100°$, which of $\angle 2$ and $\angle 4$ is the smaller?

**21.** If $m(\widehat{AB}) = 80°$, $m(\widehat{BC}) = 90°$, and $m(\widehat{DA}) = 100°$, which angle in the figure is the largest?

**22.** If $m(\angle 1) = 45°$ and $m(\angle 3) = 50°$, which of $\widehat{AB}$ and $\widehat{DC}$ is the larger?

**23.** If $m(\angle 1) = 40°$, $m(\angle 2) = 120°$, and $m(\angle 3) = 70°$, which of $\widehat{AD}$ and $\widehat{BC}$ is the larger?

**24.** If $m(\angle 1) = 40°$, $m(\widehat{AD}) = 100°$, and $m(\widehat{DC}) = 80°$, which of $\widehat{BC}$ and $\widehat{DC}$ is the larger?

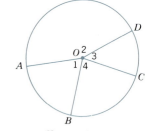

**Illustration 2**

In Review Exercises 25–26, refer to Illustration 3 in which $\overline{AB}$ is tangent to circle $O$.

**25.** Is $\overline{AB}$ or $\overline{OA}$ the longer? Explain.

**26.** Is $\angle O$ or $\angle 1$ the smaller? Explain.

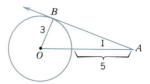

**Illustration 3**

In Review Exercises 27–30, refer to Illustration 4, in which $m(\widehat{AB}) = 88°$ and $m(\angle 2) = 86°$.

**27.** Is $\widehat{AB}$ or $\widehat{DC}$ the longer? Explain.

**28.** Is $\angle 1$ or $\angle 2$ the smaller? Explain.

**29.** Which of $\overline{AB}$ or $\overline{CD}$ is the further from the center of the circle?

**30.** If $\overline{AD}$ is further from the center than $\overline{BC}$, which of $\overline{AD}$ or $\overline{BC}$ is the smaller?

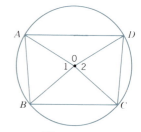

**Illustration 4**

In Review Exercises 31–34, refer to Illustration 5, in which $m(\widehat{AB}) = 72°$ and $m(\angle 1) = 65°$.

**31.** Is $\widehat{AB}$ or $\widehat{CD}$ the longer? Explain.

**32.** Is $m(\widehat{AB})$ or $m(\widehat{CD})$ the greater? Explain.

**33.** Is m(∠1) or m(∠2) the greater? Explain.

**34.** Is $\overline{AB}$ or $\overline{CD}$ further from point $O$? Explain.

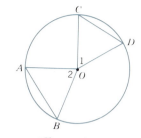

***Illustration 5***

In Review Exercises 35–36, refer to Illustration 6, in which m($\widehat{AB}$) > m($\widehat{BC}$) > m($\widehat{CD}$).

**35.** Write the following measures in increasing order.

m($\overline{AB}$), m($\overline{BC}$), m($\overline{CD}$)

**36.** Write the following measures in decreasing order.

m($\overline{OP}$), m($\overline{OQ}$), m($\overline{OR}$)

***Illustration 6***

In Review Exercises 37–38, refer to Illustration 7, in which m($\widehat{AB}$) > m($\widehat{BC}$) > m($\widehat{CD}$).

**37.** Write the following measures in decreasing order.

m($\overline{BC}$), m($\overline{AB}$), m($\overline{CD}$)

**38.** Write the following measures in increasing order.

m($\overline{OP}$), m($\overline{OR}$), m($\overline{OQ}$)

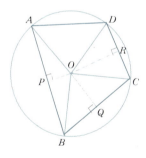

***Illustration 7***

## CHAPTER EIGHT TEST

In Exercises 1–20, classify each statement as true or false.

**1.** If $a$ does not equal $b$, then $a < b$.

**2.** The commutative law of addition states that $a + b = b + a$.

**3.** If unequal quantities are added to unequal quantities, then the sums are unequal quantities.

**4.** If $a > b$ and $b > c$, then $a > c$.

**5.** If $a > b$ and $c < d$, then $ac < bc$.

**6.** If a base angle of an isosceles triangle is less than 60°, the base is the longest side of the triangle.

**7.** The sum of the measures of the exterior angles of a triangle is greater than the sum of the measures of its interior angles.

**8.** The hypotenuse is longer than either leg in a right triangle.

**9.** The longer of two chords in a circle is nearer the center of the circle.

**10.** The greater of two arcs in a circle has the greater chord.

**11.** The greatest angle of an isosceles triangle must be the vertex angle.

**12.** If $a < b$, then $ac < bc$.

**13.** If $a < b$, then $a - c < b - c$.

**14.** Refer to Illustration 1. The longest side of $\triangle ABC$ is $\overline{AB}$.

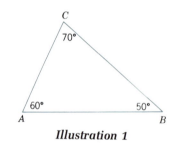

*Illustration 1*

**15.** A chord of a circle cannot be greater than a diameter of the circle.

**16.** The sum of the interior angles of a pentagon (five-sided figure) is greater than the sum of its exterior angles.

**17.** If unequal quantities are multiplied by equal quantities, then the results are unequal quantities and in the same order.

**18.** The larger of two central angles in a circle intercepts the greater arc.

**19.** The measure of an exterior angle in a triangle is greater than the difference between the measures of the nonadjacent interior angles.

**20.** If a central angle of a circle has a measure of 50°, then the chord intercepted by the angle is longer than the radius of the circle.

# 9

*Aristotle (384–322 B.C.). Aristotle, a Greek philosopher and logician, was a student of Plato and in turn a tutor of Alexander the Great. One of his greatest works was the Organum, a set of six treatises on logic. The logic of Aristotle forms the basis of the formal logic studied today. This representation is from the Cathedral of Chartres, France (New York Public Library Picture Collection).*

# Geometric Loci

Suppose that a very eager dog is tied to a stake with a rope, and that the dog always keeps the rope tight. The collection of all the points where the dog might be found is a circle with a radius equal to the length of the rope. The taut rope restricts the dog to only certain possible locations. They are the points of a circle. A **locus** is a set of points (such as the circle in this example) that are determined by some restriction or condition (such as the stake and the length of the rope).

The locus of the center of a car wheel as it rolls down a straight and level road is a line parallel to the road and above it at a distance equal to the radius of the wheel. See Figure 9.1.

*Figure 9.1*

The locus of points representing the locations of a bouncing rubber ball might be described by the curve in Figure 9.2.

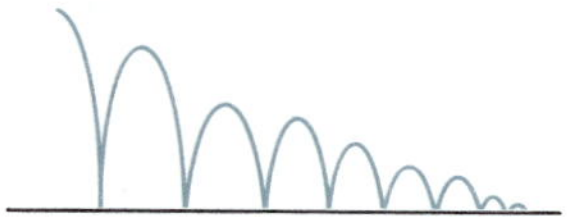

*Figure 9.2*

# 9.1 BASIC LOCUS THEOREMS

We begin with a formal definition of the concept of *locus*.

---

*Definition 9.1* A **locus** is a set of points such that

1. all points in the set satisfy a given condition, and
2. all points that satisfy the given condition are points in the set.

---

The plural of *locus* is *loci*.

A circle was defined in Chapter 5. A circle can also be defined using the concept of locus.

---

*Definition 9.2* A **circle** is the locus of points in a plane at a given distance *r* from a fixed point called the **center** of the circle. The distance *r* is called the **radius** of the circle.

---

We present the first locus theorem of this chapter without proof.

---

*Theorem 9.1* The locus of a point at a given distance from a given line is a pair of lines, one on each side of the given line, parallel to the given line and at the given distance from it.

---

EXAMPLE 1

In Figure 9.3, the points that lie on lines $l_1$ and $l_2$ lie at the same given distance from line $L$. Note that lines $l_1$ and $l_2$ are parallel to each other.

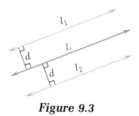

**Figure 9.3**

We now prove three basic theorems about geometric loci. Each proof has two parts: We must prove that all the points that satisfy the given condition are on the locus and that all the points on the locus satisfy the given condition.

---

**Theorem 9.2** The locus of all points equidistant from two points is the perpendicular bisector of the segment joining the two points.

---

PART 1

*Given*: Point $P$ is equidistant from points $A$ and $B$; that is, $\overline{AP} \cong \overline{BP}$. $M$ is the midpoint of $\overline{AB}$.

*Prove*: $\overline{PM}$ is the perpendicular bisector of $\overline{AB}$.

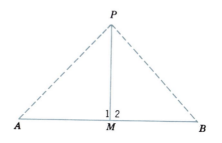

*Construction*: Draw line segments $\overline{AP}$ and $\overline{BP}$.

*Proof*:

| STATEMENTS | REASONS |
|---|---|
| **1.** $M$ is the midpoint of $\overline{AB}$. | **1.** Given. |
| **2.** $\overline{MA} \cong \overline{MB}$. | **2.** The midpoint divides the segment into two congruent segments. |

| STATEMENTS | REASONS |
|---|---|
| **3.** $\overline{PM} \cong \overline{PM}$. | **3.** A quantity is congruent to itself. |
| **4.** Point $P$ is equidistant from $A$ and $B$: $\overline{AP} \cong \overline{BP}$. | **4.** Given. |
| **5.** $\triangle AMP \cong \triangle BMP$. | **5.** SSS $\cong$ SSS. |
| **6.** $\angle 1 \cong \angle 2$. | **6.** cpctc. |
| **7.** $\angle 1$ and $\angle 2$ are adjacent angles. | **7.** Definition of adjacent angles. |
| **8.** $\overline{PM} \perp \overline{AB}$. | **8.** If lines meet and form congruent adjacent angles, then the lines are perpendicular. |
| **9.** $\overline{PM}$ is the perpendicular bisector of $\overline{AB}$. | **9.** If $\overline{PM}$ is perpendicular to $\overline{AB}$ and bisects $\overline{AB}$, then $\overline{PM}$ is the perpendicular bisector of $\overline{AB}$ (Steps 1 and 8). |

PART 2

*Given*: $\overline{PM}$ is the perpendicular bisector of $\overline{AB}$.

*Prove*: $P$ is equidistant from $A$ and $B$.

*Construction*: Draw line segments $\overline{AP}$ and $\overline{BP}$.

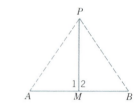

*Proof*:

| STATEMENTS | REASONS |
|---|---|
| **1.** $\overline{PM}$ is the perpendicular bisector of $\overline{AB}$. | **1.** Given. |
| **2.** $\angle 1 \cong \angle 2$. | **2.** Perpendicular lines meet to form congruent adjacent angles. |
| **3.** $\overline{AM} \cong \overline{BM}$. | **3.** A bisector of a line segment divides the segment into two congruent segments. |
| **4.** $\overline{MP} \cong \overline{MP}$. | **4.** A quantity is congruent to itself. |
| **5.** $\triangle APM \cong \triangle BPM$. | **5.** SAS $\cong$ SAS. |
| **6.** $\overline{AP} \cong \overline{BP}$. | **6.** cpctc. |
| **7.** m($\overline{AP}$) = m($\overline{BP}$). | **7.** If two segments are congruent, they have equal measures. |
| **8.** $P$ is equidistant from points $A$ and $B$. | **8.** Step 7. |

> **Theorem 9.3** The locus of all points equidistant from the sides of an angle is the angle bisector.

## PART 1

*Given*: Point $P$ is equidistant from lines $l_1$ and $l_2$, which intersect at $A$: $\overline{PB} \cong \overline{PC}$.

*Prove*: $\overrightarrow{PA}$ is the bisector of $\angle A$.

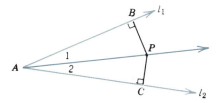

*Construction*: Construct perpendiculars $\overline{PB}$ and $\overline{PC}$ to lines $l_1$ and $l_2$ forming right triangles $ABP$ and $ACP$.

*Proof*:

| STATEMENTS | REASONS |
|---|---|
| **1.** $\overline{PB} \perp \overline{AB}$; $\overline{PC} \perp \overline{AC}$; right triangles $ABP$ and $ACP$. | **1.** By construction. |
| **2.** $P$ is equidistant from lines $l_1$ and $l_2$: $\overline{PB} \cong \overline{PC}$. | **2.** Given. |
| **3.** $\overline{AP} \cong \overline{AP}$. | **3.** A quantity is congruent to itself. |
| **4.** $\triangle ABP \cong \triangle ACP$. | **4.** $hl \cong hl$. |
| **5.** $\angle 1 \cong \angle 2$. | **5.** cpctc. |
| **6.** $\overrightarrow{PA}$ is the bisector of $\angle A$. | **6.** Definition of angle bisector. |

## PART 2

*Given*: $\overrightarrow{PA}$ is the angle bisector of $\angle A$.

*Prove*: $P$ is equidistant from lines $l_1$ and $l_2$.

*Construction*: Construct segments $\overline{PB}$ and $\overline{PC}$ perpendicular to lines $l_1$ and $l_2$ forming right triangles $ABP$ and $ACP$.

The completion of the proof of Theorem 9.3 is left as an exercise. □

---

**Theorem 9.4** The locus of the vertex of the right angle of a right triangle with fixed hypotenuse is a circle with the hypotenuse as diameter.

---

PART 1

*Given*: $\triangle ABC$ is a right triangle with $\overline{AC} \perp \overline{BC}$.
Circle $O$ with diameter $\overline{AB}$.

*Prove*: Point $C$ is on $\overparen{ADB}$.

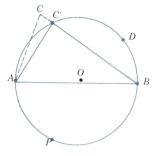

*Construction*: Let the intersection of $\overparen{ADB}$ and $\overline{BC}$ (or $\overline{BC}$ extended if $C$ were to lie inside the circle) be the point $C'$. Draw $\overline{AC'}$.

*Proof*:

| STATEMENTS | REASONS |
|---|---|
| **1.** $\triangle ABC$ is a right triangle with $\overline{AC} \perp \overline{BC}$; circle $O$ with diameter $\overline{AB}$. | **1.** Given. |
| **2.** $C'$ is on $\overparen{ADB}$. | **2.** By construction. |
| **3.** $\overparen{APB}$ is a semicircle. | **3.** If the endpoints of an arc are on a diameter, then the arc is a semicircle. |
| **4.** $m(\overparen{APB}) = 180°$. | **4.** The measure of a semicircle is 180°. |

| STATEMENTS | REASONS |
|---|---|
| **5.** m($\angle AC'B$) = 90°. | **5.** The measure of an inscribed angle is one-half the measure of its intercepted arc. |
| **6.** $\angle AC'B$ is a right angle. | **6.** If an angle measures 90°, then it is a right angle. |
| **7.** $\overline{AC'} \perp \overline{BC}$. | **7.** If two lines meet to form right angles, then they are perpendicular. |
| **8.** $\overline{AC}$ coincides with $\overline{AC'}$. | **8.** There is only one perpendicular to a line from a point not on the line. |
| **9.** $C$ coincides with $C'$. | **9.** Two lines can intersect only in one point. |

PART 2

*Given:* $\triangle ABC$; circle $O$, with diameter $\overline{AB}$; $C$ is on circle $O$.

*Prove:* $\angle ACB$ is a right angle.

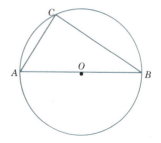

The completion of the proof of Theorem 9.4 is left as an exercise. □

We can also think of locus problems in three dimensions. For example, the locus of all points in space that are 20 cm from a fixed point is a sphere with a radius of 20 cm. See Figure 9.4a. In general, the locus of all points in space that are equidistant from a fixed point is a sphere. The fixed point is the center of the sphere, and the length of its radius is equal to the distance from its center to one of the given points in space.

As another example, the locus of all points in space that are equidistant from the planes $AB$ and $CD$ shown in Figure 9.4b is the plane $MN$.

If two planes intersect, the angles that are formed are called **dihedral angles**. Two dihedral angles, theta ($\theta$) and phi ($\phi$), are shown in Figure 9.5a. The locus of all points in space that are equidistant from the faces of the dihedral angle $ABC$ shown in Figure 9.5b is the plane $BD$ that bisects the dihedral angle.

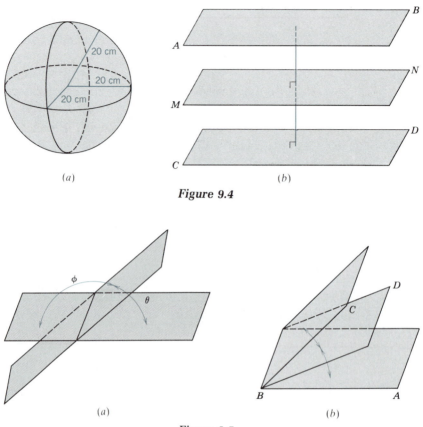

(a) (b)

**Figure 9.4**

(a) (b)

**Figure 9.5**

## EXERCISE 9.1

In Exercises 1–10, make a sketch of each locus. Assume that all points lie in the same plane.

1. Find the locus of all points in a plane that are equidistant from two parallel lines.
2. Find the locus of the vertex angle of a triangle with a fixed base if the vertex angle is always 60°.
3. Find the locus of the vertex angle of a triangle with fixed base if the vertex angle is always 45°.
4. Find the locus of all points that are equidistant from three fixed points.
5. Find the locus of all points that are equidistant from the two points $A$ and $B$ in Illustration 1 and that are also a fixed distance from point $C$. Discuss the possibilities.

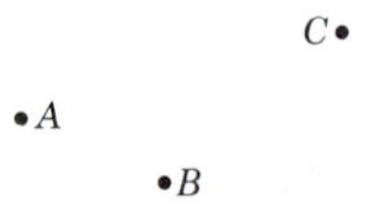

$C \bullet$

$\bullet A$

$\bullet B$

**Illustration 1**

**6.** Find the locus of all points that are equidistant from the sides of angle $A$ and that are a fixed distance from point $D$. See Illustration 2. Discuss the possibilities.

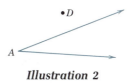

*Illustration 2*

**7.** Find the locus of all points that are equidistant from a line and a fixed point that is not on that line. This locus is called a **parabola**.

**8.** A point is marked on a circle of radius $a$. Find the locus of that point as the circle rolls along a flat surface without slipping. This locus is called a **cycloid**.

**9.** A fine thread is wrapped several times around a circle. A point is marked on the thread. Find the locus of that point if the thread is kept taut as it is unwound from the circle. This locus, called an **involute**, is important in the design of gears.

**10.** Find the locus of all points such that the *sum* of the distances to two fixed points is a constant. This locus is called an **ellipse**.

In Exercises 11–14, describe each locus. Assume that all points lie in three-dimensional space.

**11.** Find the locus of all points in space that are equidistant from a line.

**12.** Find the locus of all points in space that are equidistant from a plane.

**13.** Find the locus of all points in space that are equidistant from two points.

**14.** One side and the vertex of a 30° angle are given. Find the locus of all points in space that could lie on the other side of the angle.

**15.** Devise a procedure for constructing an isosceles right triangle with a given hypotenuse.

**16.** How could you use a carpenter's square (see Illustration 3) to draw a circle with a diameter of 10 in.?

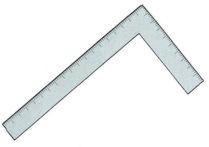

*Illustration 3*

**17.** Complete the proof of Theorem 9.3.

**18.** Complete the proof of Theorem 9.4.

## 9.2 CONCURRENCY THEOREMS

It is not surprising to find that two lines intersect at a point, but it is unusual to find that *three* lines intersect at one point. Three or more lines that intersect at a point are called **concurrent lines**. The theorems of this section deal with certain concurrent lines associated with any triangle.

---

***Theorem 9.5***  The perpendicular bisectors of the sides of a triangle meet at a point that is equally distant from the vertices of the triangle.

---

*Given*:  $\triangle ABC$ with $\overline{HD}$ perpendicular bisector of $\overline{AC}$ and $\overline{HE}$ perpendicular bisector of $\overline{AB}$. $F$ is midpoint of $\overline{BC}$.

*Prove*:  The perpendicular bisector of $\overline{BC}$ passes through point $H$, and $H$ is equidistant from $A$, $B$, and $C$.

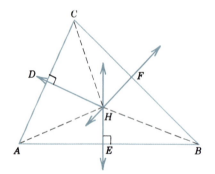

*Construction*:  Draw segments $\overline{AH}$, $\overline{CH}$, and $\overline{BH}$.

*Proof*:

| STATEMENTS | REASONS |
|---|---|
| **1.** $\overline{HD}$ is the perpendicular bisector of $\overline{AC}$, and $\overline{HE}$ is the perpendicular bisector of $\overline{AB}$. | **1.** Given. |
| **2.** m($\overline{HC}$) = m($\overline{HA}$); m($\overline{HA}$) = m($\overline{HB}$). | **2.** The locus of points equidistant from two points is the perpendicular bisector of the segment joining those points. |

| STATEMENTS | REASONS |
|---|---|
| **3.** m($\overline{HC}$) = m($\overline{HB}$). | **3.** Transitive law. |
| **4.** $H$ is equidistant from $A$, $B$, and $C$. | **4.** If a point is the same distance from three points, then it is equidistant from those points (Steps 2 and 3). |
| **5.** $F$ is the midpoint of $\overline{BC}$. | **5.** Given. |
| **6.** $F$ is on the perpendicular bisector of $\overline{BC}$. | **6.** The midpoint of $\overline{BC}$ lies on *any* bisector of $\overline{BC}$. |
| **7.** $H$ is on the perpendicular bisector of $\overline{BC}$. | **7.** If a point is equidistant from two points, then it is on the perpendicular bisector of the line segment joining those points (Theorem 9.2). |
| **8.** $\overline{FH}$ is the perpendicular bisector of $\overline{BC}$. | **8.** Two points ($F$ and $H$) determine a line (in this case the perpendicular bisector of $\overline{BC}$). |
| **9.** The perpendicular bisector of $\overline{BC}$ passes through point $H$. | **9.** $H$ is a point on the perpendicular bisector of $\overline{BC}$.    □ |

The point of intersection of the perpendicular bisectors of the sides of a triangle is called the **circumcenter** of the triangle. Because the circumcenter is equidistant from the three vertices of the triangle, it is the center of a circle that passes through those three vertices. Such a circle is called a **circumscribed circle**.

*EXAMPLE 1*

Construct the circle that passes through the three vertices of $\triangle ABC$ in Figure 9.6.

*Solution*

First, construct the perpendicular bisectors of two of the sides of a given triangle, as in Figure 9.6. By Theorem 9.5, they will meet at a point $P$ that

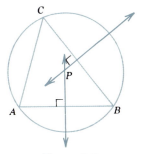

**Figure 9.6**

is equidistant from the three vertices $A$, $B$, and $C$. Finally, choose $\overline{AP}$ as the radius, and draw a circle with $P$ as center. The circle will also pass through vertices $B$ and $C$. ▪

---

**Theorem 9.6** The angle bisectors of a triangle meet in a point that is equidistant from the sides of the triangle.

---

*Given*: $\triangle ABC$ with $\overline{AP}$ and $\overline{BP}$ bisectors of $\angle A$ and $\angle B$, respectively.

*Prove*: $\overline{CP}$ bisects $\angle C$, and $P$ is equidistant from sides $\overline{AB}$, $\overline{BC}$, and $\overline{CA}$.

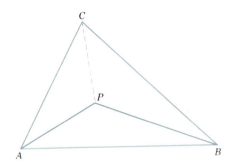

*Proof*:

| STATEMENTS | REASONS |
|---|---|
| **1.** $\overline{AP}$ bisects $\angle A$; $\overline{BP}$ bisects $\angle B$. | **1.** Given. |
| **2.** $P$ is equidistant from $\overline{AB}$ and $\overline{AC}$. | **2.** Since $P$ is on the bisector of $\angle A$, it is equidistant from the sides of $\angle A$ (Theorem 9.3). |
| **3.** $P$ is equidistant from $\overline{AB}$ and $\overline{BC}$. | **3.** Since $P$ is on the bisector of $\angle B$, it is equidistant from the sides of $\angle B$ (Theorem 9.3). |
| **4.** $P$ is equidistant from $\overline{AC}$ and $\overline{BC}$. | **4.** If $P$ is the same distance from $\overline{AB}$ as from $\overline{AC}$, and $P$ is the same distance from $\overline{BC}$ as from $\overline{AB}$, then point $P$ is equidistant from $\overline{AC}$ and $\overline{BC}$. |
| **5.** $P$ lies on the bisector of $\angle C$. | **5.** If a point is equidistant from the sides of an angle, then it lies on the bisector of the angle (Theorem 9.3). |
| **6.** $\overline{CP}$ bisects $\angle C$. | **6.** If $P$ lies on the bisector of $\angle C$, then segment $\overline{CP}$ bisects $\angle C$. |

| STATEMENTS | REASONS |
|---|---|

**7.** *P* is equidistant from sides $\overline{AB}$, **7.** Statements 2 and 3.
  $\overline{BC}$, and $\overline{CA}$.                                                    □

The point of intersection of the angle bisectors of a triangle is called the **incenter of the triangle**. Because the incenter is equidistant from the three sides of the triangle, it is the center of a circle that is tangent to all three sides. Such a circle is called an **inscribed circle**.

*EXAMPLE 2*    Construct the inscribed circle of △*ABC* in Figure 9.7.

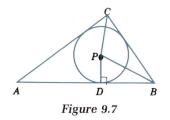

**Figure 9.7**

*Solution*    To construct the inscribed circle of a given triangle, first construct the angle bisectors of two of its angles. See Figure 9.7. These bisectors will meet at the incenter, point *P*. Then construct the perpendicular from *P* to one side of the triangle, say $\overline{AB}$, and call the point of intersection point *D*. Finally, draw a circle with center *P* and radius $\overline{PD}$. This circle is the inscribed circle.   ∎

---

**Theorem 9.7** The altitudes of a triangle (or their extensions) meet in a point.

---

*Given*: △*ABC* with altitudes $\overline{AE}$, $\overline{CD}$, and $\overline{BF}$.

*Prove*: $\overline{AE}$, $\overline{CD}$, and $\overline{BF}$ meet in a point.

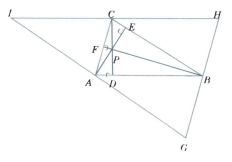

*Plan*: Construct $\overline{IG} \parallel \overline{CB}$, $\overline{IH} \parallel \overline{AB}$, and $\overline{GH} \parallel \overline{AC}$, and show that $\overline{AE}$, $\overline{CD}$, and $\overline{BF}$ are the perpendicular bisectors of the sides of $\triangle IGH$. They would then be concurrent by Theorem 9.5. To do so, show that $\overline{AE}$, $\overline{CD}$, and $\overline{BF}$ are perpendicular to $\overline{IG}$, $\overline{IH}$, and $\overline{GH}$, respectively. Then show that point $A$ is the midpoint of $\overline{IG}$ by considering parallelograms $ABCI$ and $AGBC$. In a similar fashion, points $B$ and $C$ can be shown to be midpoints.

The details of the proof of Theorem 9.7 are left as an exercise.

The point of intersection of the altitudes of a triangle is called the **orthocenter of the triangle**.

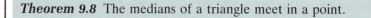

**Theorem 9.8**  The medians of a triangle meet in a point.

*Given*: $\triangle ABC$ with medians $\overline{NA}$ and $\overline{MB}$, intersecting at point $P$.

*Prove*: $\overline{CPQ}$ is a median of $\triangle ABC$.

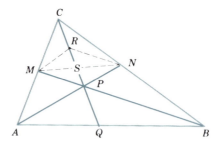

*Plan*: Locate the midpoint of $\overline{CP}$ and call it point $R$. Draw segments $\overline{MR}$, $\overline{RN}$, and $\overline{MN}$. Show that $MRNP$ is a parallelogram. Then $\overline{MS} \cong \overline{SN}$ because the diagonals of a parallelogram bisect each other. Since $\triangle ACQ \sim \triangle MCS$ and $m(\overline{AC}) = 2[m(\overline{MC})]$, we have $m(\overline{AQ}) = 2[m(\overline{MS})]$. Similarly $m(\overline{QB}) = 2[m(\overline{SN})]$ and $\overline{AQ} \cong \overline{QB}$ and, therefore, $CPQ$ is a median of $\triangle ABC$.

The details of the proof of Theorem 9.8 are left as an exercise.

The point of intersection of the medians of a triangle is called the **centroid of the triangle**.

**Theorem 9.9**  The point of intersection of the medians of a triangle is two-thirds of the way from the vertex to the midpoint of the opposite side.

*Given:* $\triangle ABC$ with medians $\overline{AN}$, $\overline{BM}$, and $\overline{CQ}$.

*Prove:* $m(\overline{AP}) = \dfrac{2}{3}[m(\overline{AN})]$; $m(\overline{BP}) = \dfrac{2}{3}[m(\overline{BM})]$; and

$\qquad m(\overline{CP}) = \dfrac{2}{3}[m(\overline{CQ})]$.

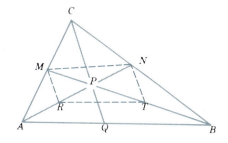

*Plan:* Locate the midpoints of segments $\overline{AP}$ and $\overline{BP}$ and call them $R$ and $T$, respectively. Draw segments $\overline{MR}$, $\overline{RT}$, $\overline{MN}$, and $\overline{TN}$. Show that $MRTN$ is a parallelogram and that $\overline{RP} \cong \overline{PN}$ and $\overline{TP} \cong \overline{PM}$. Thus, $m(\overline{AR}) = m(\overline{RP}) = m(\overline{PN})$ and $m(\overline{BT}) = m(\overline{TP}) = m(\overline{PM})$ and $m(\overline{AP}) = \dfrac{2}{3}[m(\overline{AN})]$ and $m(\overline{BP}) = \dfrac{2}{3}[m(\overline{BM})]$. To show $m(\overline{CP}) = \dfrac{2}{3}[m(\overline{CQ})]$, consider the midpoints of segments $\overline{PC}$ and $\overline{AP}$ and proceed with a similar argument.

The details of the proof Theorem 9.9 are left as an exercise.

## EXERCISE 9.2

1. Draw a triangle and locate its circumcenter.
2. Draw a triangle and locate its incenter.
3. Draw a triangle and locate its orthocenter.
4. Draw a triangle and locate its centroid.
5. Draw a triangle and construct its circumscribed circle.
6. Draw a triangle and construct its inscribed circle.
7. Draw three points (not in a straight line) and construct a circle that passes through them.
8. Can a circle be drawn through any *four* points? Why or why not?
9. Construct a square and circumscribe a circle around it.
10. Construct a square, and inscribe a circle within it.
11. Can the orthocenter, incenter, circumcenter, and centroid be concurrent in a triangle?
12. If the altitude of an equilateral triangle is 4 in., find the radius of its inscribed circle.

13. If the altitude of an equilateral triangle is 4 in., find the radius of its circumscribed circle.

14. The distance between the centroid of an equilateral triangle and one of the vertices is 6 in. How long is a median?

15. Cut out a large triangle from a sheet of stiff cardboard, and carefully locate its centroid. Try to balance the triangle on the tip of your finger. How close to the centroid does your finger touch? The point at which the triangle balances is called the **center of mass**. For a uniform triangle, the center of mass and the centroid coincide.

16. *Complete the proof of Theorem 9.7*: The altitudes of a triangle meet in a point.

17. *Complete the proof of Theorem 9.8*: The medians of a triangle meet in a point.

18. *Complete the proof of Theorem 9.9*: The point of intersection of the medians of a triangle is two-thirds of the way from the vertex to the midpoint of the opposite side.

## CHAPTER NINE REVIEW EXERCISES

In Review Exercises 1–4, sketch each locus.

1. The locus of all points in a plane equidistant from a given line segment $\overline{AB}$.
2. The locus of all points in a plane equidistant from three points in the plane.
3. The locus of all points in a plane that are a fixed distance from a circle.
4. The locus of all points in a plane that are equidistant from the vertices of a square.

In Review Exercises 5–6, describe each locus.

5. The locus of all points in space that are equidistant from a circle.
6. The locus of all points in space that are equidistant from three points in a plane.

In Review Exercises 7–10, perform each construction.

7. Construct the circumcenter of $\triangle ABC$ shown in Illustration 1.
8. Construct the incenter of $\triangle DEF$ shown in Illustration 2.

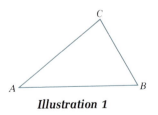

*Illustration 1*

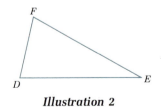

*Illustration 2*

9. Construct the orthocenter of $\triangle GHI$ shown in Illustration 3.
10. Construct the centroid of $\triangle JKL$ shown in Illustration 4.

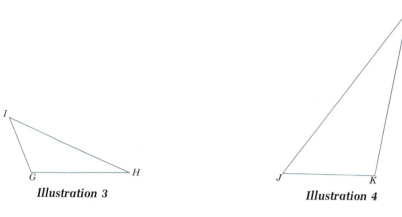

**Illustration 3**                    **Illustration 4**

## CHAPTER NINE TEST

In Exercises 1–20, classify each statement as true or false. Unless stated otherwise, assume that each locus is a two-dimensional locus.

**1.** The altitudes of a triangle meet in a point that is in the interior of the triangle.

**2.** The angle bisectors of the three angles of a triangle meet in a point that is in the interior of the triangle.

**3.** The medians of a triangle meet in a point that is in the interior of the triangle.

**4.** The locus of points equidistant from two points is the perpendicular bisector of the line segment joining the two points.

**5.** The locus of points equidistant from the sides of an angle is the bisector of the angle.

In Statements 6–10, refer to Illustration 1.

**6.** If $M$, $N$, and $P$ are midpoints of the three sides of $\triangle ABC$ and m($\overline{OM}$) = 2 in., then m($\overline{ON}$) = 2 in.

**7.** If $M$, $N$, and $P$ are midpoints of the three sides of $\triangle ABC$ and m($\overline{ON}$) = 2 in., then m($\overline{AO}$) = 4 in.

**8.** If $M$, $N$, and $P$ are midpoints of the three sides of $\triangle ABC$ and m($\overline{CM}$) = 3 m, then m($\overline{CO}$) = 2 m.

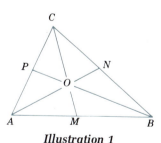

**9.** If $\overline{CM}$, $\overline{AN}$, and $\overline{BP}$ are bisectors of the angles of $\triangle ABC$, then point $O$ is the midpoint of $\overline{CM}$.

**Illustration 1**

**10.** If $\overline{CM}$, $\overline{AN}$, and $\overline{BP}$ are bisectors of the angles of $\triangle ABC$, then m($\overline{OP}$) = m($\overline{ON}$) = m($\overline{OM}$).

**11.** The locus of points that are equidistant from the sides of an angle and at a given distance from the vertex of the angle contains exactly one point.

**12.** The locus of points equidistant from two parallel lines is a line parallel to the given lines and midway between them.

13. The locus of the vertex of the right angle of a right triangle with fixed hypotenuse is a circle minus the endpoints of the hypotenuse.

14. The point of intersection of the altitudes of a triangle is called the orthocenter of the triangle.

15. The point of intersection of the medians of a triangle is called the centroid of the triangle.

16. The point of intersection of the angle bisectors of a triangle is called the circumcenter of the triangle.

17. It is always possible to construct a circle passing through three points if the points are not colinear.

18. In three space, the locus of points equidistant from a line is a pair of parallel lines.

19. In three space, the locus of points equidistant from two fixed points is a single point midway between them.

20. In three space, the locus of points equidistant from three fixed noncolinear points is a line.

# 10

*Carl Friedrich Gauss (1777–1855). Gauss is acknowledged to be one of the greatest mathematicians of all time. In addition to his work in geometry, he made significant contributions to the fields of arithmetic, number theory, analysis, and algebra as well as astronomy and physics. Gauss considered as one of his greatest discoveries a method for constructing a regular polygon of 17 sides. (Deutsches Museum, Munich).*

# Regular Polygons and Circles

Recall that a regular polygon is a polygon with all sides and angles congruent. In this chapter, we shall prove several theorems about such polygons.

## 10.1 CIRCLES CIRCUMSCRIBED ABOUT AND INSCRIBED WITHIN REGULAR POLYGONS

Recall that if a polygon has all of its vertices on a circle, it is said to be **inscribed within the circle**. An equivalent statement is to say that the circle is **circumscribed about the polygon**. In Figure 10.1, polygon *ABCDE* is inscribed within circle *O*, and circle *O* is circumscribed about the polygon.

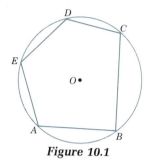

**Figure 10.1**

Also recall that if each side of a polygon is tangent to a circle, the polygon is said to be **circumscribed about the circle**. An equivalent statement is to say that the circle is **inscribed within the polygon**. In Figure 10.2, polygon *ABCDE* is circumscribed about circle *O*, and circle *O* is inscribed within the polygon.

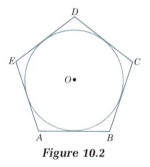

**Figure 10.2**

---

**Theorem 10.1** If a circle is divided into *n* congruent arcs, then the chords of the arcs form a regular polygon (*n* > 2).

---

*Given*: $\overset{\frown}{AB} \cong \overset{\frown}{BC} \cong \overset{\frown}{CD} \cong \overset{\frown}{DE} \cong \overset{\frown}{EA}$.

*Prove*: *ABCDE* is a regular polygon.

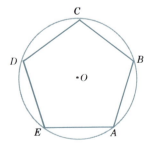

*Proof*: The circle is divided into $n$ congruent arcs. For the sake of argument, we take $n = 5$. Since $\overset{\frown}{AB}$, $\overset{\frown}{BC}$, $\overset{\frown}{CD}$, $\overset{\frown}{DE}$, and $\overset{\frown}{EA}$ are all congruent arcs and since in the same circle congruent arcs have congruent chords, we have $\overline{AB} \cong \overline{BC} \cong \overline{CD} \cong \overline{DE} \cong \overline{EA}$. Thus, all sides of the polygon are congruent. Each angle formed by two chords of the circle, such as $\angle A$, is an inscribed angle that intercepts an arc three times the measure of a single arc. Hence, all angles of the inscribed polygon are congruent. Because $ABCDE$ has all sides congruent and all angles congruent, it is a regular polygon. □

The preceding argument can be generalized to any number of congruent arcs.

---

**Theorem 10.2** If a circle is divided into $n$ congruent arcs and tangents are drawn to the circle at the endpoints of these arcs, then the figure formed by the tangents will be a regular polygon ($n > 2$).

---

*Given*: $\overset{\frown}{AB} \cong \overset{\frown}{BC} \cong \overset{\frown}{CD} \cong \overset{\frown}{DE} \cong \overset{\frown}{EA}$. $\overline{MN}$, $\overline{NP}$, $\overline{PQ}$, $\overline{QR}$, and $\overline{RM}$ are all tangents to the circle.

*Prove*: $MNPQR$ is a regular polygon.

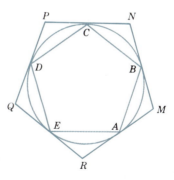

*Proof*: The circle is divided into $n$ congruent arcs. For the sake of argument, we take $n = 5$. Because all of the arcs are congruent, all of the chords are congruent as well. Because tangents drawn to a circle from a point outside the circle are congruent, triangles $AMB$, $BNC$, $CPD$, $DQE$, and $ERA$ are all isosceles triangles. Angles formed by a tangent and a chord have a measure equal to one-half their intercepted arc, and since all five arcs are congruent, the base angles of the above triangles are congruent also. These five triangles are all congruent by ASA $\cong$ ASA. Because they are corresponding parts of congruent triangles, angles *M*, *N*, *P*, *Q*, and *R* are congruent. Thus, all angles of the polygon are congruent. Also, $\overline{MN}$ ≅ $\overline{NP}$ ≅ $\overline{PQ}$ ≅ $\overline{QR}$ ≅ $\overline{RM}$, and all sides of the polygon are congruent. Can you explain why? Since *MNPQR* has all sides congruent and all angles congruent, it is a regular polygon.    □

The preceding argument can be generalized to any number of congruent arcs.

---

**Theorem 10.3** A circle can be circumscribed about any regular polygon.

---

We will first give a construction for circumscribing a circle about a regular polygon and then prove that the construction works.

Construct the perpendicular bisectors $\overline{FP}$ and $\overline{GP}$ of two sides of a regular polygon such as the one shown in Figure 10.3. Call their intersection point *P*. Using point *P* as center and segment $\overline{PA}$ as radius, draw circle *P*. Circle *P* is circumscribed about the regular polygon *ABCDE*. This construction works for all regular polygons, regardless of the number of sides.

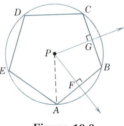

**Figure 10.3**

Here is the proof of Theorem 10.3.

*Given*: Regular polygon *ABCDE*.

*Prove*: A circle can be circumscribed about *ABCDE*.

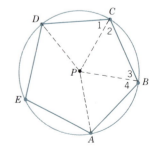

*Construction*: Construct the circle *P* that passes through vertices *A*, *B* and *C*, and show that a fourth vertex lies on the circle. Draw radii $\overline{PA}$, $\overline{PB}$, $\overline{PC}$, and $\overline{PD}$.

*Proof*:

| STATEMENTS | REASONS |
|---|---|
| **1.** $ABCDE$ is a regular polygon. | **1.** Given. |
| **2.** $\overline{AB} \cong \overline{CD}$. | **2.** All sides of a regular polygon are congruent. |
| **3.** $\overline{PB} \cong \overline{PC}$. | **3.** All radii of the same circle are congruent. |
| **4.** $\angle DCB \cong \angle CBA$. | **4.** All angles of a regular polygon are congruent. |
| **5.** m($\angle DCB$) = m($\angle CBA$). | **5.** If two angles are congruent, they have equal measures. |
| **6.** m($\angle DCB$) = m($\angle 1$) + m($\angle 2$). | **6.** Postulate 1.11. |
| **7.** m($\angle CBA$) = m($\angle 3$) + m($\angle 4$). | **7.** Postulate 1.11. |
| **8.** m($\angle 1$) + m($\angle 2$) = m($\angle 3$) + m($\angle 4$). | **8.** Substitution. |
| **9.** $\angle 2 \cong \angle 3$. | **9.** In a triangle, if two sides are congruent, then the angles opposite those sides are congruent. |
| **10.** m($\angle 2$) = m($\angle 3$). | **10.** If two angles are congruent, they have equal measures. |
| **11.** m($\angle 1$) = m($\angle 4$). | **11.** Equal subtracted from equals are equal. |
| **12.** $\angle 1 \cong \angle 4$. | **12.** If two angles have equal measures, they are congruent. |
| **13.** $\triangle ABP \cong \triangle CDP$. | **13.** SAS $\cong$ SAS. |
| **14.** $\overline{PA} \cong \overline{PD}$. | **14.** cpctc. |
| **15.** Point $D$ lies on the circle. | **15.** If a point is an endpoint of the radius $\overline{PD}$ of a circle, the point lies on the circle. |

In a similar way, we can show that any vertex of the regular polygon lies on the circle. This proof is general and applies to all regular polygons regardless of the number of sides. □

---

**Theorem 10.4** A circle can be inscribed in any regular polygon.

*Given*: Regular polygon *ABCDE*.

*Prove*: A circle can be inscribed in regular polygon *ABCDE*.

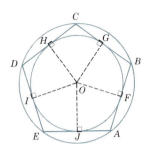

*Construction*: Construct circle *O* so that circle *O* is circumscribed about regular polygon *ABCDE*. Construct perpendiculars from point *O* to sides $\overline{AB}$, $\overline{BC}$, $\overline{CD}$, $\overline{DE}$, and $\overline{EA}$.

*Proof*:

| STATEMENTS | REASONS |
| --- | --- |
| 1. *ABCDE* is a regular polygon. | 1. Given. |
| 2. A circle *O* can be circumscribed about any regular polygon. | 2. Theorem 10.3. |
| 3. $\overline{OF} \perp \overline{AB}$; $\overline{OG} \perp \overline{BC}$; $\overline{OH} \perp \overline{CD}$; $\overline{OI} \perp \overline{DE}$; and $\overline{OJ} \perp \overline{EA}$. | 3. A perpendicular can be drawn from a point to a line. |
| 4. $\overline{AB} \cong \overline{BC} \cong \overline{CD} \cong \overline{DE} \cong \overline{EA}$. | 4. All sides of a regular polygon are congruent. |
| 5. $\overline{OF} \cong \overline{OG} \cong \overline{OH} \cong \overline{OI} \cong \overline{OJ}$. | 5. In a circle, congruent chords are the same distance from the center. |
| 6. A circle *O* exists with radii of $\overline{OF}$, $\overline{OG}$, $\overline{OH}$, $\overline{OI}$, and $\overline{OJ}$. | 6. Definition of a circle. |
| 7. $\overline{AB}$, $\overline{BC}$, $\overline{CD}$, $\overline{DE}$, and $\overline{EA}$ are all tangents to circle *O*. | 7. If a line is perpendicular to a radius at its intersection with its circle, then the line is tangent to the circle. |
| 8. Circle *O* is inscribed within regular polygon *ABCDE*. | 8. If each side of a polygon is tangent to a circle, then the circle is inscribed within the polygon. |

        □

This proof is general and applies to all regular polygons, regardless of the number of sides.

Note that in the proof of Theorem 10.4, the center of the circumscribed circle is also the center of the inscribed circle. This fact is stated as Theorem 10.5.

**Theorem 10.5** The center of a circle that is circumscribed about a regular polygon is the center of the circle that is inscribed within the regular polygon.

**EXAMPLE 1**

Construct a square and circumscribe a circle about it.

Solution

Draw a line segment $\overline{AB}$ as in Figure 10.4. Construct a perpendicular to segment $\overline{AB}$ at point $A$. Copy side $\overline{AB}$ on the perpendicular to locate point $D$ such that $\overline{AB} \cong \overline{AD}$. Construct arcs of radius $\overline{AB}$ using points $D$ and $B$ as centers. These arcs will intersect at point $C$ such that quadrilateral $ABCD$ is a square.

To circumscribe a circle about the square, construct the perpendicular bisectors of two adjacent sides of the square, say sides $\overline{AB}$ and $\overline{BC}$. These perpendicular bisectors will intersect at point $O$, the center of the desired circle. Draw circle $O$ using segment $\overline{OB}$ as a radius.

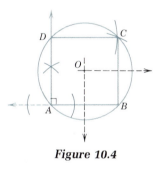

**Figure 10.4**                    ■

**Definition 10.1** The **center of a regular polygon** is the center of the circle inscribed within the polygon.

In the exercises, you will be asked to explain why the center of a regular polygon is also the center of the circle circumscribed about the polygon.

**Definition 10.2** The **radius of a regular polygon** is a line segment drawn from the center of the polygon to one of its vertices.

In the exercises, you will be asked to show that all radii of the same regular polygon are equal.

> **Definition 10.3** The **central angle of a regular polygon** is the angle formed by radii drawn to two consecutive vertices.

In the exercises, you will be asked to show that all central angles of the same regular polygon are congruent. Because of this fact, we have the following theorem.

> **Theorem 10.6** The measure $a°$ of a central angle of a regular polygon is given by the formula
>
> $$a° = \frac{360°}{n}$$
>
> where $n$ is the number of sides of the polygon.

*Given*: $ABCDE$ is a regular polygon.

*Prove*: $a° = \dfrac{360°}{n}$

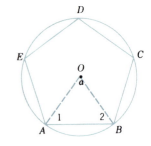

*Proof*: There are $n$ equal central angles whose sum is 360°. Hence, one such central angle must contain $360°/n$.    □

## EXERCISE 10.1

1. Construct a square and inscribe a circle within it.
2. Construct an equilateral triangle and inscribe a circle within it.
3. Construct an equilateral triangle and circumscribe a circle about it.
4. Construct a circle. Inscribe an equilateral triangle within it.
5. Construct a circle. Inscribe a square within it.
6. Construct a regular hexagon and circumscribe a circle about it.
7. Construct a regular octagon and inscribe a circle within it.
8. Can a circle be inscribed within a rhombus that is not a square? Explain your answer.
9. Can a circle be circumscribed about a rectangle that is not a square? Explain your answer.

**10.** Can a circle be circumscribed about a trapezoid that is not isosceles? Explain your answer.

**11.** Find the measure of a central angle of a regular octagon.

**12.** Find the measure of a central angle of a regular decagon.

**13.** Find the radius of a regular hexagon with a side of length 10 cm.

**14.** A square has a radius 6 in. long. Find the length of a side.

**15.** Explain why the center of a regular polygon is also the center of the circle circumscribed about the polygon.

**16.** *Prove*: All radii of the same regular polygon are congruent.

**17.** *Prove*: All central angles of the same regular polygon are congruent.

**18.** *Prove*: A radius of a regular polygon bisects an interior angle of the polygon.

## 10.2 AREAS OF REGULAR POLYGONS

We have previously discussed how to find areas of certain polygons. We now discuss how to find areas of regular polygons.

> **Definition 10.4** An **apothem of a regular polygon** is a line segment drawn from the center of the polygon and perpendicular to one of its sides.

> **Theorem 10.7** An apothem of a regular polygon bisects its respective side.

*Given*: $ABCDE$ is a regular polygon.
$\overline{OM}$ is an apothem.

*Prove*: $\overline{OM}$ bisects $\overline{EA}$.

The proof of Theorem 10.7 is left as an exercise.

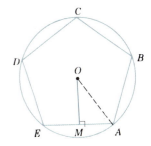

In the exercises, you will be asked to prove that all apothems of the same regular polygon are the same length. Because this is true, it is common to say that any apothem of a regular polygon is *the* apothem of that polygon.

> **Theorem 10.8** The area of a regular polygon is equal to the product of one-half the length of its apothem and its perimeter ($A = \frac{1}{2}ap$).

*Given*: $ABCDE$ is a regular polygon.
$\overline{OM}$ is an apothem.

*Prove*: $A(ABCDE) = \dfrac{1}{2}\{m(\overline{OM})[m(\overline{AB}) + m(\overline{BC}) + m(\overline{CD}) + m(\overline{DE}) + m(\overline{EA})]\}.$

*Proof*: We find the areas of triangles $ABO$, $BCO$, $CDO$, $DEO$, and $EAO$ and add them together to find the area of the regular polygon $ABCDE$. The details of the proof are left as an exercise. ☐

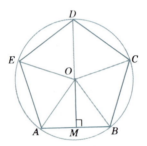

**EXAMPLE 1**

A side of the regular hexagon shown in Figure 10.5 is 8 cm long. Find its area.

Solution

The perimeter $p$ of the regular hexagon is the sum of the lengths of its six sides. Thus, we have

$$p = 6(8 \text{ cm})$$
$$= 48 \text{ cm}$$

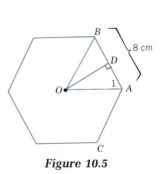

**Figure 10.5**

Because each interior angle of a regular hexagon is 120°, and $\overline{OA}$ bisects $\angle BAC$ (see Exercise 18 on page 323), we know that m($\angle 1$) = 60°. Because $\overline{OD}$ is perpendicular to side $\overline{AB}$, it is an apothem of the regular hexagon. Thus, it bisects side $\overline{AB}$, and segment $\overline{AD}$ is 4 centimeters in length. By Theorem 5.8, the length of apothem $\overline{OD}$ is $4\sqrt{3}$ cm. The area of the regular hexagon can be found by substituting $4\sqrt{3}$ for $a$ and 48 for $p$ in the formula

$$A = \frac{1}{2} ap$$

and simplifying.

$$A = \frac{1}{2} ap$$
$$= \frac{1}{2} (4\sqrt{3}) (48)$$
$$= 96\sqrt{3}$$

The area of the regular hexagon is $96\sqrt{3}$ cm². ∎

We recall a definition from Chapter 5.

**Definition 5.4** The **circumference of a circle** is the distance around a circle.

We also recall that the formula used to compute the circumference of a circle is

$$C = 2\pi r$$

where $r$ is the length of the radius of the circle, and $\pi$ = 3.14159 . . . .

. We can use the above information to motivate the formula used for finding the area of a circle. Suppose that we are given a circle and we inscribe a square within it. See Figure 10.6. The area of the square will be

*Figure 10.6*

an approximation of the area of the circumscribed circle. Now suppose that we bisect each arc associated with a side of the square and form a regular octagon using the four midpoints of the arcs and the vertices of the square. The area of this octagon will be a much better approximation of the area of the circle. We could then repeat the process again and construct a regular polygon of 16 sides. The area of this polygon would be a very good approximation of the area of the circumscribed circle. In general, the more sides the regular polygon has, the better the approximation will be.

The area of each regular polygon generated is given by the formula

$$A = \frac{1}{2} ap$$

where $a$ is the apothem, and $p$ is the perimeter of the regular polygon. In this limiting process, the apothem approaches the length of a radius and the perimeter of the polygon approaches the circumference of the circle. Thus, the area $A$ approaches

$$A = \frac{1}{2} rC = \frac{1}{2} r(2\pi r) = \pi r^2$$

This suggests the following theorem.

---

**Theorem 10.9** The **area of a circle** is given by the formula

$$A = \pi r^2$$

where $r$ is the radius of the circle, and $\pi = 3.14159 \ldots$ .

---

*EXAMPLE 2*

A circle has an area of $64\pi$ in.$^2$. Find its diameter.

Solution

Because the circle has an area of $64\pi$ in.$^2$, we have

$$
\begin{aligned}
A &= \pi r^2 \\
64\pi &= \pi r^2 \\
64 &= r^2 \qquad \text{Divide both sides by } \pi. \\
8 &= r \qquad \text{Take the square root of both sides.}
\end{aligned}
$$

Because a diameter is twice as long as radius, a diameter of the circle has length of 16 in.  ■

---

**Definition 10.5** A **sector of a circle** is the set of points between two radii and their intercepted arc.

Area $OAB$ is a sector in the circle $O$ shown in Figure 10.7.

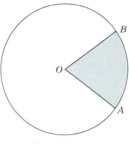

**Figure 10.7**

---

**Postulate 10.1**   The area of a sector of a circle divided by the area
of the circle is equal to the measure of its central angle divided by
360°.

---

**EXAMPLE 3**

Find the area of the sector $AOB$ shown in Figure 10.8.

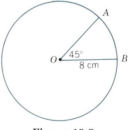

**Figure 10.8**

Solution

The area of circle $O$ is $\pi 8^2$ cm², or $64\pi$ cm². By Postulate 10.1, we can set
up the following proportion, substitute values for the area of the circle and
the measure of the central angle, and solve for the area of the sector.

$$\frac{A_{\text{sector}}}{A_{\text{circle}}} = \frac{\text{central angle of the sector}}{360°}$$

$$\frac{A}{64\pi} = \frac{45°}{360°}$$

$$A = 64\pi \cdot \frac{45°}{360°}$$

$$= 8\pi$$

The area of the sector is $8\pi$ cm².    ∎

> **Definition 10.6** A **segment of a circle** is the area enclosed by an arc of a circle and its respective chord.

**EXAMPLE 4**

Find the area of the segment of the circle shown in Figure 10.9.

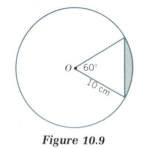

**Figure 10.9**

Solution

Find the area of the sector of the circle as follows.

$$\frac{A_{sector}}{A_{circle}} = \frac{\text{central angle of the sector}}{360°}$$

$$\frac{A}{\pi 10^2} = \frac{60°}{360°}$$

$$A = \frac{50}{3}\pi$$

The area of the equilateral triangle shown in Figure 10.9 is found as follows.

$$A = \frac{1}{2} bh$$

$$= \frac{1}{2}(10)(5\sqrt{3}) \qquad 5\sqrt{3} \text{ is the height of the equilateral triangle.}$$

$$= 25\sqrt{3}$$

The area of the segment is found by computing the difference of the above two areas.

$$A_{segment} = \frac{50}{3}\pi - 25\sqrt{3} \approx 9.06$$

The area of the segment is approximately 9 cm². ∎

If the central angle in Example 4 had not been 60°, the triangle would not have been equilateral, and computing its area would have required other techniques.

## EXERCISE 10.2

**1.** A regular polygon has a perimeter of 160 cm and an apothem of 20 cm. Find its area.

**2.** A regular polygon has a perimeter of $24\sqrt{3}$ in. and an apothem of 4 in. Find its area.

**3.** A side of a regular hexagon is 10 m long. Find its area.

**4.** The area of a regular hexagon is $216\sqrt{3}$ yd². Find the length of one of its sides.

**5.** Estimate the area of a 16-sided regular polygon with an apothem of length 15 cm. (*Hint*: The area of a circle is $\pi r^2$.)

**6.** Estimate the circumference of a 16-sided regular polygon with an apothem of length 15 cm. (*Hint*: The circumference of a circle is $\pi D$.)

**7.** Find the area of a sector of a circle if the circle has a radius of 5 ft and the sector has an arc of 30°.

**8.** The area of a sector of a circle is $12\pi$ square units. The sector has an arc of 120°. Find the diameter of the circle.

**9.** In Illustration 1, $ABC$ is an equilateral triangle inscribed in circle $O$ with a radius of 8 in. Find the area of the triangle.

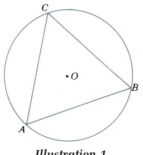

*Illustration 1*

**10.** Find the area of a regular decagon with sides of length 10 in. and an apothem of length 15 in.

**11.** A circle has a radius of 7 in. Find the area of a sector of the circle with a central angle of 50°.

**12.** Find the radius of a circle if its circumference is numerically equal to its area.

**13.** Two circles are called **concentric circles** if they have the same point as center. Prove that the area between two concentric circles with radii of $R$ and $r$ is given by the formula

$$A = \pi(R^2 - r^2)$$

**14.** A football stadium has a circular track. The longest distance that a person can run in a straight line and still remain on the track is 100 yd. Find the area of the track. (*Hint*: See Exercise 13.)

**15.** In Illustration 2, $m(\angle O) = 90°$ and $m(\overline{OA}) = 6\sqrt{2}$ cm. Find the area of the indicated segment of the circle.

**16.** In Illustration 2, $m(\angle O) = 120°$ and $m(\overline{AB}) = 20\sqrt{3}$ cm. Find the area of the indicated segment of the circle.

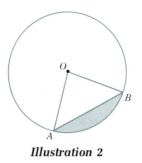

*Illustration 2*

**17.** *Prove*: All apothems of the same regular polygon are the same length.

**18.** *Prove*: If two regular polygons are similar, then the ratio of their perimeters is equal to the ratio of their apothems.

**19.** *Prove Theorem 10.7*: An apothem of a regular polygon bisects its respective side.

**20.** *Complete the proof of Theorem 10.8*: The area of a regular polygon is equal to the product of one-half its apothem and its perimeter.


## CHAPTER TEN REVIEW EXERCISES

**1.** Draw a triangle and circumscribe a circle around it.

**2.** Draw a triangle and inscribe a circle within it.

**3.** Construct a regular hexagon and inscribe a circle within it.

**4.** Can a circle be circumscribed about an isosceles trapezoid. Explain.

**5.** Find the measure of a central angle of a regular hexagon.

**6.** Find the measure of a central angle of a regular 20-sided figure.

**7.** Find the radius of a regular hexagon with a side 12 m long.

**8.** A regular octagon has a radius of 4 cm. Find its perimeter. (*Hint*: Use trigonometry.)

**9.** Find the area of a regular octagon with a side of length 8 ft. (*Hint*: Use trigonometry.)

**10.** A regular polygon has a perimeter of $8\sqrt{3}$ cm and an apothem that is 2 cm long. Find its area.

**11.** The area of a regular octagon with an apothem of 9.6 m is $216\sqrt{2}$ m². Find the length of one of its sides.

**12.** Estimate the area of a 20-sided regular polygon with an apothem that is 12 in. long.

**13.** A circle has a radius of 10 ft. If a sector of the circle has an arc of 60°, find the area of the sector.

**14.** The area of a sector of a circle is $12\pi$ cm² and the central angle intercepts an arc of 120°. Find the radius of the circle.

**15.** Find the area of a regular hexagon with sides of length 12 in. and an apothem of 15 in.

**16.** A circle has a diameter of 12 m. Find the area of a sector of the circle if the sector has a central angle of 40°.

**17.** A rectangle has dimensions of $\pi$ in. by 9 in. Find the radius of a circle with the same area.

**18.** A rectangle has dimensions of $3\sqrt{\pi}$ ft by $12\sqrt{\pi}$ ft. Find the radius of a circle with the same area.

**19.** A circle with a radius 1 m long has an area equal to that of a certain square. Find the perimeter of the square.

**20.** In Illustration 1, the sides of right triangle $ABC$ are diameters of the semicircles shown in the illustration. Prove that the sum of the areas of the semicircles drawn on the triangle's legs is equal to the area of the semicircle drawn on the hypotenuse.

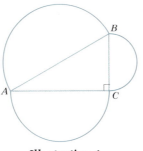

*Illustration 1*

## CHAPTER TEN TEST

In Exercises 1–20, classify each statement as true or false.

**1.** A rhombus is a regular polygon.

**2.** If a regular polygon is inscribed in a circle, then the polygon will divide the circle into equal arcs.

**3.** Any rectangle can be inscribed within a circle.

**4.** The center of an inscribed polygon coincides with the center of its circumscribed circle.

**5.** An apothem is the radius of an inscribed regular polygon.

**6.** The perimeter of an inscribed polygon is equal to one-half the apothem multiplied by the area of the polygon.

**7.** The circumference of a circle is the distance around the circle.

**8.** A central angle of a regular pentagon is 72°.

**9.** The area of a circle with diameter of 14 ft is 49 ft².

**10.** The circumference of a circle with radius of 6 in. is $12\pi$ in.

**11.** A sector of a circle is a triangle.

**12.** If a polygon is inscribed in a circle, then it is a regular polygon.

**13.** If a regular 12-sided figure is inscribed in a circle, each of the 12 arcs formed will equal 30°.

**14.** The circumference of a circle with radius of 1 in. is equal to the circle's area.

**15.** In the same circle, sectors with equal central angles always have equal areas.

**16.** The area of a segment of a circle is less than the area of every sector of a circle.

**17.** If the central angle of a sector of a circle is doubled, the area of the sector is doubled.

**18.** If the central angle of a segment of a circle is doubled, the area of the segment is doubled.

**19.** If the length of a radius of a sector of a circle is doubled, the area of the sector is doubled.

**20.** In congruent circles, sectors with the same area have congruent central angles.

# 11

*Felix Klein was a German mathematician who first developed criteria for classifying the different geometries. In his* Erlangen Program, *Klein defined geometry as the study of invariants in a transformation group. This led to a unification of the ideas of Euclidean geometry, non-Euclidean geometries, projective geometry, and finally a rather new kind of geometry called topology (Courtesy Dover Publications, Inc., publishers of Felix Klein's* Elementary Mathematics from an Advanced Standpoint).

# Logic

The old saying that "life is full of surprises" is not entirely true. Unexpected events do occur, but much of life is expected and predictable. When an expected event occurs, we can say "I thought so" because we are able to *reason*. We are able to take facts and observations, and on the basis of past experience and common sense, predict certain results. When events occur as expected, we are not surprised.

In this chapter, we will study **mathematical logic**, the mathematical principles behind the common sense of forming conclusions.

## 11.1 SIMPLE AND COMPOUND STATEMENTS

In English, some sentences state a fact, while others ask a question, make an exclamation, or give a command. In mathematics, we are concerned with declarative sentences, called **statements**, that can be classified as *true* or *false*, but not both true and false. For example, the sentences *How are you feeling?* and *Do your homework!* are not statements; the first is a question, the second a command, and neither can be classified as true or false. The sentence *Roosters lay eggs* is a statement, although it is a false statement.

When we refer to statements, we will use **statement variables** such as *p*, *q*, *r*, and *s*. Here are more examples of statements:

*p*:   The Kiwi bird is extinct.
*q*:   A triangle is a rectangle.
*r*:   $3 + 7 = 9$

Statement *p* is true, and statements *q* and *r* are false.

The **negation** of a statement has a truth value that is always opposite that of the statement itself. For example, the negation of the true statement *A triangle has three sides*, is the false statement *A triangle does **not** have three sides*. The negation of the false statement *London is in France* is the true statement *London is **not** in France*.

The symbol ~ is used to indicate negation. The symbol ~*p* is read "the negation of *p*," "not *p*," or "it is not true that *p*." Because the negation of a true statement is false, and the negation of a false statement is true, the properties of negation can be summarized in the following table, called a **truth table**, which lists all possible truth values for the logical variable *p* (T for *true,* and F for *false,*) and the corresponding truth values for the logical expression ~*p*.

---

**Definition 11.1** The Truth Table for Negation.

| *p* | ~*p* |
|-----|------|
| T   | F    |
| F   | T    |

---

**EXAMPLE 1**

Let *p* be the statement *The bluebird is a sign of happiness* and let *q* be the statement *Poets are impractical*. Find (a) ~*p*, and (b) ~*q*.

Solution

   (a)   The negation of *The bluebird is a sign of happiness* is the statement *The bluebird is **not** a sign of happiness*.

   (b)   The negation of *Poets are impractical* is the statement *Poets are not impractical*, or more briefly, *Poets are practical*.   ∎

Simple statements can be combined to form more complex statements, called **compound statements**. For example, if *p* is the simple statement *John played golf* and *q* is the statement *Betty played tennis*, then the statement

*s*:   *John played golf and Betty played tennis.*

is a compound statement made up of two simple statements joined by the
**connective** *and*. The truth of statement *s* can be determined if we know
the truth of the two statements it contains. Statement *s* asserts that *both*
*p* and *q* are true. If either *p* or *q* were false, then *s* would be false, also.

The compound statement *s* is called a **conjunction**. It is written symbolically as $p \wedge q$, where the symbol $\wedge$ represents the connective *and*.
Because $p \wedge q$ involves two variables and each variable can be either true
or false, the truth table for conjunction must consider four possible com-
binations of truth values. Only when *p* and *q* are both true is the conjunction
$p \wedge q$ true. If either *p* or *q* is false, the conjunction $p \wedge q$ is also false.

---

**Definition 11.2** The Truth Table for Conjunction.

| $p$ | $q$ | $p \wedge q$ |
|:---:|:---:|:---:|
| T | T | T |
| T | F | F |
| F | T | F |
| F | F | F |

---

**EXAMPLE 2**

If *p* is a true statement and *q* is a false statement, find the truth value of
(a) $p \wedge q$, (b) $p \wedge \sim q$, and (c) $\sim(p \wedge q)$.

**Solution**

(a)  For the conjunction $p \wedge q$ to be true, *both* *p* and *q* must be true.
     Because you are given that *q* is false, the conjunction $p \wedge q$ is
     false, also. You can also determine this by referring to the second
     line of the truth table for conjunction: there *p* is true, *q* is false,
     and the result $p \wedge q$ is false.

(b)  To determine the truth value of $p \wedge \sim q$, you must first determine
     the truth value of its two terms: *p* and $\sim q$. You are given that *p*
     is true. Because you are given that *q* is false, its negation $\sim q$ is
     true. Thus $p \wedge \sim q$ is the conjunction of two true statements. By
     the *first* line of the truth table for conjunction, $p \wedge \sim q$ is true.

(c)  The truth value of $\sim(p \wedge q)$ is opposite that of $p \wedge q$. In part a.
     you found that $p \wedge q$ is false. Therefore, $\sim(p \wedge q)$ is true.  ∎

Two simple statements joined by the connective *or* form a compound
statement called a **disjunction**. If *p* is the statement *Jim has measles* and
*q* is the statement *Jenny has mumps*, then the compound statement

*Jim has measles or Jenny has mumps.*

would be written as $p \vee q$, where the symbol $\vee$ represents the connective
*or*. The only way that the disjunction $p \vee q$ can be false is that *both* *p* and
*q* be false. If either or both is true, the disjunction is true.

---

**Definition 11.3** The Truth Table for Disjunction.

| $p$ | $q$ | $p \lor q$ |
|:---:|:---:|:---:|
| T | T | T |
| T | F | T |
| F | T | T |
| F | F | F |

---

**EXAMPLE 3**

If $p$ is false and $q$ is true, find the truth value of (a) $p \lor q$, (b) $p \lor \sim q$, and (c) $\sim(p \lor q)$.

**Solution**

(a)  A disjunction $p \lor q$ is true if either of its terms is true. Because $q$ is given to be true, $p \lor q$ is true. You can also determine this by looking at the third line of the truth table for disjunction: there $p$ is false, $q$ is true, and the disjunction $p \lor q$ is true.

(b)  To find the truth value of $p \lor \sim q$, first determine the truth values of its two terms, $p$ and $\sim q$. Because $q$ is true, its negation, $\sim q$, is false. $p$ is given as false. Therefore, $p \lor \sim q$ is the disjunction of two false statements. By the last line of the truth table for disjunction, $p \lor \sim q$ is false.

(c)  The truth value of $\sim(p \lor \sim q)$ is opposite that of $p \lor \sim q$, which you determined to be false in part (b). Thus, $\sim(p \lor \sim q)$ is true.    ∎

Many statements in mathematics are *If . . . then . . .* sentences:

*If* a closed figure has three sides, *then* it is a triangle.
*If* a triangle is isosceles, *then* its base angles are congruent.
*If* $x = y$, then $y = x$.

An *If . . . then . . .* statement is called an **implication** or a **conditional statement**. The conditional statement *If p*, *then q* is written symbolically as $p \Rightarrow q$. To determine the truth table for the conditional, we consider an example. Let $p$ be the statement *It is sunny* and let $q$ be the statement *We will go on a picnic*. The conditional statement $p \Rightarrow q$ is a promise:

**If** it is sunny, **then** we will go on a picnic.

The conditional $p \Rightarrow q$ will be false only if the promise is broken: the day *is* sunny, and the picnic is cancelled. That is, $p \Rightarrow q$ is false only when $p$ is true (the conditions are satisfied) and $q$ is false (the promise is not kept.) Under all other conditions the promise is not broken, and the conditional statement is true. Note that if the day is *not* sunny, we might still go on the picnic; the promise is not broken because it does not cover actions on such days. The truth table for the for the conditional follows.

**Definition 11.4** The Truth Table for the Conditional.

| $p$ | $q$ | $p \Rightarrow q$ |
|---|---|---|
| T | T | T |
| T | F | F |
| F | T | T |
| F | F | T |

**EXAMPLE 4**

If $p$ is false and $q$ is true, find the truth value of (a) $p \Rightarrow q$ and (b) $p \Rightarrow \sim q$.

Solution

(a) Refer to the third line of the truth table for the conditional to determine that under the given conditions, $p \Rightarrow q$ is true.

(b) To determine the truth value of $p \Rightarrow \sim q$ determine that under the given conditions, $p$ and $\sim q$ are both false. Then refer to the last line of the truth table for the conditional to determine that $p \Rightarrow \sim q$ is true. ∎

**EXAMPLE 5**

Compound sentences can be written symbolically. Consider the following statements:

$p$: It is raining.
$q$: The sun is shining.
$r$: The grass is wet.

Then study each of the following sentences and its equivalent symbolic form.

| SENTENCE | SYMBOLIC FORM |
|---|---|
| It is raining and the sun is shining. | $p \wedge q$ |
| It is raining and the sun is not shining. | $p \wedge \sim q$ |
| It is raining or the grass is not wet. | $p \vee \sim r$ |
| It is not raining or the sun is not shining or the grass is wet. | $\sim p \vee \sim q \vee r$ |
| It is raining and the sun is shining, or the grass is wet. | $(p \wedge q) \vee r$ |
| It is raining, and the sun is shining or the grass is wet. | $p \vee (q \vee r)$ |
| If it is raining, then the grass is wet. | $p \Rightarrow r$ |
| If the grass is not wet, then the sun is not shining. | $\sim r \Rightarrow (\sim q)$ ∎ |

## CONVERSE, INVERSE, AND CONTRAPOSITIVE

In the *If . . . then . . .* statement $p \Rightarrow q$, the *if* clause, $p$, is called the **antecedent**, and the *then* clause, $q$, is called the **consequent**. Other conditional statements are related to the *If . . . then . . .* statement $p \Rightarrow q$. One of these is the **converse** of $p \Rightarrow q$, formed by interchanging the antecedent and consequent. The converse of $p \Rightarrow q$ is $q \Rightarrow p$. Another is the **inverse**, formed by negating both the antecedent and the consequent. The inverse of $p \Rightarrow q$ is $\sim p \Rightarrow \sim q$. Finally, the **contrapositive** of $p \Rightarrow q$ is $\sim q \Rightarrow \sim p$, formed by interchanging the antecedent and the consequent, and negating them both.

---

**Definition 11.5** The Converse, Inverse, and Contrapositive.

| | |
|---|---|
| $p \Rightarrow q$ | Statement |
| $q \Rightarrow p$ | Converse |
| $\sim p \Rightarrow \sim q$ | Inverse |
| $\sim q \Rightarrow \sim p$ | Contrapositive |

---

**EXAMPLE 6**

Find the converse, inverse, and contrapositive of *If I have measles, then I am sick.*

Solution

*If I have measles, then I am sick* is a conditional statement of the form $p \Rightarrow q$, where the antecedent $p$ is *I have measles*, and the consequent $q$ is *I am sick*. By the definition of converse, inverse, and contrapositive, you have

| | | |
|---|---|---|
| $p \Rightarrow q$ | Statement | *If I have measles, then I am sick.* |
| $q \Rightarrow p$ | Converse | *If I am sick, then I have measles.* |
| $\sim p \Rightarrow \sim q$ | Inverse | *If I don't have measles, then I am not sick.* |
| $\sim q \Rightarrow \sim p$ | Contrapositive | *If I am not sick, then I don't have measles.* |

■

By examining the results of Example 6, we see that a statement and its converse do not necessarily agree: a statement can be true and its converse false. We will see in the next section, however, that a statement and its contrapositive always agree: they are either both true, or both false.

We consider one more logical connective, known as the **biconditional** and indicated by the symbol $\Leftrightarrow$. The expression $p \Leftrightarrow q$ is read as ***p if and only if q*** or as ***p is equivalent to q***. The biconditional is logic's equal sign: $p \Leftrightarrow q$ is true when $p$ and $q$ have the *same* truth value (both true or both false). If $p$ and $q$ have opposite truth values, $p \Leftrightarrow q$ is false.

---

**Definition 11.6** The Truth Table for the Biconditional.

| $p$ | $q$ | $p \Leftrightarrow q$ |
|---|---|---|
| T | T | T |
| T | F | F |
| F | T | F |
| F | F | T |

---

## EXERCISE 11.1

In Exercises 1–8, tell whether the expression is a statement.

**1.** $3 + 7 = 10$                  **2.** Think Snow!

**3.** $5 + 7 = 10$                  **4.** Are you sure of that?

**5.** You can't fire me, I quit!    **6.** Play it again, Sam.

**7.** It was a dark and stormy night.    **8.** If $x = 3$, then $y = 2$.


In Exercises 9–18, write the logical negation of each statement.

**9.** Today is Tuesday.             **10.** The crops will not survive.

**11.** The earth is round.          **12.** The moon is made of green cheese.

**13.** $4 = 7$                      **14.** $7 \neq 7$

**15.** Spaghetti is slimy.          **16.** Crime does not pay.

**17.** Bill and Sue are friends.    **18.** Bill is older than Sue.


In Exercises 19–30 assume that $p$ is true and that $q$ is false. Then determine the truth value of each expression.

**19.** $\sim p$          **20.** $\sim q$          **21.** $p \wedge q$          **22.** $p \vee q$
**23.** $p \Rightarrow q$  **24.** $q \Rightarrow p$  **25.** $\sim p \wedge q$      **26.** $\sim p \vee q$
**27.** $p \wedge \sim q$  **28.** $p \wedge \sim q$  **29.** $p \Leftrightarrow q$  **30.** $p \Leftrightarrow \sim q$


In Exercises 31–42, assume that $p$ is false and that $q$ is true. Then determine the truth value of each expression.

**31.** $\sim p$          **32.** $\sim q$          **33.** $p \wedge q$          **34.** $p \vee q$
**35.** $p \Rightarrow q$  **36.** $q \Rightarrow p$  **37.** $\sim p \wedge q$      **38.** $\sim p \vee q$
**39.** $p \vee \sim q$    **40.** $p \wedge \sim q$  **41.** $p \Leftrightarrow q$  **42.** $p \Leftrightarrow \sim q$


In Exercises 43–54, write each symbolic expression as an English sentence. Assume that the variables represent the following statements.

$p$:  Bob caught a fish.
$q$:  Jim caught a cold.

**43.** $\sim p$          **44.** $\sim q$          **45.** $p \vee q$          **46.** $p \wedge q$
**47.** $p \Rightarrow q$  **48.** $q \Rightarrow p$  **49.** $p \wedge \sim q$    **50.** $\sim q \vee p$
**51.** $\sim p \Rightarrow q$  **52.** $p \Rightarrow \sim q$  **53.** $p \Leftrightarrow q$  **54.** $\sim p \Leftrightarrow \sim q$

In Exercises 55–64, write each symbolic expression as an English sentence. Assume that the variables represent the following statements.

- $p$: A triangle has three sides.
- $q$: A quadrilateral has four sides.
- $r$: A pentagon has five sides.

**55.** $\sim p \wedge \sim r$      **56.** $\sim q \vee \sim r$      **57.** $p \wedge q \wedge r$      **58.** $p \vee q \vee r$
**59.** $p \Rightarrow (r \wedge q)$      **60.** $r \wedge (p \vee q)$      **61.** $(p \Rightarrow r) \wedge q$      **62.** $(r \wedge p) \vee q$
**63.** $(p \wedge q) \Rightarrow r$                      **64.** $p \wedge (q \Rightarrow r)$

In Exercises 65–78, write each English sentence as a symbolic expression. Assume the following:

- $p$: It is snowing.
- $q$: Travel is difficult.
- $r$: The guests will be late.

**65.** It is snowing and travel is difficult.

**66.** It is snowing or the guests will be late.

**67.** It is not snowing or the guests will be late.

**68.** If it is snowing, then travel is difficult.

**69.** If travel is not difficult, then the guests will not be late.

**70.** If it is snowing, then either travel is difficult or the guests will be late.

**71.** If it is snowing or travel is difficult, then the guests will be late.

**72.** It is not snowing and travel is not difficult, or the guests will be late.

**73.** If the guests will be late, then not only is it snowing but also travel is difficult. (*Hint: not only . . . but also . . . is a fancy way of saying and.*)

**74.** If it is not snowing or travel is not difficult, then the guests will not be late.

**75.** It is not snowing or travel is not difficult, then the guests will not be late.

**76.** If it is not snowing and travel is not difficult, then the guests will not be late.

**77.** It is snowing if and only if travel is difficult.

**78.** It is snowing is equivalent to travel is difficult.

In Exercises 79–92, write the converse, inverse, and contrapositive of each statement.

**79.** If you study, then you will pass.

**80.** If Sue is not invited, then she will be disappointed.

**81.** If a triangle is equilateral, then it is equiangular.

**82.** If you are a full-time student, then you will take five classes.

**83.** If you do not like the weather, then just wait a minute.

**84.** If today is Saturday, then I missed Friday's exam.

**85.** If you water the plants, then they will grow.

**86.** If you think you can, you can.

**87.** If it's a mild spring, then we will have a hot summer.

**88.** If you don't finish your vegetables, then you can't have dessert.

**89.** You would do better if you did your homework. (*Hint*: What is the *if* clause?)

**90.** You won't succeed if you don't try.

**91.** If you can keep your head while others around you are losing theirs, then you don't understand the situation.

**92.** If you do not tell the truth, then people will learn not to trust you.

**93.** Use your imagination to create a true *If . . . then . . .* statement that has a false converse.

**94.** Use your imagination to create a false *If . . . then . . .* statement that has a true converse.

**95.** Use your imagination to create a true *If . . . then . . .* statement that has a true inverse.

**96.** Use your imagination to create a false *If . . . then . . .* statement that has a false inverse.

**97.** Find the contrapositive of the converse of *If you like mosquitoes, you'll love camping.*

**98.** Find the converse of the contrapositive of *Sam will stay if Sue stays.*

# 11.2 LONGER STATEMENT FORMS AND TAUTOLOGIES

As we have seen, many compound statements can be broken into simpler statements and written symbolically. For example, if *p* and *q* represent the two simple statements

    *p*:  *Jim fell.*

and

    *q*:  *Jim broke his arm.*

Then the compound statement

    *Jim fell and did not break his arm.*

can be expressed symbolically as

$$p \wedge \sim q$$

Before we can use the properties of conjunction to determine the value of $p \wedge \sim q$, we must know the truth values of the two terms, $p$ and $\sim q$. Thus, we must form a truth table with an extra column headed $\sim q$, as an aid to determining the truth value of the given expression, $p \wedge \sim q$. The entries in this extra column will be the negations of the corresponding values found in the column headed by $q$.

| $p$ | $q$ | $\sim q$ | $p \wedge \sim q$ |
|---|---|---|---|
| T | T | F | F |
| T | F | T | T |
| F | T | F | F |
| F | F | T | F |

The truth values in the last column are obtained from the values found in the first and third columns, using the definition of $\wedge$.

**EXAMPLE 1**    Construct a truth table for $\sim p \wedge (p \vee q)$.

Solution    The expression $\sim p \wedge (p \vee q)$ is the conjunction of two terms: $\sim p$ and ($p \vee q$). The truth value of these two terms must be known before you can determine the truth value of the conjunction. Therefore, include two extra columns in your truth table to record these intermediate results.

In the first line of your table, both $p$ and $q$ are true. Because $p$ is true, $\sim p$ is false; place an F in the column headed by $\sim p$. Because $p$ and $q$ are both true, the

| $p$ | $q$ | $\sim p$ | $p \vee q$ | $\sim p \wedge (p \vee q)$ |
|---|---|---|---|---|
| T | T | F | T | F |

A false and a true, joined by **and**, is false.

disjunction $p \vee q$ is true; place a T in the column headed by $p \vee q$. Finally, look at these two intermediate results, and combine them with the operation $\wedge$. By the definition of $\wedge$, the conjunction of a *false* statement and a *true* statement is *false*; place an F in the final column headed by $\sim p \wedge (p \vee q)$. Complete the table by performing a similar analysis of the remaining possibilities. Here is the complete truth table:

| $p$ | $q$ | $\sim p$ | $p \vee q$ | $\sim p \wedge (p \vee q)$ |
|---|---|---|---|---|
| T | T | F | T | F |
| T | F | F | T | F |
| F | T | T | T | T |
| F | F | T | F | F |

## TAUTOLOGIES

We have seen that the negation of a simple statement such as *I am tired* is *I am **not** tired.* If this negation were negated again, it would read

*It is not the case that I am not tired.*

which is a complicated way of saying, simply,

*I am tired.*

Thus, the **double negation** of a statement appears to be equivalent to the statement itself. To show that $p$ and $\sim(\sim p)$ are equivalent, we use a truth table to show that

$$p \Leftrightarrow \sim(\sim p)$$

is always true. We include two extra columns for the intermediate results $\sim p$ and $\sim(\sim p)$.

| $p$ | $\sim p$ | $\sim(\sim p)$ | $p \Leftrightarrow \sim(\sim p)$ |
|-----|----------|----------------|----------------------------------|
| T   | F        | T              | T                                |
| F   | T        | F              | T                                |

⇑――――⇑

Because the results in these columns agree, the expressions are logically equivalent.

A logical expression that is true for *all* values of its variables is called a **tautology**. Because the last column of the previous truth table contains only T's, the expression $p \Leftrightarrow \sim(\sim p)$, known as the **law of double negation**, is a tautology.

---

**Theorem 11.1** The Law of Double Negation. The statements $p$ and $\sim(\sim p)$ are equivalent. That is,

$$p \Leftrightarrow \sim(\sim p)$$

is a tautology.

---

The negation of a compound statement such as the conjunction

*I walked to work* **and** *I am tired.*

is the disjunction

*I did **not** walk to work or I am **not** tired.*

To show this, we must prove that the negation of the conjunction $p \wedge q$ is equivalent to the disjunction $\sim p \vee \sim q$. That is, we must prove that

$$\sim(p \wedge q) \Leftrightarrow (\sim p \vee \sim q)$$

is a tautology. Example 2 provides the proof.

**EXAMPLE 2**

Show that the expression $\sim(p \wedge q) \Leftrightarrow (\sim p \vee \sim q)$ is a tautology.

Solution

Create a truth table for $\sim(p \wedge q) \Leftrightarrow (\sim p \vee \sim q)$, and verify that the expression is always true. To determine truth values for the left-hand side of the biconditional, include extra columns for $(p \wedge q)$ and $\sim(p \wedge q)$. To determine the truth value of the right-hand side, include three extra columns for the intermediate results $\sim p$, $\sim q$, and $\sim p \vee \sim q$. The complete truth table follows.

| $p$ | $q$ | $p \wedge q$ | $\sim(p \wedge q)$ | $\sim p$ | $\sim q$ | $\sim p \vee \sim q$ | $\sim(p \wedge q) \Leftrightarrow (\sim p \vee \sim q)$ |
|---|---|---|---|---|---|---|---|
| T | T | T | **F** | F | F | **F** | T |
| T | F | F | **T** | F | T | **T** | T |
| F | T | F | **T** | T | F | **T** | T |
| F | F | F | **T** | T | T | **T** | T |

⇑————————————⇑

Because the results in these columns agree, the expressions are logically equivalent.

Because the last column of the truth table of Example 2 consists entirely of T's, the expression

$$\sim(p \wedge q) \Leftrightarrow (\sim p \vee \sim q)$$

is a tautology. It, together with another tautology,

$$\sim(p \vee q) \Leftrightarrow (\sim p \vee \sim q)$$

are known as **DeMorgan's laws**, named after the English mathematician Augustus DeMorgan (1806–1871). DeMorgan's laws provide methods for determining the negations of conjunctions and disjunctions.

> **Theorem 11.2 DeMorgan's Laws.**   The following are tautologies:
>
> $$\sim(p \wedge q) \Leftrightarrow (\sim p \vee \sim q)$$
> $$\sim(p \vee q) \Leftrightarrow (\sim p \wedge \sim q)$$

**EXAMPLE 3**

Write the negation of the sentence *Dave will get a letter, or he will not be pleased.*

Solution

The given statement is a disjunction of the form $p \vee \sim q$, where

    *p*:  *Dave will get a letter.*
    *q*:  *He will be pleased.*

The negation of $p \vee \sim q$ is $\sim(p \vee \sim q)$. To simplify this negation, proceed as follows:

$$\sim(p \vee \sim q) \Leftrightarrow (\sim p \wedge \sim(\sim q)) \qquad \text{Use DeMorgan's law.}$$
$$\Leftrightarrow (\sim p \wedge q) \qquad \text{Use the Law of Double Negation:}$$
$$\sim(\sim q) \Leftrightarrow q$$

Thus, the negation of the given statement is $\sim p \wedge q$, or

*Dave will not get a letter, **and** he will be pleased.*

**EXAMPLE 4**

Show that an *If . . . then . . .* statement is equivalent to its contrapositive.

Solution

An *If . . . then . . .* statement has the form $p \Rightarrow q$. Its contrapositive is $\sim q \Rightarrow \sim p$. To show that the statement is equivalent to its contrapositive, use a truth table to show that

$$(p \Rightarrow q) \Leftrightarrow (\sim q \Rightarrow \sim p)$$

is a tautology. Several columns are needed for intermediate results.

| $p$ | $q$ | $p \Rightarrow q$ | $\sim q$ | $\sim p$ | $\sim q \Rightarrow \sim p$ | $(p \Rightarrow q) \Leftrightarrow (\sim q \Rightarrow \sim p)$ |
|---|---|---|---|---|---|---|
| T | T | T | F | F | T | T |
| T | F | F | T | F | F | T |
| F | T | T | F | T | T | T |
| F | F | T | T | T | T | T |

Because the results in these columns agree, the expressions are logically equivalent. ∎

---

**Theorem 11.3** The equivalence of a statement and its contrapositive. The following is a tautology:

$$(p \Rightarrow q) \Leftrightarrow (\sim q \Rightarrow \sim p)$$

---

Because a statement and its contrapositive are logically equivalent, they convey exactly the same information. If we prove a certain *If . . . then . . .* statement, it is unnecessary to prove the contrapositive as well. Proving a theorem and its contrapositive is just duplicated effort; either proof alone is enough. However, sometimes we have done second proofs to illustrate the indirect method of proof. For example, we proved Theorem 3.18, although it is the contrapositive of Theorem 3.5. Similarily, Theorem 3.19 is the contrapositive of Theorem 3.6.

**EXERCISE 11.2**

In Exercises 1–14, construct a truth table for each expression.

**1.** $\sim p \vee p$

**2.** $p \wedge \sim p$

**3.** $\sim p \vee q$

**4.** $p \wedge \sim q$

**5.** $\sim p \Leftrightarrow q$

**6.** $p \Leftrightarrow \sim q$

**7.** $\sim (p \Rightarrow q)$

**8.** $\sim (p \Leftrightarrow q)$

**9.** $p \wedge (p \vee q)$

**10.** $p \vee (p \wedge q)$

**11.** $p \Rightarrow (p \vee q)$

**12.** $(p \wedge q) \Rightarrow q$

**13.** $\sim p \vee (p \Leftrightarrow q)$

**14.** $(p \Leftrightarrow \sim p) \Rightarrow q$

In Exercises 15–28, construct truth tables to verify that each expression is a tautology.

**15.** $\sim p \lor p$

**16.** $\sim(p \land \sim p)$

**17.** $(p \lor p) \Leftrightarrow p$

**18.** $p \Leftrightarrow (p \land p)$

**19.** $p \Rightarrow (p \lor q)$

**20.** $(p \land q) \Rightarrow p$

**21.** $(p \lor q) \Leftrightarrow (q \lor p)$

**22.** $(p \land q) \Leftrightarrow (q \land p)$

**23.** $p \land (p \lor q) \Leftrightarrow p$

**24.** $p \lor (p \land q) \Leftrightarrow p$

**25.** $(p \Rightarrow q) \Leftrightarrow (\sim p \lor q)$

**26.** $\sim(p \Rightarrow q) \Leftrightarrow (p \land \sim q)$

**27.** $\sim(p \Leftrightarrow q) \Leftrightarrow (p \Leftrightarrow \sim q)$

**28.** $(p \Leftrightarrow q) \Leftrightarrow [(p \Rightarrow q) \land (q \Rightarrow p)]$

Exercises 29–32 are properties of logic that are similar to properties of algebra. In each exercise, construct a truth table to verify that the expression is a tautology. Because each of the three variables can be either *true* or *false*, each truth table will contain eight lines.

**29.** The associative property of $\land$: $(p \land q) \land r \Leftrightarrow p \land (q \land r)$

**30.** The associative property of $\lor$: $(p \lor q) \lor r \Leftrightarrow p \lor (q \lor r)$

**31.** A distributive property: $p \lor (q \land r) \Leftrightarrow (p \lor q) \land (p \lor r)$

**32.** A distributive property: $p \land (q \lor r) \Leftrightarrow (p \land q) \lor (p \land r)$

In Exercises 33–42, write the negation of each statement.

**33.** It is hot and humid.

**34.** The moon is full and the kids are wild.

**35.** We are late and my watch is broken.

**36.** I will watch the late movie or I will go to bed.

**37.** I heard a voice or I am imagining things.

**38.** Tim is fat and forty.

**39.** If you don't buy me a ring, then I'll break the engagement. (*Hint*: Use the result of Exercise 26 to find the negation of an *If . . . then . . .* statement.)

**40.** If wishes were horses, then beggars would ride.

**41.** If it tastes good, then it's fattening.

**42.** If you make a mess, then clean it up or say you're sorry.

## 11.3 VALID ARGUMENTS

An **argument** in logic is a set of statements, called **premises**, and a **conclusion** that may or may not be a consequence of the premises. The argument is **valid** if it is impossible that the premises be true and the conclusion false. For example, the two premises

*If I add 3 and 4, then my answer is 7.*
*I add 3 and 4.*

and the conclusion

*My answer is 7.*

provide a valid argument, because it is not possible for the premises to be true, and not get an answer of seven.

The form of this example of valid argument is

$$p \Rightarrow q$$
$$\underline{\phantom{xxx}p\phantom{xxx}}$$
$$q$$

where the horizontal line separates the premises from the conclusion. The two premises together imply the conclusion. Using the notation of Section 11.1, this form of valid argument may be written

$$[(p \Rightarrow q) \wedge p] \Rightarrow q$$

The next example will show this expression to be a tautology.

**EXAMPLE 1**   Show that $[(p \Rightarrow q) \wedge p] \Rightarrow q$ is a tautology.

Solution   Use a truth table to show that

$$[(p \Rightarrow q) \wedge p] \Rightarrow q$$

is a tautology. Several columns are needed for intermediate results.

| $p$ | $q$ | $p \Rightarrow q$ | $(p \Rightarrow q) \wedge p$ | $[(p \Rightarrow q) \wedge p] \Rightarrow q$ |
|-----|-----|-------------------|------------------------------|----------------------------------------------|
| T | T | T | T | T |
| T | F | F | F | T |
| F | T | T | F | T |
| F | F | T | F | T |

Because the last column of the truth table contains only T's, the given expression is a tautology.   ∎

As another example, the two premises

*If I add 3 and 4, then my answer is 7.*
*My answer is 7.*

and the conclusion

*I added 3 and 4.*

do *not* provide a valid argument, for it is quite possible for the two premises to be true and the conclusion false: I could have added 5 and 2, for example. An argument which is not valid is called an **invalid argument**. The structure of this invalid argument is

$$[(p \Rightarrow q) \wedge q] \Rightarrow p$$

This expression is not a tautology, as the next example shows.

**EXAMPLE 2**

Show that $[(p \Rightarrow q) \wedge q] \Rightarrow p$ is not a tautology.

*Solution*

Use a truth table to show that

$$[(p \Rightarrow q) \wedge q] \Rightarrow p$$

is not a tautology by showing that the expression could be false.

| $p$ | $q$ | $p \Rightarrow q$ | $(p \Rightarrow q) \wedge q$ | $[(p \Rightarrow q) \wedge q] \Rightarrow p$ |
|---|---|---|---|---|
| T | T | T | T | T |
| T | F | F | F | T |
| F | T | T | T | F |
| F | F | T | F | T |

Because the last column of the truth table contains an F, the given expression is not a tautology. ∎

The important fact of valid reasoning illustrated by the previous discussion is this:

> If the argument is valid, then the form of the argument must be a tautology.

The form of valid argument represented by $[(p \Rightarrow q) \wedge p] \Rightarrow q$ or by

$$p \Rightarrow q$$
$$\underline{p}$$
$$q$$

is known as **modus ponens**, or as **the law of detatchment**. In the law of detatchment, one premise is a conditional, and the other premise asserts that the antecedent is true. The conclusion asserts that the consequent must be true, also.

**EXAMPLE 3**

Is the following argument valid?

*If Jenny has a rash, then she has measles.*
*Jenny has a rash.*

*She has measles.*

*Solution*    Determine if the argument is valid by assigning letters to the various phrases, and seeing if its form is a tautology. Let

    $p$:  *Jenny has a rash.*
    $q$:  *She has measles.*

Then the form of the argument is

$$p \Rightarrow q$$
$$\underline{\phantom{aa} p \phantom{aa}}$$
$$q$$

This is the law of detatchment, a tautology. The argument is valid. ■

Perhaps you disagree with the first premise of the previous example, thinking "If she has a rash, she could have chicken pox, or poison ivy." You are correct; the first premise is false. Even so, the argument is valid, because validity is determined by the *form* of the argument. The previous example illustrates that it is possible for a valid argument with *false* premises to have a conclusion that is either true or false. However, in a valid argument, true premises guarantee a true conclusion. The points to remember are these:

---

If the premises of a valid argument are *true*, then the conclusion must be true, also.
If the premises of a valid argument are *false*, however, the conclusion *doesn't have to be false*!

---

Another valid argument, known as **modus tolens**, has the form

$$\boldsymbol{p \Rightarrow q}$$
$$\underline{\phantom{aa} \boldsymbol{\sim q} \phantom{aa}} \qquad \text{or} \qquad \boldsymbol{[(p \Rightarrow q) \wedge \sim q] \Rightarrow \sim p}$$
$$\boldsymbol{\sim p}$$

To show that this form of argument is valid, we could use a truth table to show that

$$[(p \Rightarrow q) \wedge \sim q] \Rightarrow \sim p$$

is a tautology. However, the next example shows an easier way.

*EXAMPLE 4*    Show that the argument represented by

$$p \Rightarrow q$$
$$\underline{\phantom{aa} \sim q \phantom{aa}}$$
$$\sim p$$

is valid.

**Solution**

You already know by the law of detatchment that the form

$$\sim q \Rightarrow \sim p$$
$$\underline{\sim q}$$
$$\sim p$$

is valid because the second premise agrees with the antecedent of the first premise, and the conclusion asserts the consequent of the first premise.

Because a conditional statement is equivalent to its contrapositive, replace the first premises with its equivalent, $p \Rightarrow q$ to get

$$p \Rightarrow q$$
$$\underline{\sim q}$$
$$\sim p$$

Thus, the given form of argument is valid. In this form of valid argument, one premise is a conditional, and the other premise is a *denial* of the consequent of the conditional. The conclusion is the denial of the antecedent. ■

**EXAMPLE 5**

Find, if possible, a valid conclusion to the premises:

> *If Sue is invited, then the party will be a success.*
> *Sue is not invited.*
> ─────────────────────────────────────────
> *?? Therefore ??*

**Solution**

Determine if a valid conclusion is possible by assigning letters to the various phrases, and seeing if a logical conclusion can be drawn. Let

$p$:   *Sue is invited.*
$q$:   *The party will be a success.*

Then the form of the argument is

$$p \Rightarrow q$$
$$\underline{\sim p}$$
$$???$$

The first premise is a conditional. The second premise does not agree with the antecedent, so the form is not modus ponens. The second premise does not deny the consequent, so the form is not modus tolens. No valid conclusion is possible. ■

Other forms of argument are valid. Two of them, together with the two we have just studied, are stated here for reference. You will be asked in the exercises to show that these new forms are valid.

**Theorem 11.4** Valid Forms of Argument.

$$p \Rightarrow q$$
$$\underline{p}$$
$$q$$

The Law of Detatchment
or Modus Ponens

$$p \Rightarrow q$$
$$\underline{\sim q}$$
$$\sim p$$

Modus Tolens

$$p \vee q \qquad\qquad p \vee q$$
$$\underline{\sim p} \qquad\qquad \underline{\sim q}$$
$$q \qquad\qquad\quad p$$

Denial of the Alternative

$$p \Rightarrow q$$
$$\underline{q \Rightarrow r}$$
$$p \Rightarrow r$$

Law of the Syllogism

## EXERCISE 11.3

In Exercises 1–50, determine the valid conclusion if one is warranted. If no valid conclusion is possible, write "no conclusion."

**1.** If I am tired, then I shouldn't drive.
I am tired.

**2.** If I am tired, then I shouldn't drive.
I should drive.

**3.** If I am tired, then I shouldn't drive.
I am not tired.

**4.** If I am tired, then I shouldn't drive.
I should not drive.

**5.** If I come home early, then I'll take a nap.
I will come home early.

**6.** If I come home early, then I'll take a nap.
I will take a nap.

**7.** If I come home early, then I'll take a nap.
I will not come home early.

**8.** If I come home early, then I'll take a nap.
I will not take a nap.

**9.** If my dog is scratching, then she has fleas.
My dog is scratching.

**10.** If my dog has fleas, then she should be bathed.
My dog should be bathed.

**11.** If my dog is scratching, then she has fleas.
My dog has fleas.

**12.** If my dog has fleas, then she should be bathed.
My dog has fleas.

**13.** If today is Friday, then tomorrow is a holiday.
Today is Friday.

**14.** If today is Friday, then tomorrow is a holiday.
Tomorrow is a holiday.

**15.** If today is Friday, then tomorrow is a holiday.
Today is not Friday.

**16.** If today is Friday, then tomorrow is a holiday.
Tomorrow is not a holiday.

**17.** If it is raining, then it is cloudy.
It is not cloudy.

**18.** If it is raining, then it is cloudy.
It is cloudy.

**19.** If it is raining, then it is cloudy.
It is raining.

**20.** If it is raining, then it is cloudy.
It is not raining.

**21.** If it is cloudy, then it is raining.
It is cloudy.

**22.** If it is cloudy, then it is raining.
It is raining.

**23.** If it is cloudy, then it is raining.
It is not cloudy.

**24.** If it is hot and humid, then it will rain.
It is humid.

**25.** If it is hot and humid, then it will rain.
It is hot and humid.

**26.** If wishes were horses, then beggars would ride.
Beggars do not ride.

**27.** If it is hot and humid, then it will rain.
It will rain.

**28.** If it is hot and humid, then it will rain.
It will not rain.

**29.** If I study hard, then I will pass geometry.
I study very hard.

**30.** If I study hard, then I will pass geometry.
I don't study hard, at all!

**31.** If red is green, then purple is not yellow.
Red is green.

**32.** If red is green, then purple is not yellow.
Purple is yellow.

**33.** I will pass geometry if I study hard.
I study very hard. (*Hint*: What is the *if* clause?)

**34.** The mail is late if it is snowing.
It is not snowing.

**35.** The mail is late or it is not snowing.
It is snowing.

**36.** I am not sick, or I have measles.
I am sick.

**37.** I am sick, or I don't have measles.
I have measles.

**38.** It is not Friday, or it is payday.
Unfortunately, it is not payday.

**39.** It's Friday, or it's not payday.
It is payday.

**40.** Geometry is fun, or I'm a monkey's uncle.
I'm not a monkey's uncle.

**41.** It is not cloudy, or it is raining.
It is not raining.

**42.** Red is green, or purple is not yellow.
Red is not green.

**43.** Red is purple, or green is yellow.
Green is not yellow.

**44.** If my parakeet gets lonesome, then I'll buy him a friend.
If I buy him a friend, then he will sing.

**45.** If I don't get a diamond, then our engagement is off.
If our engagement is off, then my mother will be annoyed.

**46.** If red is green, then purple is yellow.
If purple is yellow, then I'll adjust the TV.

**47.** If Frank quits the game, then he will take his bat with him.

If he takes his bat with him, the game is over.

**49.** If the door is unlocked, then someone can steal the jewels.

If security is good, then no one can steal the jewels.

(*Hint*: Replace the second premise with its contrapositive.)

**51.** Show that denial of the alternative is a valid form of argument by showing that $[(p \lor q) \land \sim p] \Rightarrow q$ is a tautology.

**48.** If Jim gets home early, then he will go to a movie.

If Jim gets home early, then he will not watch TV.

**50.** If the temperatures have been below zero, then it is safe to skate.

If I don't invite my friends, then it is not safe to skate.

**52.** Show that the law of the syllogism is a valid form of argument by showing that $[(p \Rightarrow q) \land (q \Rightarrow r)] \Rightarrow (p \Rightarrow r)$ is a tautology.

# 11.4 EULER CIRCLES

There is another way of looking at the various forms of logical argument. Consider the example *If an animal is a collie, then it is a dog*. Without changing its meaning, we can write that sentence as *All collies are dogs*. In this form, the sentence relates two sets of animals—animals that are collies, and animals that are dogs. The sentence indicates that all animals in the first collection are also in the second; that is, the first group is a *subset* of the second. A picture of this relationship of inclusion is provided by the **Euler circles** of Figure 11.1. Euler circles are named after the eighteenth-century Swiss mathematician Leonhard Euler.

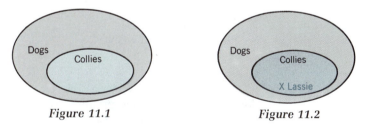

*Figure 11.1*          *Figure 11.2*

Another statement, *Lassie is a collie*, can be represented in the diagram by placing an X (to represent Lassie) within the circle of collies. This new diagram appears in Figure 11.2. Because X represents an element of the smaller set, and the smaller set is contained in the larger set, X must also be a member of the larger collection. Thus, the two premises

*All collies are dogs.*
*Lassie is a collie.*

support the valid conclusion

*Lassie is a dog.*

Find, if possible, a valid conclusion to the statements

All friendly animals are cuddly.
My teddy bear is cuddly.

Solution

The statement *All friendly animals are cuddly* indicates the relationship between two sets of animals diagrammed in Figure 11.3.

Where shall you place an X to represent the teddy bear? Either position indicated in Figure 11.3 conveys the meaning of the last premise, *My teddy bear is cuddly*, because each position appears within the boundary of the larger circle. May you conclude that "My teddy bear *is* friendly?" No, for the X need not be within the smaller circle. May you conclude that "My

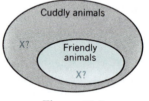

**Figure 11.3**

teddy bear is *not* friendly?" No, for the X need not be outside of the smaller circle. No conclusion is possible. ∎

Find, if possible, a valid conclusion to the statements

All hippopotomi are expensive.
No turtles are expensive
Tillie is a turtle.

Solution

Diagram the first sentence as one circle contained within another, as in Figure 11.4. The second premise says that the set of turtles and the set of expensive things are completely separate; there is no overlap at all. Represent the second premise as a circle completely separate from the others.

**Figure 11.4**

Finally, diagram the third premise by placing an X (representing Tillie) inside the circle representing the set of turtles.

The conclusion to be drawn from the three statements is "Tillie is not a hippopotomus." ■

## EXERCISE 11.4

In Exercises 1–27, draw Euler's circles to help find a valid conclusion if one is possible. If no conclusion is possible, indicate "no conclusion." Exercises 23–27 are inspired by Lewis Carroll (C. L. Dodgson), a nineteenth-century mathematics teacher best known as the author of *Alice in Wonderland* and *Through the Looking-Glass*.

1. All who study hard will pass geometry.
   Jean studies hard.

2. All who study hard will pass geometry.
   Euclid will pass geometry.

3. All lazy students will flunk.
   Bob is lazy.

4. All lazy students will flunk.
   Jeanne will flunk.

5. No lazy students will succeed.
   Jim is lazy.

6. No lazy students will succeed.
   George will succeed.

7. No successful students are lazy.
   All my friends are lazy.

8. No successful students are lazy.
   All my friends are successful.

9. All foreign cars are good.
   All foreign cars are expensive.

10. All foreign cars are good.
    All good cars are expensive.

11. All foreign cars are good.
    All expensive cars are good.

12. Mathematics teachers are dull.
    Dull people are rich.

13. No ostriches can vote.
    All voters are intelligent.

14. No ostriches can vote.
    All who are literate can vote.

15. All alligators are introverts.
    No introverts are good teachers.

16. All alligators are introverts.
    All introverts can be helped.

17. All aardvarks can whistle.
    No animal that can whistle is a good companion.

18. No reptiles make good pets.
    All kittens make good pets.

19. No pterodactyls are now living.
    All pterodactyls can fly.

20. All petty officers are enlisted sailors.
    All gunners' mates are petty officers.

21. All cats are mammals.
    All mammals are warm blooded.
    No reptiles are warm blooded.

22. No incompetents are inspiring teachers.
    All biologists are inspiring teachers.
    All babies are incompetent.

23. All my sisters have colds.
    No one can sing who has a cold.

24. No bald creature needs a hairbrush.
    No lizards have hair.

25. Babies are illogical.
    Nobody is despised who can manage a crocodile.
    Illogical people are despised.

26. All unripe fruit is unwholesome.
    All these apples are wholesome.
    No fruit grown in the shade is ripe.

**27.** Trustworthy people keep their promises.
Wine drinkers are very communicative.
All men who keep their promises are honest.
All pawnbrokers are wine drinkers.
One can always trust a very communicative person.

# 11.5 NON-EUCLIDEAN GEOMETRIES (OPTIONAL)

Euclid's geometry is based on 10 postulates he thought were self-evident truths that described the physical world. From these postulates he proved over 450 theorems, building this structure of theorems so carefully, logically, and clearly, that his *Elements* has been studied for centuries as the model of deductive reasoning. Euclid's geometry, however, was not just an exercise in proving theorems; Euclid and many of his students through the centuries never questioned that the content of geometry provided *true* information about the real world. For example, the theorem

*The sum of the measures of the angles of a triangle is 180°.*

was not just the logical result of previous theorems; it was a truth so imbedded in the fabric of the universe that no conceivable triangle could behave differently. For centuries the development of geometry continued, but no one thought to question the truth of Euclid's postulates or his theorems.

In the centuries that followed, study of Euclid's work revealed several flaws and subtle gaps in his reasoning. One perceived flaw in the structure was this: Some of Euclid's postulates seemed to be more self-evidently true than others. For example, the truth of the postulate *Two points determine a line* seemed obvious, but even Euclid felt that his fifth postulate, known as the **parallel postulate**, sounded more like a theorem. Although Euclid's version was worded differently, we have this equivalent form:

> *The Parallel Postulate.* If a point does not lie on a given line, then through that point *one* line and *only one* line can be drawn that is parallel to the given line.

As well as we can know the thoughts of one who lived so long ago, it appears that Euclid did not doubt that the parallel postulate was true, but was uncomfortable including it as a postulate if it were really a theorem. He did not introduce the postulate until he had proved as many theorems as he could without it.

Concern over the parallel postulate spans 2300 years. Later mathematicians felt similar discomfort, thinking that if the postulate *could* be proved, it *should* be proved. Many tried to prove it using Euclid's other nine postulates. They had no success. Others tried discarding the postulate and replacing it with another. These postulates, however, seemed no more self-evident than Euclid's, and the discomfort remained. In 1759, the mathematician Jean-le-Rond d'Alembert (1717–1783) called the difficulty of the parallel postulate "the scandal of geometry."

It was not until the nineteenth century that real progress was made. It was then that mathematicians discovered that the parallel postulate was not a theorem. They did so by replacing the parallel postulate with its logical negation, and then applying a simple logical tautology. We begin by showing how Euclid's parallel postulate can be negated, and then we continue by discussing a tautology which shows that the postulate is not a theorem.

The postulate makes two statements: first, that through a point not on a line, a parallel line *exists*, and second, that it is *unique* (there is *just one* such line.) Symbolically, if we let

*p*:   Through a point not on a line, a parallel line *exists*.

and

*q*:   Through a point not on a line, the parallel line is *unique*.

then the parallel postulate can be written as $p \wedge q$.

We can use DeMorgan's law to find $\sim(p \wedge q)$, the negation of the parallel postulate:

$$\sim(p \wedge q) \Leftrightarrow \sim p \vee \sim q$$

Thus, there are two ways to negate the parallel postulate: either use the negation of *p*, or use the negation of *q*. The first, $\sim p$, denies that parallel lines exist:

*Through a point not on a line, no line can be drawn parallel to the given line.*

The other, $\sim q$, denies that the parallel line is unique:

*Through a point not on a line, several lines can be drawn parallel to the given line.*

The mathematicians Carl Friedrich Gauss (1777–1855), Nikolai Ivanovich Lobachevsky (1793–1856), and John Bolyai (1802–1860) chose the second form of the negation. By assuming that several parallel lines are possible, they proved many strange theorems. The most surprising is this:

- The sum of the measures of the angles of a triangle is *less than* 180°, and as the area of the triangle increases, the sum of the angles decreases.

Because the measure of the angles decreases as the triangle grows larger, two triangles cannot be similar unless they are also congruent! Gauss was so surprised by this result that he and two friends tried an experiment. They stationed themselves on three neighboring mountain peaks, and measured the angles of the triangle they formed. Because their calculations were within the expected errors of measurement, the experiment proved nothing. Perhaps, they thought, even that triangle was not big enough.

The mathematician Bernhard Riemann (1826–1866) chose the first method of negating Euclid's parallel postulate: he assumed that *no* parallel lines exist. This assumption also led to strange results, including these:

- Two points do not necessarily determine one line; sometimes two points determine more than one line.
- All the perpendiculars to a line meet at one point.
- The sum of the measures of the angles of a triangle is *greater than* 180°, and as the area of the triangle increases, the sum of the angles increases.

Although these strange results contradict theorems of Euclid, they helped prove that Euclid's parallel postulate is not a theorem. The reasoning uses this tautology:

$$(p \wedge \sim q) \Rightarrow \sim(p \Rightarrow q)$$

In the exercises, you will be asked to prove that this is a tautology.

If we let

$p$:  the conjunction of Euclid's nine postulates (not including the parallel postulate)

and

$q$:  Euclid's parallel postulate,

then the tautology $(p \wedge \sim q) \Rightarrow \sim(p \Rightarrow q)$ can be understood in this way: If it is possible to find an interpretation of $p \wedge \sim q$, that is, a real-world situation in which Euclid's nine postulates hold along with the negation of his parallel postulate, then it is *not* true that those nine postulates can be used to prove the parallel postulate.

Strange as the results of these nineteenth-century mathematicians appear, both Lobachevsky's and Riemann's geometry do have understandable interpretations. We will consider an interpretation of Riemannian geometry.

On the surface of a sphere, the shortest distance between two points $A$ and $B$ (see Figure 11.5a) is an arc of a **great circle**, a circle that girdles the sphere as the equator does the earth. The center of a great circle must be the center of the sphere; the circle passing through point $P$ in Figure 11.5a is *not* a great circle. If we consider a *plane* to be the *surface of a sphere*, and a *line* to be a *great circle*, then it is true that in the plane *no* lines can be parallel, for two great circles will always intersect at the endpoints of a diameter. See Figure 11.5b. The two points A and B in Figure 11.5a determine only *one* line (great circle), but the points C and D in Figure 11.5b determine *many* lines. The two lines x and y in Figure 11.5c are each perpendicular to line z, yet intersect at point P. The angle between lines x and y is given to be 30°. The sum of the measures of the angles of the triangle they form is greater than 180°, because 90° + 90° + 30° is 210°. If the triangle were quite small compared to the sphere, however, the sum of the measures of the angles would be so slightly greater than 180° as to be unnoticeable.

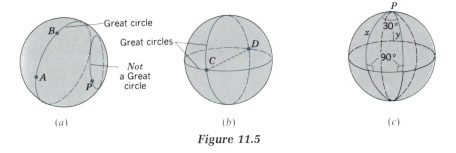

**Figure 11.5**

Most surprising and unexpected is the fact that although these geometries (Euclid's, Lobachevsky's, and Riemann's) contradict each other, each alone is logical and internally consistent. It is ironic that attempts to improve the structure of Euclidean geometry and clean up a few of its flaws have led to the discovery that Euclid's geometry does *not* have any special claim to truth, and almost certainly is *not* a description of the real world. Rather, Euclid's is only one of many geometries. Any geometry based on axioms that contradict the axioms of Euclid is called a **non-Euclidean geometry**. Lobachevsky's is a non-Euclidean geometry known as **hyperbolic geometry**, and Riemann's is known as **elliptic geometry**. Mathematicians have created many more. Which geometry, then, is the *true* description of the real world? Perhaps none of them! Each, however, has proven to be useful: navigation on the earth's surface and Einstein's theory of relativity both depend on variations of Riemann's non-Euclidean geometry.

## EXERCISE 11.5

In Exercises 1–8, investigate the properties of Riemannian geometry by sketching with a marker on the surface of a spherical balloon. Write a brief comment on each of the following.

**1.** Draw a line (great circle) on the balloon, and mark a point P not on that line. Can you draw a line (another great circle) through P that is parallel to the first line? Why or why not?

**2.** Do two points determine a single line segment (an arc of a great circle)?

**3.** Is it possible to have isosceles triangles in Riemannian geometry?

**4.** Are the base angles of an isosceles triangle congruent?

**5.** Draw a large triangle and a small triangle, and use a protractor to determine the measures of the angles. Are the totals different for each triangle?

**6.** What is the greatest total of the measures of the angles of a triangle that you can find?

**7.** Can you draw an equilateral triangle that is not equiangular?

**8.** Does the Pythagorean theorem still hold?

**9.** Many mathematicians contributed to the development of non-Euclidean geometry. Write a paragraph or two on the contribution of one of them, Gerolamo Saccheri (1667–1733).

**10.** Use a truth table to show that $(p \wedge \sim q) \Rightarrow \sim(p \Rightarrow q)$ is a tautology.

## CHAPTER ELEVEN REVIEW EXERCISES

In Exercises 1–6, write the logical negation of the statement.

**1.** Today is Wednesday.

**2.** $8 \neq 9$

**3.** I am going to a store or to a movie.

**4.** Jim is shopping and Jill is bored.

**5.** If I take a nap, then I will go out tonight.

**6.** If you have the time, call me.

In Exercises 7–10, assume that $p$ and $q$ are both false. Then determine the truth value of each expression.

**7.** $\sim p$

**8.** $\sim q \wedge p$

**9.** $p \vee \sim q$

**10.** $p \Rightarrow q$

In Exercises 11–14, assume that $p$ is false and that $q$ is true. Then determine the truth value of each expression.

**11.** $p \Rightarrow \sim q$

**12.** $\sim q \Rightarrow p$

**13.** $\sim p \vee \sim q$

**14.** $\sim p \wedge \sim q$

In Exercises 15–20, write each symbolic expression as an English sentence. Assume that the variables represent the following statements.

$p$:  Fred is fat.
$q$:  Ted is tired.
$r$:  Sam is sloppy.

**15.** $\sim(p \wedge r)$

**16.** $\sim p \vee \sim q$

**17.** $(p \vee \sim q) \wedge r$

**18.** $p \vee (\sim q \wedge r)$

**19.** $(p \wedge q) \Rightarrow r$

**20.** $p \wedge (q \Rightarrow r)$

In Exercises 21–24, write each English sentence as a symbolic expression. Assume the following:

$p$:  Bill will be transferred to Omaha.
$q$:  Bill will be fired.
$r$:  Bill's company is losing money.

**21.** Bill will be fired or transferred to Omaha.

**22.** If Bill's company is losing money, then Bill will be fired.

**23.** If Bill will be fired, then Bill will not be transferred to Omaha.

**24.** If Bill's company is not losing money and Bill will not be fired, then Bill will not be transferred to Omaha.

In Exercises 25–28, write the converse, inverse, and contrapositive of each statement.

**25.** If you apply the brakes, the car will stop.

**26.** If you do not pay for that, you are shoplifting.

**27.** If a triangle is isosceles, then it has two congruent angles.

**28.** You would understand me better if you would just listen.

In Exercises 29–36, construct a truth table for each expression.

**29.** $p \wedge q$

**30.** $p \vee q$

**31.** $p \Rightarrow q$

**32.** $p \Leftrightarrow q$

**33.** $\sim p \wedge (p \vee q)$

**34.** $p \vee (\sim p \wedge \sim q)$

**35.** $p \vee (p \Leftrightarrow q)$

**36.** $(p \Leftrightarrow p) \Rightarrow q$

In Exercises 37–40, construct truth tables to verify that each expression is a tautology.

**37.** $(p \wedge q) \Leftrightarrow (q \wedge p)$

**38.** $(p \wedge q) \Rightarrow p$

**39.** $(p \wedge q) \Rightarrow (p \vee q)$

**40.** $(p \Leftrightarrow q) \Leftrightarrow [(p \Rightarrow q) \wedge (q \Rightarrow p)]$

In Exercises 41–54, determine the valid conclusion if one is warranted. If no valid conclusion is possible, write "no conclusion."

**41.** If I come home early, then I'll watch a movie.
I will come home early.

**42.** If I come home early, then I'll watch a movie.
I will watch a movie.

**43.** If it is snowing, then it is cold.
It is not cold.

**44.** If it is snowing, then it is cold.
It is cold.

**45.** If I work hard, then I will succeed.
I do not work hard.

**46.** If I work hard, then I will succeed.
I will not succeed.

**47.** It is not snowing, or we will ski.
It is snowing.

**48.** It is not snowing, or we will ski.
We will not ski.

**49.** If triangles are congruent, then corresponding parts are congruent.
These triangles are congruent.

**50.** If two lines are parallel, then they lie in the same plane.
These lines lie in the same plane.

**51.** If two angles of a triangle are congruent, then two sides are congruent.
     If two sides are congruent, then their measures are equal.

**52.** If two angles are complements of the same angle, then they are congruent.
     If two angles are congruent, then their measures are equal.

**53.** If I don't pay the fine, then I will be suspended from school.
     If I pay the fine, then I will not go to jail.

**54.** If the barn burns, then the fire trucks will come.
     If the cow is a careless smoker, then the barn will burn.

In Exercises 55–62, use Euler circles to determine, if possible, a valid conclusion to the given premises.

**55.** Reading books is a waste of time.
     Whatever wastes time is not a worthy activity.

**56.** All aardvarks are animals.
     No apples are animals.

**57.** All beagles are hunting dogs.
     All hunting dogs are trusted friends.

**58.** All parallelograms are quadrilaterals.
     All rectangles are quadrilaterals.

**59.** All sophomores are upperclassmen.
     No freshmen are sophomores.

**60.** All sophomores are upperclassmen.
     No freshmen are upperclassmen.

**61.** No criminals are trustworthy.
     All mathematics professors are trustworthy.
     All who drive too fast are criminals.

**62.** No poisonous foods are good to eat.
     All blueberries are good to eat.
     Purple mushrooms are poisonous.

## CHAPTER ELEVEN TEST

In Exercises 1–16, classify each statement as true or false.

**1.** *Be quick about it!* is a statement.

**2.** The negation of *He and I are friends* is *He and I are enemies.*

**3.** If $p$ is true and $q$ is false, then $p \land q$ is false.

**4.** If $p$ is false and $q$ is true, then $p \Rightarrow q$ is false.

**5.** If $p$ and $q$ are both true, then $\sim p \land \sim q$ is false.

**6.** If $p$ and $q$ are both false, then $\sim p \Rightarrow q$ is false.

**7.** The converse of $p \Rightarrow q$ is $q \Rightarrow p$.

**8.** The inverse of $p \Rightarrow q$ is $\sim q \Rightarrow \sim p$.

**9.** The contrapositive of $p \Rightarrow q$ is $\sim p \Rightarrow \sim q$.

**10.** The contrapositive of $p \Rightarrow \sim q$ is $q \Rightarrow \sim p$.

**11.** The contrapositive of a true statement is necessarily true.

**12.** The converse of a false statement must be true.

**13.** The contrapositive of any statement is necessarily true.

**14.** If the premises of an argument are true and the reasoning is valid, then the conclusion must be true also.

**15.** If the conclusion of a valid argument is false, then the premises must be false.

**16.** If the conclusion of a valid argument is true, then the premises must be true.

In Exercises 17–25, answer "V" if the argument is valid, and "I" if it is invalid.

**17.** If he is a doctor, then he plays golf on Wednesdays; Jim does not play golf on Wednesdays; *Therefore*, Jim is not a doctor.

**18.** The soccer team will either lose or go on to State championships; The team will lose; *Therefore*, The team will not go on to State championships.

**19.** If Sue is a good citizen, then she will pay taxes; If Sue is not a good citizen, then Sue will not vote; *Therefore*, If Sue will vote, then Sue will pay taxes.

**20.** If a number is even, then it is divisible by 2; If a number is odd, then it is not divisible by 2; *Therefore*, If a number is even, then it is not odd.

**21.** If all roses are red and all violets are blue, then no roses are violets.

**22.** If no anteaters are redheads, then no redheads are anteaters.

**23.** If all geometry students are smart and Sally is smart, then Sally is a geometry student.

**24.** If logic is interesting and difficult subjects are interesting, then logic is difficult.

**25.** If all flying birds are feathered and no bats are feathered, then no bats are flying birds.

# 12

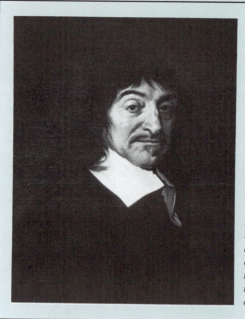

*René Descartes (1596–1650). René Descartes was a French mathematician, theologian, and philosopher, who is credited with applying algebraic methods to geometry. This branch of mathematics is now known as analytic geometry. Descartes died of pneumonia as a result of regular tutoring sessions with Queen Christina of Sweden in a cold room at 5 o'clock in the morning.*

# Analytic Geometry

The French mathematician, theologian, and philosopher René Descartes (1596–1650) is credited with joining algebra and geometry into a branch of mathematics called **analytic geometry**. At the heart of his discovery lies the **Cartesian coordinate system**, which associates pairs of real numbers (an idea of algebra) with points in a plane (an idea of geometry.) Analytic geometry allows mathematicians to study geometric figures by looking at certain equations and to understand the behavior of equations by studying geometric figures.

## 12.1 THE CARTESIAN COORDINATE SYSTEM

The Cartesian coordinate system is based on two perpendicular number lines, called the **x-axis** and the **y-axis**, which divide the plane into four

**quadrants** numbered as in Figure 12.1. These number lines intersect at a point called the **origin**, which is the zero point on each number line. The positive direction on the $x$-axis is to the right, the positive direction of the $y$-axis is upward, and the unit distance on each axis is the same.

Each point on the plane is associated with a pair of real numbers, one indicating its position left or right, and the other its position up or down. To **plot** the point associated with the pair of real numbers (2, 3), for example, we start at the origin and count 2 units to the right, and then 3 units up. See Figure 12.2. This determines point $P$, which is called the **graph** of the pair (2, 3). The pair (2, 3) gives the **coordinates** of point $P$. To plot point $Q$ with coordinates $(-3, 1)$, we start at the origin and count 3 units to the left, and then 1 unit up. Point $P$ lies in the first quadrant. Point $Q$ lies in the second quadrant. Point $R$ with coordinates $(1, -3)$ lies in the fourth quadrant.

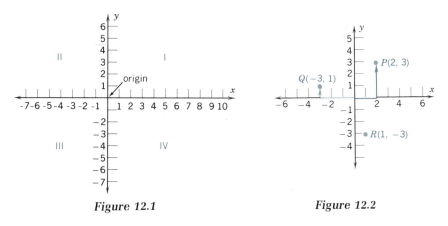

*Figure 12.1*          *Figure 12.2*

Note that the two pairs $(-3, 1)$ and $(1, -3)$ represent different points. Because the order of the numbers in each pair is important, such pairs are called **ordered pairs**. The first coordinate, $a$, in the ordered pair $(a, b)$ is called the **$x$-coordinate** or the **abscissa**, and the second coordinate, $b$, is called the **$y$-coordinate**, or the **ordinate**.

**EXAMPLE 1**     Graph the points $A(-2, -7)$, $B(0, -3)$, $C(1, -1)$, and $D(3, 3)$.

Solution     Draw a coordinate system as in Figure 12.3 shown on page 366, and plot each ordered pair.     ∎

**EXAMPLE 2**     Determine the coordinates of four points that mark the vertices of a square.

Solution     There are many solutions to this problem. The points $O(0, 0)$, $A(4, 0)$, $B(4, 4)$, and $C(0, 4)$ are the vertices of the square in Figure 12.4a. The

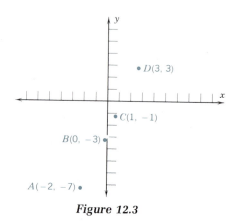

**Figure 12.3**

points $E(5, 0)$, $F(0, 5)$, $G(-5, 0)$, and $H(0, -5)$ are the vertices of the square in Figure 12.4*b*. The points $J(4, 2)$, $K(-2, 4)$, $L(-4, -2)$, and $M(2, -4)$ are the vertices of the square in Figure 12.4*c*. Can you find other answers? ■

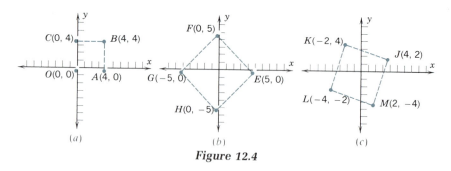

**Figure 12.4**

A line drawn on the Cartesian coordinate system is **horizontal** if, and only if, it is parallel to the *x*-axis. Because parallel lines are the same distance apart, all of the points on a horizontal line have the same *y*-coordinate. Similarly, a line is **vertical** if, and only if, it is parallel to the *y*-axis. All of the points on a vertical line have the same *x*-coordinate. In Figure 12.5, the *y*-coordinates of the points on the horizontal line $\overleftrightarrow{AB}$ are equal, and the *x*-coordinates of the points on the vertical line $\overleftrightarrow{CD}$ are equal.

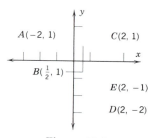

**Figure 12.5**

To find the distance between points on a horizontal line, we first subtract the x-coordinates of the points. Because a distance cannot be negative, we then find the absolute value of that difference.

---

**Definition 12.1** The **distance between two points** *on a horizontal line* is the absolute value of the difference of the x-coordinates of the two points.

---

Similarly, we can find the distance between two points on a vertical line.

---

**Definition 12.2** The **distance between two points** *on a vertical line* is the absolute value of the difference of the y-coordinates of the two points.

---

**EXAMPLE 3**

Refer to Figure 12.6. The length of horizontal line segment $\overline{AB}$ is the distance between the two points $A$ and $B$. That distance is the absolute value of the difference of the x-coordinates of points $A$ and $B$.

$$\begin{aligned}
m(\overline{AB}) &= |x\text{-coordinate of } A - x\text{-coordinate of } B| \\
&= |1 - 6| \\
&= |-5| \\
&= 5
\end{aligned}$$

Similarly, the length of the vertical line segment $\overline{CD}$ is the absolute value of the difference of the y-coordinates of points $C$ and $D$.

$$\begin{aligned}
m(\overline{CD}) &= |y\text{-coordinate of } C - y\text{-coordinate of } D| \\
&= |3 - (-2)| \\
&= |3 + 2| \\
&= |5| \\
&= 5
\end{aligned}$$

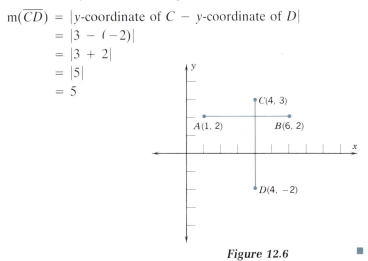

*Figure 12.6*

## THE DISTANCE FORMULA

There is a formula, based on the Pythagorean theorem, for finding the length of a line segment that is not necessarily parallel to a coordinate axis. In this formula, we use **subscript notation** to represent the coordinates of points. One point might be denoted as $P(x_1, y_1)$, read as "point $P$ with coordinates of $x$ sub 1 and $y$ sub 1." Another point might be $Q(x_2, y_2)$, and so on.

If $P(x_1, y_1)$ and $Q(x_2, y_2)$ are two points in Figure 12.7, then right triangle $PQR$ can be formed, with $R$ having coordinates $(x_2, y_1)$. Because the line segment $\overline{RQ}$ is vertical, the square of its length is $(y_2 - y_1)^2$. Because $\overline{PR}$ is horizontal, the square of its length is $(x_2 - x_1)^2$. By the Pythagorean theorem, we know that the square of the hypotenuse of right triangle $PQR$ is equal to the sum of the squares of the lengths of its two

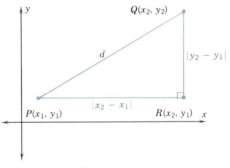

**Figure 12.7**

legs. Thus, we have

$$[m(\overline{PQ})]^2 = [m(\overline{PR})]^2 + [m(\overline{QR})]^2$$
$$d^2 = (x_2 - x_1)^2 + (y_2 - y_1)^2$$

Because equal positive numbers have equal square roots, we can take the positive square root of both sides of this equation to obtain a formula called the **distance formula**.

---

**Theorem 12.1  The Distance Formula.**    The distance $d = m(\overline{PQ})$ between points $P(x_1, y_1)$ and $Q(x_2, y_2)$ is given by

$$d = \sqrt{(x_2 - x_1)^2 + (y_2 - y_1)^2}$$

---

**EXAMPLE 4**

Find the length, $d$, of the line segment joining $P(3, -2)$ and $Q(7, 1)$.

Solution

Refer to Figure 12.8. Let $P(x_1, y_1)$ be $P(3, -2)$ and let $Q(x_2, y_2)$ be $Q(7, 1)$. That is, let

$$x_1 = 3, \quad y_1 = -2, \quad x_2 = 7, \quad \text{and} \quad y_2 = 1$$

Then, substitute these values into the distance formula, and simplify.

$$d = \sqrt{(x_2 - x_1)^2 - (y_2 - y_1)^2}$$
$$= \sqrt{(7 - 3)^2 + [1 - (-2)]^2}$$
$$= \sqrt{4^2 + 3^2}$$
$$= \sqrt{16 + 9}$$
$$= \sqrt{25}$$
$$= 5$$

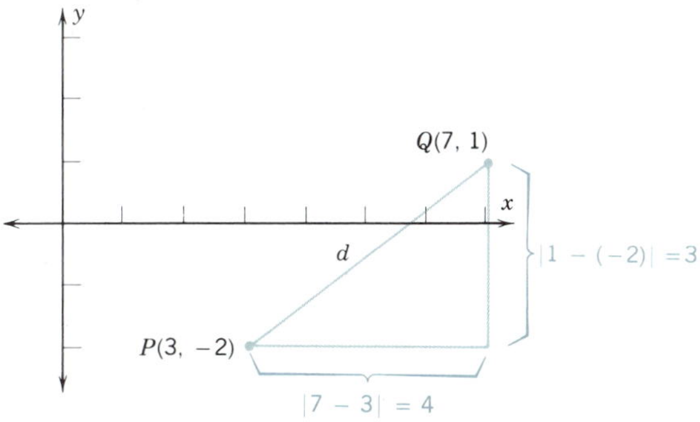

**Figure 12.8**

Thus, points $P$ and $Q$ are 5 units apart. ∎

## THE MIDPOINT FORMULA

Analytic geometry provides an easy way of finding the midpoint of a line segment drawn in a Cartesian coordinate system. We state this theorem without proof.

---

**Theorem 12.2 The Midpoint Formula.** The midpoint of the line segment with endpoints $A(x_1, y_1)$ and $B(x_2, y_2)$ has the coordinates

$$\left( \frac{x_1 + x_2}{2}, \frac{y_1 + y_2}{2} \right)$$

---

Theorem 12.2 points out that the $x$-coordinate of the midpoint of a line segment is the average of the $x$-coordinates of the endpoints of the segment, and the $y$-coordinate of the midpoint is the average of the $y$-coordinates of the endpoints.

**EXAMPLE 5**

Find the midpoint of the line segment that joins $A(2, -3)$ and $B(4, 7)$.

*Solution*

Let $A$ be the point $(x_1, y_1)$ and let $B$ be $(x_2, y_2)$. Then, we have

$$x_1 = 2, \quad y_1 = -3, \quad \text{and} \quad x_2 = 4, \quad y_2 = 7$$

Substitute these values into the formula for the coordinates of the midpoint

$$\left(\frac{x_1 + x_2}{2}, \frac{y_1 + y_2}{2}\right)$$    The coordinates of the midpoint.

$$\left(\frac{2 + 4}{2}, \frac{-3 + 7}{2}\right)$$    Substitute values for the variables.

$$\left(\frac{6}{2}, \frac{4}{2}\right)$$    Combine terms.

$$(3, 2)$$    Simplify.

Thus, the coordinates of the midpoint of the given segment are (3, 2). ■

**EXAMPLE 6**

The midpoint of the line segment joining $P(-3, 2)$ and $Q(x_2, y_2)$ is $M(1, 4)$. Find the coordinates of $Q$.

*Solution*

Let $P(x_1, y_1)$ be $P(-3, 2)$ and let $M(x_M, y_M)$ be $M(1, 4)$. That is, let

$$x_1 = -3, \quad y_1 = 2, \quad x_M = 1, \quad \text{and} \quad y_M = 4$$

To find the coordinates $x_2$ and $y_2$ of point $Q$, solve two equations, as follows.

$$x_M = \frac{x_1 + x_2}{2} \qquad \qquad y_M = \frac{y_1 + y_2}{2}$$

$$1 = \frac{-3 + x_2}{2} \qquad \qquad 4 = \frac{2 + y_2}{2}$$    Substitute values for the indicated variables.

$$2 = -3 + x_2 \qquad \qquad 8 = 2 + y_2$$    Multiply both sides by 2.

$$5 = x_2 \qquad \qquad 6 = y_2$$

The coordinates of point $Q$ are (5, 6).    ■

## EXERCISE 12.1

In Exercises 1–6, indicate the quadrant in which the graph of each ordered pair lies.

**1.** $(1, 5)$    **2.** $(-1, 5)$    **3.** $(5, -2)$    **4.** $(-2, 2)$    **5.** $(-1, -2)$    **6.** $(-2, -7)$

In Exercises 7–20, plot each point. The points form the shape of a pine tree.

**7.** $(0, 3)$       **8.** $(-1, -1)$     **9.** $(-1, 2)$      **10.** $(2, 1)$
**11.** $(0, 1)$      **12.** $(2, 0)$      **13.** $(-3, 0)$     **14.** $(1, 0)$
**15.** $(1, -1)$     **16.** $(-1, 0)$     **17.** $(-2, 1)$     **18.** $(3, 0)$
**19.** $(1, 2)$                           **20.** $(-2, 0)$

In Exercises 21–26, find the length of the line segment $\overline{AB}$, and indicate whether the line segment is vertical or horizontal.

**21.** $A(3, 5)$, $B(3, 7)$

**22.** $A(5, -4)$, $B(2, -4)$

**23.** $A(-3, 8)$, $B(3, 8)$

**24.** $A(3, 9)$, $B(3, -5)$

**25.** $A(-2, -4)$, $B(-5, -4)$

**26.** $A(-7, 8)$, $B(-3, 8)$

In Exercises 27–40, use the distance formula to find the length of the line segment $\overline{PQ}$.

**27.** $P(1, 3)$; $Q(4, 7)$

**28.** $P(2, -1)$; $Q(5, 3)$

**29.** $P(3, 3)$; $Q(3, 7)$

**30.** $P(0, -1)$; $Q(5, 11)$

**31.** $P(-5, 3)$; $Q(1, 11)$

**32.** $P(-1, -3)$; $Q(-7, 5)$

**33.** $P(0, 3)$; $Q(3, 0)$

**34.** $P(6, -2)$; $Q(2, -2)$

**35.** $P(-12, -8)$; $Q(0, 8)$

**36.** $P(0, 0)$; $Q(24, -10)$

**37.** $P\left(-\frac{1}{2}, -\frac{1}{3}\right)$; $Q\left(\frac{5}{2}, \frac{11}{3}\right)$

**38.** $P\left(-\frac{7}{3}, \frac{31}{5}\right)$; $Q\left(\frac{8}{3}, -\frac{29}{5}\right)$

**39.** $P(-2, 7)$; $Q(-3, 2)$

**40.** $P(-7, 2)$; $Q(2, -7)$

In Exercises 41–48, find the midpoint of the line segment $\overline{AB}$.

**41.** $A(6, 4)$, $B(5, 8)$

**42.** $A(-3, 6)$, $B(6, -3)$

**43.** $A(-7, 9)$, $B(7, 9)$

**44.** $A(-5, -2)$, $B(-11, -23)$

**45.** $A(14, -14)$, $B(0, 2)$

**46.** $A(-4, -21)$, $B(-16, 13)$

**47.** $A\left(\frac{1}{2}, \frac{2}{3}\right)$, $B\left(\frac{3}{2}, \frac{5}{3}\right)$

**48.** $A\left(\frac{4}{7}, -\frac{3}{5}\right)$, $B\left(-\frac{11}{7}, \frac{8}{5}\right)$

In Exercises 49–54, one endpoint $P$ and the midpoint $M$ of line segment $\overline{PQ}$ are given. Find the coordinates of the other endpoint, $Q$.

**49.** $P(2, 4)$; $M(5, 4)$

**50.** $P(3, -8)$; $M(3, -4)$

**51.** $P(-3, 2)$; $M(-2, 3)$

**52.** $P(-1, 7)$; $M(-3, -3)$

**53.** $P(-1, -4)$; $M(0, 0)$

**54.** $P(0, 0)$; $M(-5, 4)$

**55.** The endpoints of the base of an isosceles triangle are $(-1, 2)$ and $(5, 2)$. What are the possible coordinates of the vertex?

**56.** The endpoints of a leg of a right triangle are $(4, -1)$ and $(4, 3)$. Find possible coordinates for a third vertex.

**57.** Three vertices of a square are the points $(-2, -1)$, $(1, -4)$, and $(4, -1)$. What are the coordinates of the other vertex?

**58.** The endpoints of the base of an equilateral triangle are $(0, 0)$ and $(0, 2)$. What are two possible coordinates of the vertex?

**59.** Show that a triangle with vertices at $(13, -2)$, $(9, -8)$, and $(5, -2)$ is isosceles, by using the distance formula to show that two of its sides have equal length.

**60.** Show that a triangle with vertices at $(-1, 2)$, $(3, 1)$, and $(4, 5)$ is isosceles.

**61.** Show that the points $(-3, -1)$, $(3, 1)$, $(1, 7)$, and $(-5, 5)$ are the vertices of a rhombus.

**62.** Use the distance formula to find the radius of a circle with center at $(4, -2)$ and passing through $(6, 9)$.

**63.** Find the area of a square with the endpoints of a diagonal at $(3, -17)$ and $(-2, -5)$.

**64.** Find the coordinates of the two points on the $y$-axis that are $\sqrt{74}$ units from the point $(5, -3)$.

## 12.2 THE STRAIGHT LINE

Analytic geometry is a marriage of algebra and geometry. We illustrate this by showing that certain equations of algebra are related to an idea of geometry, the line. It is possible to draw a picture or a graph of an equation in two variables. The **graph of an equation** in the two variables $x$ and $y$ is the collection of all points on a Cartesian coordinate system with coordinates $(x, y)$ that satisfy the equation.

*EXAMPLE 1*

Graph the equation $3x + 2y = 6$.

*Solution*

Pick some arbitrary values for either $x$ or $y$, substitute those values in the equation, and solve for the other variable. For example, if $x = 4$, you can find $y$ as follows:

$$3x + 2y = 6$$
$$3(4) + 2y = 6 \qquad \text{Substitute 4 for } x.$$
$$12 + 2y = 6 \qquad \text{Simplify.}$$
$$2y = -6 \qquad \text{Add } -12 \text{ to both sides.}$$
$$y = -3 \qquad \text{Divide both sides by 2.}$$

Thus, one ordered pair that satisfies the equation is $(4, -3)$.
   If $y = 2$, you have

$$3x + 2y = 6$$
$$3x + 2(2) = 6 \qquad \text{Substitute 2 for } y.$$
$$3x + 4 = 6 \qquad \text{Simplify.}$$
$$3x = 2 \qquad \text{Add } -4 \text{ to both sides.}$$
$$x = \frac{2}{3} \qquad \text{Divide both sides by 3.}$$

Thus, another ordered pair that satisfies the equation is $(\frac{2}{3}, 2)$.
   The ordered pairs $(4, -3)$ and $(\frac{2}{3}, 2)$, and others that satisfy the equation $3x + 2y = 6$ are shown in the table of values in Figure 12.9. Plot each of these ordered pairs on a Cartesian coordinate system, as in the figure. Note that the resulting points appear to lie on a line. It can be shown that all

points with coordinates satisfying the equation $3x + 2y = 6$ do, in fact, lie on a line. Draw the line that joins the points. This line is the graph of the equation.

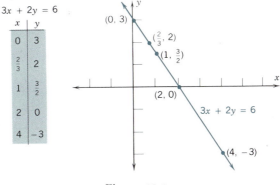

**Figure 12.9**

Because the line in Example 1 intersects the y-axis at the point $(0, 3)$, the number 3 is called the **y-intercept** of the graph of the line. Likewise, 2 is called the **x-intercept**.

Any equation that can be written in the form $Ax + By = C$ is called a **linear equation**. It can be shown that all linear equations have graphs that are straight lines.

**EXAMPLE 2**

Use the x- and y-intercepts to graph the equation $2x + 5y = 10$.

Solution

Begin by finding the x- and y-intercepts. To find the y-intercept, substitute 0 for x and solve for y:

$$2x + 5y = 10$$
$$2(\mathbf{0}) + 5y = 10 \qquad \text{Substitute 0 for } x.$$
$$5y = 10 \qquad \text{Simplify.}$$
$$y = 2 \qquad \text{Divide both sides by 5.}$$

Because the y-intercept is 2, the line intersects the y-axis at the point $(0, 2)$. To find the x-intercept, substitute 0 for y and solve for x:

$$2x + 5y = 10$$
$$2x + 5(\mathbf{0}) = 10 \qquad \text{Substitute 0 for } y.$$
$$2x = 10 \qquad \text{Simplify.}$$
$$x = 5 \qquad \text{Divide both sides by 2.}$$

Because the x-intercept is 5, the line intersects the x-axis at the point $(5, 0)$.

Although these two points, if calculated correctly, are sufficient to draw the line, it is a good idea to find and plot a third ordered pair to act as a check. To find the coordinates of a third point, substitute any convenient number, such as 1, for $y$ and solve for $x$:

$$2x + 5y = 10$$
$$2x + 5(\mathbf{1}) = 10 \qquad \text{Substitute 1 for } y.$$
$$2x + 5 = 10 \qquad \text{Simplify.}$$
$$2x = 5 \qquad \text{Add } -5 \text{ to both sides.}$$
$$x = \frac{5}{2} \qquad \text{Divide both sides by 2.}$$

Thus, the line passes through the point $(\frac{5}{2}, 1)$.

A table of values and the graph of the equation $2x + 5y = 10$ are shown in Figure 12.10.

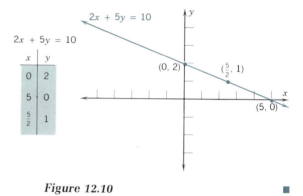

Figure 12.10

## SLOPE OF A NONVERTICAL LINE

A measure of the steepness or tilt of a line is called the **slope** of the line.

**Definition 12.3** The **slope**, $m$, **of a nonvertical line** passing through points $P(x_1, y_1)$ and $Q(x_2, y_2)$ is given by the formula

$$m = \frac{y_2 - y_1}{x_2 - x_1}$$

Note that the slope of a line passing through two points is the ratio of the difference of the $y$-coordinates of the points to the difference of the $x$-coordinates, *taken in the same order.*

Since the $x$-coordinates of a vertical line are all equal, the difference of the $x$-coordinates is zero. Because a fraction with a denominator of zero is meaningless, a vertical line has no defined slope.

**EXAMPLE 3**

Find the slope of the line passing through the points $P(-3, 2)$ and $Q(2, -5)$.

Solution

Let $P(x_1, y_1) = P(-3, 2)$ and let $Q(x_2, y_2) = Q(2, -5)$. Then you have

$$x_1 = -3, \quad y_1 = 2, \quad \text{and} \quad x_2 = 2, \quad y_2 = -5$$

Substitute these values into the formula for the slope of a line and simplify, as follows:

$$m = \frac{y_2 - y_1}{x_2 - x_1} \qquad \text{The formula for the slope of a line.}$$

$$m = \frac{(-5) - 2}{2 - (-3)} \qquad \text{Substitute values for the variables.}$$

$$m = \frac{-7}{5} \qquad \text{Simplify.}$$

The slope of the line passing through $P(-3, 2)$ and $Q(2, -5)$ is $-\frac{7}{5}$. Note that if you let $(x_1, y_1) = (2, -5)$ and $(x_2, y_2) = (-3, 2)$, then the slope is still $-\frac{7}{5}$. ∎

With the definition of slope and the following postulate, it is possible to write the equation of a nonvertical line that passes through two given points.

---

**Postulate 12.1** All segments of a nonvertical line have the same slope.

---

**EXAMPLE 4**

Find the equation of the line that passes through the points $P(2, -5)$ and $Q(4, 3)$.

Solution

Refer to Figure 12.11 and pick any third point on the line and call it $Z(x, y)$. Points $P$, $Q$, and $Z$ are all on the required line, so by Postulate 12.1, the slope determined by $P$ and $Q$ must equal the slope determined by $P$ and $Z$. Calculate the two slopes, set them equal to each other, and simplify. The equation that results is the equation of the required line. Proceed as follows.

First, use the coordinates of $P$ and $Q$ to determine the slope of the line.

$$\text{Slope of } \overleftrightarrow{PQ} = \frac{3 - (-5)}{4 - 2}$$

$$= \frac{8}{2}$$

$$= 4 \qquad \text{Simplify.}$$

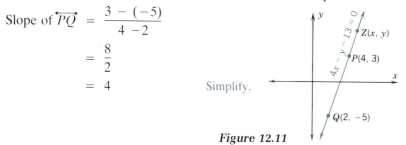

*Figure 12.11*

Then, use the coordinates of $P$ and $Z$ to determine the slope of the line.

$$\text{Slope of } \overleftrightarrow{PZ} = \frac{y - 3}{x - 4}$$

Finally, set these two quantities equal to each other and simplify.

$$\text{Slope of } \overleftrightarrow{PQ} = \text{Slope of } \overline{PZ}$$

$$4 = \frac{y - 3}{x - 4}$$

| | | |
|---|---|---|
| $4(x - 4) = y - 3$ | Multiply both sides by $x - 4$. |
| $4x - 16 = y - 3$ | Remove parentheses. |
| $4x - y = 13$ | Add $-y$ and 16 to both sides. |

The equation of the required line is $4x - y = 13$.    ■

## POINT-SLOPE FORM OF THE EQUATION OF A LINE

Suppose we want to write the equation of a line that passes through some point $P(x_1, y_1)$ and has a slope of $m$. Following the lead of Example 4, we pick another point $Z(x, y)$ that also lies on the line, and use $P$ and $Z$ to find the slope of the line.

$$\text{Slope of } \overleftrightarrow{PZ} = \frac{y - y_1}{x - x_1}$$

Because this slope must also equal $m$, we can set the two expressions equal to each other.

$$\frac{y - y_1}{x - x_1} = m$$

We then multiply both sides of this equation by $(x - x_1)$ to obtain **point-slope form** of the equation of a line.

---

**Theorem 12.3 Point-Slope Form of the Equation of a Line.** The equation of a line passing through the point $P(x_1, y_1)$ and having a slope of $m$ is given by

$$y - y_1 = m(x - x_1)$$

---

**EXAMPLE 5**

Find the equation of the line passing through the point $P(-2, 5)$ and having a slope of 3.

*Solution*

Substitute 3 for $m$ and the coordinates of $P$ for $x_1$ and $y_1$, and simplify.

$$y - y_1 = m(x - x_1)$$     The point-slope form of the equation of a line.
$$y - 5 = 3[x - (-2)]$$     Substitute values for $m$, $x_1$, and $y_1$.
$$y - 5 = 3(x + 2)$$     Simplify.
$$y - 5 = 3x + 6$$     Remove parentheses.
$$-11 = 3x - y$$     Add $-y$ and $-6$ to both sides.

The equation of the line is $3x - y = -11$.     ∎

### SLOPE-INTERCEPT FORM OF THE EQUATION OF A LINE

If the $y$-intercept of the line $l$ with slope $m$ shown in Figure 12.12 is $b$, the line intersects the $y$-axis at $P(0, b)$. We can write the equation of this line by substituting 0 for $x_1$ and $b$ for $y_1$ in the point-slope form of the equation, and simplifying:

$$y - y_1 = m(x - x_1)$$
$$y - b = m(x - 0)$$
$$y - b = mx$$     Simplify and remove parentheses.
$$y = mx + b$$     Add $b$ to both sides.

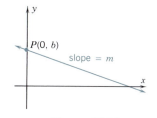

*Figure 12.12*

Because the equation $y = mx + b$ displays both the slope and the $y$-intercept of the line, it is called the **slope-intercept form** of the equation of the line.

---

**Theorem 12.4 Slope-Intercept Form of the Equation of a Line.** The equation of the line with slope of $m$ and $y$-intercept $b$ is

$$y = mx + b$$

---

*EXAMPLE 6*

Use the slope-intercept form to write the equation of a line with a slope of 4 and a $y$-intercept of $-3$.

Solution

You are given that $m = 4$ and $b = -3$. Substitute 4 for $m$ and $-3$ for $b$ in the slope-intercept form of the equation of a line, and simplify:

$$y = mx + b$$
$$y = 4x + (-3)$$
$$y = 4x - 3$$

The desired equation is $y = 4x - 3$, or (by rearranging terms) $4x - y = 3$. ∎

The slopes of lines are related to the geometric concepts of parallel and perpendicular lines. It can be shown that lines with the same slope never meet, and are therefore parallel. We will accept that fact and state it as a postulate.

> **Postulate 12.2** Nonvertical lines are parallel if, and only if, they have the same slope. Vertical lines are parallel.

**EXAMPLE 7**

Is the line passing through $P(-2, 5)$ and $Q(3, 9)$ parallel to the line passing through the origin and the point $R(5, 4)$?

Solution

Find the slope of each line:

$$\text{Slope of } \overleftrightarrow{PQ} = \frac{y_2 - y_1}{x_2 - x_1} = \frac{9 - 5}{3 - (-2)} = \frac{4}{5}$$

$$\text{Slope of } \overleftrightarrow{OR} = \frac{y_2 - y_1}{x_2 - x_1} = \frac{4 - 0}{5 - 0} = \frac{4}{5}$$

Because the slopes are equal, the lines are parallel. ∎

If the product of two numbers is 1, the numbers are said to be **reciprocals** of each other. If the product of the two numbers is $-1$, the numbers are **negative reciprocals** of each other. It can be shown that the two nonvertical lines are perpendicular if, and only if, their slopes are negative reciprocals. We accept this fact as a postulate.

> **Postulate 12.3** Nonvertical lines are perpendicular if, and only if, their slopes are negative reciprocals.
> A vertical line is perpendicular to a horizontal line.

You can verify that the lines with slopes of $\frac{2}{3}$ and $-\frac{3}{2}$, for example, are perpendicular by showing that their slopes are negative reciprocals:

$$\frac{2}{3}\left(-\frac{3}{2}\right) = -\frac{2}{3} \cdot \frac{3}{2} = -\frac{6}{6} = -1$$

**EXAMPLE 8**

Two lines intersect at point $P(3, -4)$. One passes through the origin $O$ and the other passes through point $Q(9, 4)$. Are the lines perpendicular?

*Solution*

Find the slope of lines $\overleftrightarrow{OP}$ and $\overleftrightarrow{PQ}$:

$$\text{Slope of } \overleftrightarrow{OP} = \frac{y_2 - y_1}{x_2 - x_1} = \frac{-4 - 0}{3 - 0} = -\frac{4}{3}$$

$$\text{Slope of } \overleftrightarrow{PQ} = \frac{y_2 - y_1}{x_2 - x_1} = \frac{4 - (-4)}{9 - 3} = \frac{8}{6} = \frac{4}{3}$$

Because the slopes are not negative reciprocals, the lines are not perpendicular. Incidentally, because the slopes are not equal, the lines are not parallel, either.

**EXAMPLE 9**

Show that the lines represented by $4x + 8y = 10$ and $2x = 12 - 4y$ are parallel.

*Solution*

Solve each equation for $y$ and observe that the lines are distinct and that their slopes are equal.

$$
\begin{array}{c|c}
\begin{aligned}
4x + 8y &= 10 \\
8y &= -4x + 10 \\
y &= -\frac{1}{2}x + \frac{5}{4}
\end{aligned}
&
\begin{aligned}
2x &= 12 - 4y \\
4y &= -2x + 12 \\
y &= -\frac{1}{2}x + 3
\end{aligned}
\end{array}
$$

Compare each equation with the slope-intercept form of the equation of a line, $y = mx + b$. The $y$-intercepts of the lines are $\frac{5}{4}$ and 3. Because these are different, the lines are distinct. The slope of each line is $-\frac{1}{2}$. Because these slopes are equal, the lines are parallel. ∎

**EXAMPLE 10**

Show that the lines represented by $2x + 4y = 7$ and $4x - 2y = 21$ are perpendicular.

*Solution*

Solve each equation for $y$ to write the equations in slope-intercept form. Then determine their slopes. If the slopes are negative reciprocals, then the lines are perpendicular.

$$
\begin{array}{c|c}
\begin{aligned}
2x + 4y &= 7 \\
4y &= -2x + 7 \\
y &= -\frac{1}{2}x + \frac{7}{4}
\end{aligned}
&
\begin{aligned}
4x - 2y &= 21 \\
-2y &= -4x + 21 \\
y &= 2x - \frac{21}{2}
\end{aligned}
\end{array}
$$

Compare each equation with the slope-intercept form of the equation of a line, $y = mx + b$. Because the slopes of the lines are the negative reciprocals $-\frac{1}{2}$ and 2, the lines are perpendicular. ∎

*EXAMPLE 11*    Use the slope-intercept form to write the equation of the line passing through the point $P(-2, 5)$ and parallel to the line $y = 8x - 7$.

Solution    Compare the given equation with the slope-intercept form of the equation of a line,

$$y = 8x - 7$$
$$y = mx + b$$

and determine that slope, $m$, is 8. The line represented by the desired equation must also have a slope of 8 because it is to be parallel to a line with slope of 8. Substitute $-2$ for $x$, 5 for $y$, and 8 for $m$ in the slope-intercept form and solve for $b$:

$$y = mx + b$$
$$5 = 8(-2) + b$$
$$5 = -16 + b$$
$$b = 21$$

Because $m = 8$ and $b = 21$, the desired equation is $y = 8x + 21$.    ■

*EXAMPLE 12*    Use the point-slope form to write the equation of the line passing through the point $P(-2, 5)$ and perpendicular to the line $y = 8x - 2$.

Solution    The slope of the given line is 8. Thus, the slope of the desired line must be $-\frac{1}{8}$, the negative reciprocal of 8. Substitute $-2$ for $x_1$, 5 for $y_1$, and $-\frac{1}{8}$ for $m$ into the point-slope form. Then simplify.

$$y - y_1 = m(x - x_1)$$
$$y - 5 = -\frac{1}{8}[x - (-2)]$$
$$y - 5 = -\frac{1}{8}(x + 2)$$

| | |
|---|---|
| $8y - 40 = -(x + 2)$ | Multiply both sides by 8. |
| $8y - 40 = -x - 2$ | Remove parentheses. |
| $x + 8y = 38$ | Add $x$ and 40 to both sides. |

The equation of the line is $x + 8y = 38$.    ■

## EXERCISE 12.2

In Exercises 1–14, graph each linear equation.

**1.** $y = x - 2$

**2.** $y = 5 + x$

**3.** $y = x$

**4.** $y = -x$

**5.** $y = 2x + 1$

**6.** $y = 4 - 2x$

**7.** $2x + 4y = 8$

**8.** $2x - 3y = 6$

**9.** $x - 5y = 10$

**10.** $3x + 7y = 21$

**11.** $-3x + 2y = 9$

**12.** $-2x + 4y = 3$

**13.** $\dfrac{1}{3}x + y = 2$

**14.** $2x - \dfrac{2}{3}y = 5$

In Exercises 15–20, graph each equation. To write each equation in a useful form, remove parentheses, combine terms, and so on.

**15.** $3(x + 2) + 2(y - 3) = 6$

**16.** $2(x - 1) + 4(y + 2) = -6$

**17.** $3x + 2 + 5(y - 3) = 6x$

**18.** $2(x - 2) + 4y = -(x - 1)$

**19.** $\dfrac{x - 3}{2} + \dfrac{y + 1}{5} = 0$

**20.** $\dfrac{2x + 1}{3} = 3(y - 2)$

In Exercises 21–30, find the slope of the line passing through the given points. If the slope is not defined, so indicate.

**21.** $P(3, 2);\ Q(2, 3)$

**22.** $P(-2, 0);\ Q(-3, 3)$

**23.** $P(5, 2);\ Q(7, 2)$

**24.** $P(-3, 7);\ Q(4, -7)$

**25.** $P(5, -5);\ Q(5, 3)$

**26.** $P(5, -1);\ Q(0, -6)$

**27.** $P(7, -1);\ Q(-3, 5)$

**28.** $P(-3, -2);\ Q(11, 5)$

**29.** $P\left(\dfrac{3}{2}, 0\right);\ Q\left(\dfrac{1}{2}, \dfrac{5}{4}\right)$

**30.** $P\left(\dfrac{3}{5}, -\dfrac{1}{3}\right);\ Q\left(-\dfrac{2}{5}, \dfrac{5}{3}\right)$

In Exercises 31–54, find the equation of each line.

**31.** through $(2, 3)$ and $(3, 4)$

**32.** through $(0, -1)$ and $(2, 3)$

**33.** through $(-1, 3)$ and $(5, 9)$

**34.** through $(3, 0)$ and $(0, -3)$

**35.** through $(8, 8)$ and $(-8, -8)$

**36.** through $(5, 3)$ and $(3, -2)$

**37.** through $\left(-\dfrac{1}{2}, \dfrac{2}{3}\right)$ and $\left(-\dfrac{3}{2}, \dfrac{5}{3}\right)$

**38.** through $\left(\dfrac{3}{5}, \dfrac{5}{3}\right)$ and $\left(-\dfrac{2}{5}, \dfrac{11}{3}\right)$

**39.** through $(2, 3)$ with a slope $3$

**40.** through $(3, 2)$ with a slope $-3$

**41.** through $(0, -1)$ with a slope $4$

**42.** through $(-2, -3)$ with a slope $-4$

**43.** through $(0, 0)$ with a slope $0$

**44.** through $(0, 0)$ with a slope $-1$

**45.** through $\left(-\dfrac{1}{2}, 0\right)$ with a slope $-\dfrac{1}{2}$

**46.** through $\left(0, \dfrac{2}{7}\right)$ with a slope $-7$

**47.** with slope $3$ and $y$-intercept $2$

**48.** with slope $-2$ and $y$-intercept $3$

**49.** with slope $-5$ and $y$-intercept $7$

**50.** with slope $7$ and $y$-intercept $-5$

**51.** with slope $\dfrac{3}{5}$ and $y$-intercept $5$

**52.** with slope $-\dfrac{5}{3}$ and $y$-intercept $-9$

**53.** with slope 0 and *y*-intercept $-\dfrac{1}{2}$

**54.** with slope $-\dfrac{1}{2}$ and *y*-intercept 0

In Exercises 55–60, determine if the line through points *P* and *Q* is parallel to or perpendicular to the line passing through $R(2, -4)$ and $S(-4\ 8)$. If it is neither parallel nor perpendicular, so indicate.

**55.** $P(3, 2)$; $Q(4, 2)$

**56.** $P(8, 5)$; $Q(6, 4)$

**57.** $P(-2, 1)$; $Q(6, 5)$

**58.** $P(3, 4)$; $Q(-3, -5)$

**59.** $P(5, 4)$; $Q(6, 6)$

**60.** $P(4, -9)$; $Q(-2, 3)$

In Exercises 61–66, write the equation of the line that passes through the given point and is parallel to the given line.

**61.** $P(1, 2)$; $y = 3x + 2$

**62.** $P(2, 1)$; $y = -2x + 3$

**63.** $P(-1, 3)$; $y = 5x + 3$

**64.** $P(-2, -2)$; $y = -2x - 2$

**65.** $P(0, 0)$; $y = \dfrac{5}{7}x + \dfrac{7}{11}$

**66.** $P(1, -1)$; $y = -\dfrac{7}{9}x + \dfrac{21}{31}$

In Exercises 67–72, write the equation of the line that passes through the given point and is perpendicular to the given line.

**67.** $P(1, 2)$; $y = 3x + 2$

**68.** $P(2, 1)$; $y = -2x + 3$

**69.** $P(-1, 3)$; $y = 5x + 3$

**70.** $P(-2, -2)$; $y = -2x - 2$

**71.** $P(0, 0)$; $y = \dfrac{5}{7}x + \dfrac{7}{11}$

**72.** $P(1, -1)$; $y = -\dfrac{7}{9}x + \dfrac{21}{31}$

In Exercises 73–80, determine whether the graphs determined by each pair of lines are parallel, perpendicular, or neither.

**73.** $y = 3x - \dfrac{1}{3}$; $y = 3x + 2$

**74.** $y = -\dfrac{1}{2}x - \dfrac{1}{3}$; $y = -2x + 3$

**75.** $y = \dfrac{5}{3}x - \dfrac{3}{5}$; $y = \dfrac{3}{5}x + \dfrac{5}{3}$

**76.** $y = -\dfrac{1}{2}x - 3$; $y = 2x + 3$

**77.** $y = 5x - 5$; $y = -\dfrac{1}{5}x + 5$

**78.** $2y = 3x - 3$; $y = 3x + 5$

**79.** $2x + 7y = 8$; $4x + 14y = 3$

**80.** $2x - 5y = 7$; $5x + 2y = 3$

**81.** Find the equation of the line with slope 2 and passing through the midpoint of the line segment joining $P(3, 7)$ and $Q(-5, 9)$.

**82.** Find the equation of the line with slope $-\dfrac{3}{5}$ and passing through the midpoint of the line segment joining $P(-4, 5)$ and $Q(-6, 7)$.

**83.** Find the equation of the line parallel to the line $y = 3x + 2$ and passing through the point $P(3, -1)$.

**84.** Find the equation of the line perpendicular to the line $y = 3x + 2$ and passing through the point $P(3, -1)$.

# 12.3 FINDING POINTS OF INTERSECTION OF LINES

We have considered linear equations such as $x + y = 3$ that contain two variables. Because the coordinates of any point on its graph satisfy the equation, there are many ordered pairs $(x, y)$ that will satisfy this equation. Likewise, there are many ordered pairs $(x, y)$ that will satisfy an equation such as $3x - y = 1$.

Although there are many ordered pairs that satisfy each of these equations, only the numbers $x = 1$ and $y = 2$ satisfy both equations at the same time. The pair of equations

$$\begin{cases} x + y = 3 \\ 3x - y = 1 \end{cases}$$

is called a **system of equations**. Because the ordered pair $(x, y) = (1, 2)$ satisfies both equations simultaneously, it is called a **simultaneous solution**, or just a **solution of the system of equations**. The simultaneous solution of the system of equations is the pair of coordinates of the point where their graphs intersect. To see this, we graph both equations on a single set of coordinate axes as in Figure 12.13.

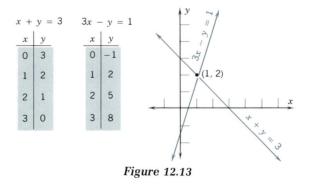

*Figure 12.13*

The point at which the lines intersect is $(1, 2)$. To verify that $x = 1$ and $y = 2$ satisfy both equations, we substitute 1 for $x$ and 2 for $y$ in each equation and verify that the pair $(1, 2)$ satisfies each equation.

$$
\begin{array}{c|c}
x + y = 3 & 3x - y = 1 \\
1 + 2 \stackrel{?}{=} 3 & 3(1) - 2 \stackrel{?}{=} 1 \\
3 = 3 & 3 - 2 \stackrel{?}{=} 1 \\
& 1 = 1
\end{array}
$$

## THE SUBSTITUTION METHOD

The graphing method for solving equations does not always provide exact solutions. For example, if the solution to a system of equations were $x = \frac{11}{97}$ and $y = \frac{13}{97}$, the graphs of the equations would not intersect at a point where we could read the solutions exactly. Fortunately, other methods exist that will determine the points of intersection exactly. We now consider one called the **substitution method**.

To solve the system

$$
\begin{cases}
y = 3x - 2 \\
2x + y = 8
\end{cases}
$$

by the substitution method, we first note the first equation, $y = 3x - 2$. Because $y = 3x - 2$, we can substitute $3x - 2$ for $y$ in the second equation $2x + y = 8$ to get

$$2x + y = 8$$
$$2x + (3x - 2) = 8$$

The resulting equation can be solved for its single variable, $x$.

$$
\begin{array}{ll}
2x + (3x - 2) = 8 & \\
2x + 3x - 2 = 8 & \text{Remove parentheses.} \\
5x - 2 = 8 & \text{Combine like terms.} \\
5x = 10 & \text{Add 2 to both sides.} \\
x = 2 & \text{Divide both sides by 5.}
\end{array}
$$

We can find the value of $y$ by substituting 2 for $x$ in either equation of the given system. Because it is already solved for $y$, it is easiest to substitute in the first equation, $y = 3x - 2$.

$$
\begin{aligned}
y &= 3x - 2 \\
y &= 3(2) - 2 \\
&= 6 - 2 \\
&= 4
\end{aligned}
$$

The solution to the given system is $x = 2$ and $y = 4$, or just $(2, 4)$.

*Check:*

| $y = 3x - 2$ | $2x + y = 8$ |
|---|---|
| $4 \overset{?}{=} 3(2) - 2$ | $2(2) + 4 \overset{?}{=} 8$ |
| $4 \overset{?}{=} 6 - 2$ | $4 + 4 \overset{?}{=} 8$ |
| $4 = 4$ | $8 = 8$ |

Because the pair of numbers $x = 2$ and $y = 4$ is a solution, the two lines represented by the equations of the given system intersect at the point $(2, 4)$.

**EXAMPLE 1**

Use the substitution method to solve the system $\begin{cases} 2x + 3y = 5 \\ 3x + 2y = 0 \end{cases}$.

*Solution*

Solve one of the equations, say the second equation, for $x$ and proceed as follows:

$$\begin{cases} 2x + 3y = 5 \\ 3x + 2y = 0 \to 3x = -2y \end{cases}$$

$$x = -\frac{2}{3}y$$

| | |
|---|---|
| $2x + 3y = 5$ | The first equation of the system. |
| $2\left(-\dfrac{2}{3}y\right) + 3y = 5$ | Substitute $-\dfrac{2}{3}y$ for $x$. |
| $-\dfrac{4}{3}y + 3y = 5$ | Remove parentheses. |
| $3\left(-\dfrac{4}{3}y\right) + 3(3y) = 3(5)$ | Multiply both sides by 3. |
| $-4y + 9y = 15$ | Remove parentheses. |
| $5y = 15$ | Combine like terms. |
| $y = 3$ | Divide both sides by 5. |

Find $x$ by substituting 3 for $y$ in the equation $x = -\dfrac{2}{3}y$.

$$\begin{aligned} x &= -\frac{2}{3}y \\ &= -\frac{2}{3}(3) \\ &= -2 \end{aligned}$$

The solution to this system is $(-2, 3)$. Check this solution in each equation. ∎

## THE ADDITION METHOD

Another method used to find the point of intersection of the lines represented by a system of equations is called the **addition method**. To use the addition method to solve the system

$$\begin{cases} x + y = 8 \\ x - y = -2 \end{cases}$$

we add the left-hand sides of the equations and the right-hand sides of the equations to eliminate the variable $y$. The resulting equation can then be solved for $x$.

$$\begin{array}{rcr} x + y & = & 8 \\ x - y & = & -2 \\ \hline 2x & = & 6 \end{array}$$

We can now solve the equation $2x = 6$ for $x$.

$$2x = 6$$
$$x = 3 \qquad \text{Divide both sides by 2.}$$

We find the value of $y$ by substituting 3 for $x$ in either of the equations of the system and solving the resulting equation for $y$. For example, substituting 3 for $x$ in the equation $x + y = 8$ and solving for $y$ gives

$$x + y = 8$$
$$3 + y = 8$$
$$y = 5 \qquad \text{Add } -3 \text{ to both sides.}$$

Thus the solution of this system of equations is $(3, 5)$. Check the solution by verifying that the pair $(3, 5)$ satisfies each equation of the system.

*EXAMPLE 2*

Use the addition method to solve the system $\begin{cases} 2x + 3y = 5 \\ 3x + 2y = 0 \end{cases}$.

Solution

In this system each equation must be adjusted so that one of the variables will be eliminated when the equations are added. To eliminate the $x$ variable, for example, multiply the first equation by 3 and the second equation by $-2$. This gives the system

$$\begin{cases} 6x + 9y = 15 \\ -6x - 4y = 0 \end{cases}$$

When the equations are added, the terms involving the variable $x$ drop out, and you obtain the following equation, which can be solved for $y$:

$$5y = 15$$
$$y = 3$$

To determine the corresponding value of $x$, substitute 3 for $y$ in either of the original equations, as you did in Example 1. The solution of the system is $x = -2$, $y = 3$. The two lines represented by the equations intersect at the point $(-2, 3)$. ∎

## EXERCISE 12.3

In Exercises 1–6, determine whether the line graphs of the two given equations intersect at the given point. Do so by checking if the given ordered pair is a solution of both equations.

**1.** $(1, 1)$; $\begin{cases} x + y = 2 \\ 2x - y = 1 \end{cases}$

**2.** $(1, 3)$; $\begin{cases} 2x + y = 5 \\ 3x - y = 0 \end{cases}$

**3.** $(3, 2)$; $\begin{cases} 2x + y = 4 \\ x + y = 1 \end{cases}$

**4.** $(-2, 4)$; $\begin{cases} 2x + 2y = 4 \\ x + 3y = 10 \end{cases}$

**5.** $(55, 39)$; $\begin{cases} 2x - 3y = -7 \\ 4x - 5y = 25 \end{cases}$

**6.** $(2, 3)$; $\begin{cases} 3x - 2y = 0 \\ 5x - 3y = -1 \end{cases}$

In Exercises 7–8, solve each system of equations by graphing. Write each answer as an ordered pair.

**7.** $\begin{cases} x + y = 7 \\ x - y = 3 \end{cases}$

**8.** $\begin{cases} y = 2x + 8 \\ y = -x - 1 \end{cases}$

In Exercises 9–14, use the substitution method to solve each system of equations. Write each answer as an ordered pair.

**9.** $\begin{cases} x = y + 5 \\ x + 2y = -1 \end{cases}$

**10.** $\begin{cases} y = 2x - 3 \\ 3x + 2y = 2 \end{cases}$

**11.** $\begin{cases} x = 3y + 9 \\ x + 2y = 9 \end{cases}$

**12.** $\begin{cases} x = 2y - 3 \\ 3x - y = 7 \end{cases}$

**13.** $\begin{cases} x = y - 15 \\ y = 2x + 12 \end{cases}$

**14.** $\begin{cases} y = 3x \\ x + 2y = 0 \end{cases}$

In Exercises 15–20, use the addition method to solve each system of equations. Write each answer as an ordered pair.

**15.** $\begin{cases} x + 3y = 2 \\ x - 3y = 0 \end{cases}$

**16.** $\begin{cases} 2x + 3y = 7 \\ 2x - 3y = 1 \end{cases}$

**17.** $\begin{cases} x - y = -7 \\ 3x + 5y = -13 \end{cases}$

**18.** $\begin{cases} 5x - 3y = 11 \\ 3x + y = 1 \end{cases}$

**19.** $\begin{cases} 7x - 8y = 0 \\ 7x + 3y = 0 \end{cases}$

**20.** $\begin{cases} 5x + 7y = 12 \\ 3x - 9y = -6 \end{cases}$

## 12.4 PROVING GEOMETRIC THEOREMS

Many geometric theorems can also be proved by the methods of analytic geometry. We have proved these theorems before by other methods, but now give different proofs using the ideas and methods of analytic geometry.

> **Theorem 4.11** The diagonals of a rectangle are congruent.

*Proof:* Because the axes of the Cartesian coordinate system are perpendicular, we can place a rectangle into the corner of the first quadrant, with the coordinates of the vertices as indicated in Figure 12.14. By the distance formula, we have

$$m(\overline{BD}) = \sqrt{(0 - a)^2 + (b - 0)^2} = \sqrt{a^2 + b^2}$$
$$m(\overline{AC}) = \sqrt{(0 - a)^2 + (0 - b)^2} = \sqrt{a^2 + b^2}$$

Hence, $\overline{AC} \cong \overline{BD}$, and the diagonals of a rectangle are congruent.

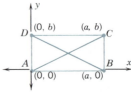

**Figure 12.14**

> **Theorem 2.4** The perpendicular bisector of the base of an isosceles triangle passes through the vertex of the triangle.

*Proof:* Since $\triangle ABC$ is isosceles, it is true that $m(\overline{AC}) = m(\overline{BC})$. If we assign the coordinates $(0, 0)$ to point $A$, $(a, 0)$ to point $B$, and $(x, y)$ to point $C$, as in Figure 12.15, we have

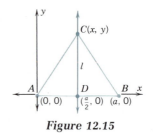

**Figure 12.15**

$$[m(\overline{AC})]^2 = (x - 0)^2 + (y - 0)^2 = x^2 + y^2$$
$$[m(\overline{BC})]^2 = (x - a)^2 + (y - 0)^2 = a^2 - 2ax + x^2 + y^2$$

Since $m(\overline{AC}) = m(\overline{BC})$, then $[m(\overline{AC})]^2 = [m(\overline{BC})]^2$ and

$$x^2 + y^2 = a^2 - 2ax + x^2 + y^2$$

$$0 = a^2 - 2ax \qquad \text{Add } -x^2 \text{ and } -y^2 \text{ to both sides.}$$

$$0 = a - 2x \qquad \text{Since } a \neq 0, \text{ divide both sides by } a.$$

$$x = \frac{a}{2}$$

and the coordinates of $C$ can be written as $\left(\dfrac{a}{2}, y\right)$.

We now consider the perpendicular bisector, $l$, of $\overline{AB}$. Since $l$ is a bisector, it must pass through $D$, the midpoint of $\overline{AB}$. The coordinates of $D$ are $\left(\dfrac{a}{2}, 0\right)$. Because line $l$ is perpendicular to the base of the triangle, it is parallel to the $y$-axis, and therefore all points with an $x$-coordinate of $\dfrac{a}{2}$ must lie on line $l$. Since $C$ has an $x$-coordinate of $\dfrac{a}{2}$, $C$ must lie on line $l$. $\qquad \square$

---

**Theorem 4.2** Opposite sides of a parallelogram are congruent.

---

*Proof*: Let $ABCD$ be a parallelogram placed in a coordinate system as in Figure 12.16. Since opposite sides of a parallelogram are parallel, we are justified in using the same $y$-coordinates for points $D$ and $C$. Since $\overline{AD} \| \overline{BC}$, we know that the slope of $\overline{AD}$ is equal to the slope of $\overline{BC}$.

**slope of $\overline{AD}$ = slope of $\overline{BC}$**

$$\frac{c - 0}{b - 0} = \frac{c - 0}{d - a}$$

$$\frac{c}{b} = \frac{c}{d - a}$$

**Figure 12.16**

We multiply both sides by $b(d - a)$ to get

$$bc = (d - a)c$$

and then divide both sides by $c$ to get

(1) $\qquad\qquad b = d - a$

Since $\overline{AB}$ and $\overline{DC}$ are horizontal line segments, we have

$$m(\overline{DC}) = d - b \qquad \text{and} \quad m(\overline{AB}) = a - 0$$
$$= d - (d - a) \qquad\qquad\qquad = a$$
$$= a$$

Hence, $m(\overline{DC}) = m(\overline{AB})$, and we have $\overline{AB} \cong \overline{DC}$.

To prove that $\overline{AD} \cong \overline{BC}$, we use the distance formula, and a substitution involving Equation (1), above:

$$m(\overline{AD}) = \sqrt{(b - 0)^2 + (c - 0)^2}$$
$$= \sqrt{b^2 + c^2}$$
$$m(\overline{BC}) = \sqrt{(d - a)^2 + (c - 0)^2}$$
$$= \sqrt{b^2 + c^2} \qquad\qquad \text{Using Equation (1), substitute } b$$
$$\text{for } d - a.$$

Thus, $m(\overline{AD}) = m(\overline{BC})$, and we have $\overline{AD} \cong \overline{BC}$. $\qquad\qquad$ □

---

**Theorem 4.14** The diagonals of a rhombus are perpendicular.

---

*Proof:* Place the rhombus $ABCD$ in a coordinate system as in Figure 12.17. Because the figure is a rhombus, we know that $\overline{AB} \cong \overline{BC} \cong \overline{CD} \cong \overline{AD}$. Because a rhombus is also a parallelogram, $\overline{BC} \| \overline{AD}$, and we are justified in using the same $y$-coordinate for points $B$ and $C$.

Since $\overline{BC} \cong \overline{AD}$, we know that $d - b = a$, and therefore that

$$(2) \qquad\qquad d = a + b.$$

Since $\overline{AB} \cong \overline{AD}$, we know that $b^2 + c^2 = a^2$, and therefore that

$$(3) \qquad\qquad c^2 = a^2 - b^2$$

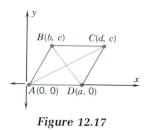

**Figure 12.17**

To prove that the diagonals $\overline{DB}$ and $\overline{AC}$ are perpendicular, we first find their slopes, and then prove that their slopes are negative reciprocals.

$$\text{Slope of } \overline{DB} = \frac{c - 0}{b - a} \qquad \text{Slope of } \overline{AC} = \frac{c - 0}{d - 0}$$

$$= \frac{c}{b - a} \qquad\qquad\qquad = \frac{c}{d}$$

To show that these slopes are negative reciprocals, we show that their product is $-1$. The product of the slopes is

$$\frac{c}{b - a} \cdot \frac{c}{d} = -\frac{c}{a - b} \cdot \frac{c}{d} \qquad -(a - b) = b - a.$$

$$= -\frac{c}{a - b} \cdot \frac{c}{a + b} \qquad \text{Use Equation (2) and substitute } a + b \text{ for } d.$$

$$= -\frac{c^2}{a^2 - b^2} \qquad \text{Multiply the fractions.}$$

$$= -\frac{c^2}{c^2} \qquad \text{Use Equation (3) and substitute } c^2 \text{ for } a^2 - b^2.$$

$$= -1 \qquad \text{Simplify.}$$

Because the product of the slopes of the diagonals is $-1$, the two diagonals are perpendicular. □

---

**Theorem 9.8** The medians of a triangle intersect at the same point.

---

*Proof:* To prove that the three medians of a triangle intersect at one point, we will show that the point of intersection of two of the medians also lies on the third median. To find that point of intersection, we will find the equations of the three medians, and solve two of them simultaneously.

We begin by placing triangle $ABC$ in a coordinate system as in Figure 12.18. We then use the midpoint formula to find the coordinates of $P$, $Q$ and $R$, the midpoints of sides $\overline{AB}$, $\overline{BC}$, and $\overline{AC}$, respectively.

$$x\text{-coordinate } P = \frac{0 + a}{2} = \frac{a}{2} \qquad y\text{-coordinate of } P = 0$$

$$x\text{-coordinate of } Q = \frac{a + b}{2} \qquad y\text{-coordinate of } Q = \frac{c + 0}{2} = \frac{c}{2}$$

and

$$x\text{-coordinate of } R = \frac{0 + b}{2} = \frac{b}{2} \qquad y\text{-coordinate of } R = \frac{c + 0}{2} = \frac{c}{2}$$

Thus,

the points $A(0, 0)$ and $Q\left(\dfrac{a + b}{2}, \dfrac{c}{2}\right)$ determine the median $\overline{AQ}$,

the points $B(a, 0)$ and $R\left(\dfrac{b}{2}, \dfrac{c}{2}\right)$ determine the median $\overline{BR}$,

and

the points $C(b, c)$ and $P\left(\dfrac{a}{2}, 0\right)$ determine the median $\overline{CP}$.

To find the equation of the median $\overline{AQ}$, we must first determine its slope.

$$
\text{Slope of } \overline{AQ} = \frac{\dfrac{c}{2} - 0}{\dfrac{a + b}{2} - 0}
$$

$$
= \frac{c}{2} \cdot \frac{2}{a + b}
$$

$$
= \frac{c}{a + b}
$$

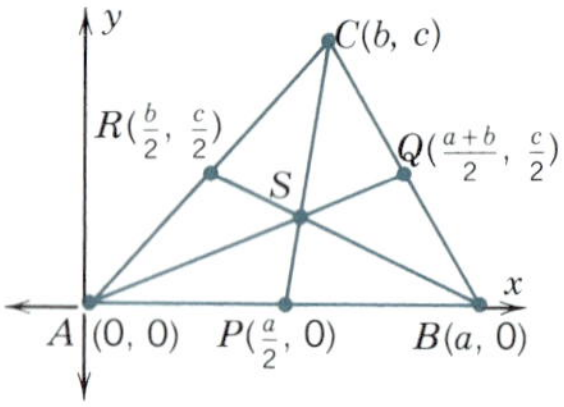

**Figure 12.18**

To find the equation of $\overline{AQ}$, we use the point-slope form of the equation of the line and let the point $(x_1, y_1)$ be the point $A(0, 0)$:

$$
y - y_1 = m(x - x_1)
$$

$$
y - 0 = \frac{c}{a + b}(x - 0) \qquad \text{Substitute 0 for } x_1 \text{ and } y_1, \text{ and } \frac{c}{a + b} \text{ for } m.
$$

We simplify to get

(4)         $y = \dfrac{c}{a + b}\, x$        The equation of $\overline{AQ}$

You will be asked in the exercises to show that the slope of $\overline{BR}$ is $\dfrac{c}{b - 2a}$, and that its equation is

(5)         $y = \dfrac{c}{b - 2a}(x - a)$        The equation of $\overline{BR}$

You will also be asked to show that the slope of $\overline{CP}$ is $\dfrac{2c}{2b - a}$, and that is equation is

(6)         $y = \dfrac{2c}{2b - a}\left(x - \dfrac{a}{2}\right)$        The equation of $\overline{CP}$

To find the point of intersection of the medians we use the method of substitution to solve Equations (4) and (5) simultaneously. Because the right-hand sides of these equations are both equal to $y$, they are equal to each other:

$$\frac{c}{a+b}x = \frac{c}{b-2a}(x-a)$$

$$(a+b)(b-2a)\frac{c}{a+b}x = \frac{c}{b-2a}(x-a)(a+b)(b-2a)$$ 
    Multiply both sides by $(a+b)(b-2a)$.

$$(b-2a)cx = c(x-a)(a+b)$$

$$bcx - 2acx = c(ax + bx - a^2 - ab)$$

$$bcx - 2acx = cax + cbx - ca^2 - cab$$ 
    Remove parentheses.

$$-2acx = cax - ca^2 - cab$$ 
    Add $-bcx$ to both sides.

$$-3acx = -ca^2 - cab$$ 
    Add $-acx$ to both sides.

$$-3acx = -ca(a+b)$$ 
    Factor.

$$x = \frac{a+b}{3}$$ 
    Divide both sides by $-3ac$.

To determine the $y$-coordinate of the point of intersection, substitute $\frac{a+b}{3}$ for $x$ in Equation (4) and simplify.

$$y = \frac{c}{a+b}x$$

$$y = \frac{c}{a+b} \cdot \frac{a+b}{3}$$

$$y = \frac{c}{3}$$

Thus, the point of intersection of the medians $\overline{AQ}$ and $\overline{BR}$ is $\left(\frac{a+b}{3}, \frac{c}{3}\right)$.

To show that this point of intersection also lies on the third median, we show that its coordinates satisfy the equation of the third median (Equation 6).

$$y = \frac{2c}{2b-a}\left(x - \frac{a}{2}\right)$$ 
    The equation of $\overline{CP}$.

$$\frac{c}{3} \overset{?}{=} \frac{2c}{2b-a}\left(\frac{a+b}{3} - \frac{a}{2}\right)$$ 
    Substitute $\frac{a+b}{3}$ for $x$ and $\frac{c}{3}$ for $y$.

$$\frac{c}{3} \overset{?}{=} \frac{2c}{2b-a}\left(\frac{2(a+b)}{6} - \frac{3a}{6}\right)$$ 
    Write as fractions with the lowest common denominator of 6.

$$\frac{c}{3} \overset{?}{=} \frac{2c}{2b - a} \left( \frac{2(a + b) - 3a}{6} \right) \qquad \text{Combine the fractions.}$$

$$\frac{c}{3} \overset{?}{=} \frac{2c}{2b - a} \left( \frac{2b - a}{6} \right) \qquad \text{Remove parentheses and combine terms.}$$

$$\frac{c}{3} = \frac{c}{3} \qquad \text{Simplify.}$$

Because the coordinates of the point of intersection satisfy the equations of all three medians, we have proven that the medians intersect at one point. □

## EXERCISE 12.4

In Exercises 1–14, use the methods of analytic geometry to prove each statement.

1. The diagonals of an isosceles trapezoid are congruent.
2. The diagonals of a parallelogram bisect each other.
3. The segments joining the midpoints of adjacent sides of any quadrilateral form a parallelogram.
4. The segments joining the midpoints of opposite sides of any quadrilateral bisect each other.
5. A median from the vertex of an isosceles triangle to the base is perpendicular to the base.
6. The point of intersection of the medians of a triangle is two-thirds of the distance from each vertex to the opposite side. (*Hint*: Use Figure 12.18 and the coordinates established in the analytic proof of Theorem 9.8)
7. A line segment joining the midpoints of two sides of a triangle is parallel to the third side.
8. The measure of a line segment joining the midpoints of two sides of a triangle is one-half the measure of the third side.
9. Two lines perpendicular to a third line are parallel to each other.
10. The line segments joining the midpoints of the sides of an equilateral triangle form an equilateral triangle.
11. If the diagonals of a parallelogram are congruent, the parallelogram is a rectangle.
12. If the diagonals of a trapezoid are congruent, the trapezoid is isosceles.
13. If the diagonals of a rhombus are congruent, the rhombus is a square.
14. If the diagonals of a parallelogram are perpendicular, the parallelogram is a rhombus.
15. Find the area of the triangle formed by the $x$-axis and the lines $y = 3x$ and $9x + 2y = 45$.
16. Find the area of the triangle formed by the $x$-axis, the $y$-axis, and the line $4x + 9y = 36$.
17. Find the area of the parallelogram formed by the $x$-axis and the lines $y = 8$, $y = 4x$ and $y = 4x - 8$.
18. Find the area of the trapezoid formed by the $x$-axis and the lines $y = 4$, $y = 4x$ and $y = 12 - 4x$.

**19.** Refer to the analytic proof of Theorem 9.8 and show that the slope of $\overline{BR}$ is $\dfrac{c}{b-2a}$, and that

the equation of $\overline{BR}$ is $y = \dfrac{c}{b-2a}(x-a)$.

**20.** Refer to the analytic proof of Theorem 9.8 and show that the slope of $\overline{CP}$ is $\dfrac{2c}{2b-a}$, and that

the equation of $\overline{CP}$ is $y = \dfrac{2c}{2b-a}\left(x - \dfrac{a}{2}\right)$.

## CHAPTER TWELVE REVIEW EXERCISES

In Exercises 1–4, plot each point and indicate the quadrant in which it lies.

**1.** $(-3, 7)$      **2.** $(5, -5)$      **3.** $\left(-2, \dfrac{1}{2}\right)$      **4.** $(3, 4)$

In Exercises 5–10, find the length of the line segment $\overline{AB}$. If the line segment is vertical or horizontal, so indicate.

**5.** $A(5, 7)$, $B(-2, 7)$            **6.** $A(-5, 3)$, $B(-5, -7)$
**7.** $A(-8, 7)$, $B(0, 1)$            **8.** $A(-17, 5)$, $B(-5, 0)$
**9.** $A(8, -1)$, $B(3, 12)$           **10.** $A(-7, -5)$, $B(2, 7)$

In Exercises 11–14, find the midpoint of the line segment $\overline{AB}$.

**11.** $A(5, 3)$, $B(5, 11)$           **12.** $A(-2, 9)$, $B(7, 9)$
**13.** $A(-11, 5)$, $B(3, -9)$        **14.** $A(-8, 7)$, $B(-12, -3)$

In Exercises 15–16, one endpoint $P$ and the midpoint $M$ of line segment $\overline{PQ}$ are given. Find the coordinates of the other endpoint, $Q$.

**15.** $P(3, -4)$; $M(7, 1)$             **16.** $P(-7, -1)$; $M(3, -2)$
**17.** The endpoints of one diagonal of a square are $(-3, 2)$ and $(2, -3)$. Find the coordinates of the other vertices.
**18.** The endpoints of the base of an equilateral triangle are $(0, 0)$ and $(4, 0)$. Find two possible coordinates of the vertex.

In Exercises 19–26, graph each linear equation.

**19.** $y = 2x - 3$      **20.** $y = 5 - 2x$      **21.** $y = -x$      **22.** $y = -\dfrac{1}{2}x$

**23.** $y = -2x + 3$      **24.** $y = 4 + \dfrac{1}{2}x$      **25.** $3x - 4y = 12$      **26.** $5x + y = 10$

In Exercises 27–30, remove parentheses, combine terms, and graph each equation.

**27.** $5(x + 1) + 3y = 15$

**28.** $7(x - 3) + 2(y + 1) = -5$

**29.** $\dfrac{x - 2}{3} + \dfrac{y - 3}{2} = 0$

**30.** $\dfrac{3x + 2}{2} = 3(y + 2)$

In Exercises 31–34, find the slope of the line passing through the given points. If the slope is not defined, so indicate.

**31.** $P(5, -2)$; $Q(-2, 5)$

**32.** $P(11, -1)$; $Q(1, 11)$

**33.** $P\left(\dfrac{5}{3}, 0\right)$; $Q\left(\dfrac{2}{3}, \dfrac{1}{3}\right)$

**34.** $P\left(\dfrac{3}{7}, \dfrac{1}{5}\right)$; $Q\left(-\dfrac{4}{7}, \dfrac{6}{5}\right)$

In Exercises 35–46, find the equation of the indicated line.

**35.** through $(5, 7)$ and $(-5, -7)$

**36.** through $(-6, 2)$ and $(0, -4)$

**37.** through $\left(\dfrac{1}{3}, \dfrac{2}{3}\right)$ and $\left(\dfrac{2}{3}, \dfrac{8}{3}\right)$

**38.** through $\left(\dfrac{3}{7}, -\dfrac{2}{3}\right)$ and $\left(-\dfrac{4}{7}, \dfrac{10}{3}\right)$

**39.** through $(1, -3)$ with a slope 2

**40.** through $(-3, -1)$ with a slope $-2$

**41.** through $(0, 1)$ with a slope 5

**42.** through $(2, -3)$ with a slope $-\dfrac{1}{2}$

**43.** with slope 2 and $y$-intercept 1

**44.** with slope $-3$ and $y$-intercept 2

**45.** with slope $-2$ and $y$-intercept $-7$

**46.** with slope $-7$ and $y$-intercept 2

In Exercises 47–50, determine if the line through points $P$ and $Q$ is parallel to, or perpendicular to, the line passing through $R(3, -5)$ and $S(-2, 5)$. If the lines are neither parallel nor perpendicular, so indicate.

**47.** $P(2, -2)$; $Q(5, 4)$

**48.** $P(5, -2)$; $Q(1, 0)$

**49.** $P(-4, 3)$; $Q(-8, 11)$

**50.** $P(4, 9)$; $Q(-10, 2)$

In Exercises 51–54, write the equation of the line that passes through the given point and is parallel to the given line.

**51.** $P(2, 5)$; $y = 5x + 3$

**52.** $P(-2, 3)$; $y = -5x - 5$

**53.** $P(1, -3)$; $y = \dfrac{2}{3}x$

**54.** $P(-3, -2)$; $y = -\dfrac{3}{2}x - 2$

In Exercises 55–58, write the equation of the line that passes through the given point and is perpendicular to the given line.

**55.** $P(2, 5)$; $y = 5x + 3$

**56.** $P(-2, 3)$; $y = -5x - 5$

**57.** $P(1, -3)$; $y = \dfrac{2}{3}x$

**58.** $P(-3, -2)$; $y = -\dfrac{3}{2}x - 2$

In Exercises 59–62, determine whether the graphs determined by each pair of lines are parallel, perpendicular, or neither.

**59.** $y = 7x - \dfrac{1}{7}$; $y = 7x + 7$

**60.** $y = -\dfrac{1}{3}x - 3$; $y = -\dfrac{1}{3}x + \dfrac{1}{3}$

**61.** $2y + 4x = 7$; $2x + y = 9$

**62.** $3x + 2y = 6$; $2x - 3y = 9$

**63.** Find the equation of the line with slope $\dfrac{2}{7}$ and passing through the midpoint of the line segment joining $P(-3, 2)$ and $Q(-2, 3)$.

**64.** Find the equation of the line perpendicular to the line $y = 7x + 2$ and passing through the origin.

In Exercises 65–70, solve each system of equations. Write each answer as an ordered pair.

**65.** $\begin{cases} 2x + y = -1 \\ x + 2y = -11 \end{cases}$

**66.** $\begin{cases} 3x + 3y = 12 \\ 2x - 3y = 3 \end{cases}$

**67.** $\begin{cases} 7x - 2y = 31 \\ 5x + 7y = -20 \end{cases}$

**68.** $\begin{cases} x - 9 = y \\ x - 1 = 2y \end{cases}$

**69.** $\begin{cases} \frac{1}{3}x - \frac{2}{5}y = 3 \\ \frac{2}{3}x + y = 8 \end{cases}$

**70.** $\begin{cases} x + \frac{5}{7}y = 1 \\ \frac{2}{9}x - y = -16 \end{cases}$

In Exercises 71–72, use analytic geometry to prove each theorem.

**71.** The area of a rhombus is one-half of the product of the length of its diagonals.

**72.** The figure formed by joining the midpoints of the sides of a rectangle is a rhombus.

## CHAPTER TWELVE TEST

In Exercises 1–28, classify each statement as true or false.

**1.** The $y$-coordinate of $(3, 7)$ is 3.

**2.** The points $(3, -2)$ and $(2, -3)$ lie in the same quadrant.

**3.** If $(x, y)$ is a point in Quadrant I, then $x$ and $y$ are both positive.

**4.** The origin of a Cartesian coordinate system has no coordinates.

**5.** The points $(3, -7)$ and $(-3, -7)$ lie on a horizontal line.

**6.** None of the points $(3, -3)$, $(1, 0)$ or $(-1, 1)$ lie on the $x$-axis.

**7.** The length of the line segment joining the point $(3, 4)$ and the origin is 5.

**8.** If the coordinates of $P$ are $(-2, 3)$ and of $Q$ are $(3, -9)$, then $m(\overline{PQ}) = 13$.

**9.** The origin is the midpoint of the line segment joining $P(-7, 12)$ and $Q(7, -12)$.

**10.** The midpoint, $M$, of the line segment joining $P(3, -5)$ and $Q(5, -13)$ is $M(8, -9)$.

**11.** The slope of the line passing through the points $P(2, -7)$ and $Q(-8, 5)$ is $-\dfrac{6}{5}$.

**12.** The graph of a line with positive slope rises up and to the right.

**13.** The slope of a vertical line is undefined.

**14.** The slope of a horizontal line is undefined.

**15.** The slope of the *x*-axis is zero.

**16.** The slope of the *y*-axis is 1.

**17.** The line passing through the origin and the point $(a, b)$ is perpendicular to the line passing through the points $(b, a)$ and $(0, 0)$.

**18.** The line passing through the points $(a, b)$ and $(-a, -b)$ is parallel to the line passing through the point $(b, a)$ and the origin.

**19.** The line $y = 2x + 1$ passes through the origin.

**20.** The lines $y + 2x = 1$ and $y + 2x = 3$ are parallel.

**21.** The lines $3x - 5y = 7$ and $5x + 3y = 9$ are perpendicular.

**22.** The graph of the equation $y = 7$ is a vertical line.

**23.** The midpoint of the line segment joining the points $P(-3, 2)$ and $Q(7, -2)$ is the point $M(2, 0)$.

**24.** The graph of the equation $x - y = 2$ passes through points in Quadrant II.

**25.** The graph of the equation $y = 3x - 7$ is a line with slope of $-7$ and a *y*-intercept of 3.

**26.** The graph of the equation $y - 3 = 2(x - 5)$ is a line with slope of 2, and passing through the point $(5, 3)$.

**27.** The solution of the system of equations $\begin{cases} 3x - 5y = 10 \\ 4x - 18 = -2y \end{cases}$ is $x = 5$ and $y = -1$.

**28.** The point $(3, -2)$ lies on the intersection of the lines given by the equations $\begin{cases} 3x - 11 = y \\ 4y + 11 = x \end{cases}$.

# A Proof of the Pythagorean Theorem

The following proof of the Pythagorean theorem has been attributed to James A. Garfield, the twentieth President of the United States.

*Given:* Right $\triangle ABC$.

*Prove:* $a^2 + b^2 = c^2$.

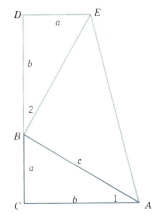

*Construction:* Extend side $\overline{BC}$ to point $D$ so that $\overline{BD} \cong \overline{CA}$. Construct segment $\overline{DE}$ so that $\overline{DE} \perp \overline{CD}$ and $\overline{DE} \cong \overline{BC}$. Draw segments $\overline{BE}$ and $\overline{AB}$.

*Proof:* Quadrilateral $ACDE$ is a trapezoid. Why? The area of the trapezoid $ACDE$ is given by

$$A = \frac{1}{2}h(b + b') = \frac{1}{2}(a + b)(a + b) = \frac{1}{2}(a^2 + 2ab + b^2)$$

The area of the trapezoid can also be found by summing the areas of triangles $ACB$, $BDE$, and $EBA$. After proving $\triangle ABC \cong \triangle BDE$ and observing that $\angle 1 \cong \angle 2$, we can show that $\angle ABE$ is a right angle. Why? Since all three of the triangles have a right angle, their areas are

$$A(\triangle ACB) = \frac{1}{2}ab$$

$$A(\triangle BDE) = \frac{1}{2}ab$$

$$A(\triangle BEA) = \frac{1}{2}c^2$$

Therefore the area of the trapezoid is

$$A = \frac{1}{2}ab + \frac{1}{2}ab + \frac{1}{2}c^2 = ab + \frac{1}{2}c^2$$

By setting the two expressions for the area equal to each other, we have

$$\frac{1}{2}(a^2 + 2ab + b^2) = ab + \frac{1}{2}c^2$$

or

$$a^2 + 2ab + b^2 = 2ab + c^2$$

or

$$a^2 + b^2 = c^2 \qquad\qquad\qquad \square$$

# Hero's Formula for the Area of a Triangle

A different way of finding the area of a triangle is credited to Heron of Alexandria (Hero), an early Greek mathematician. His formula gives the area of a triangle as a function of the lengths of its sides. Hero's formula is

$$A = \sqrt{s(s - a)(s - b)(s - c)}$$

where $A$ is the area, $s$ is one-half the perimeter, and $a$, $b$, and $c$ are the lengths of the sides of the triangle.

**EXAMPLE**

Find the area of a triangle with sides of length 4 in., 5 in., and 7 in.

*Solution*

One-half the perimeter is

$$s = \frac{1}{2}(4 + 5 + 7) = 8$$

and therefore the area is

$$A = \sqrt{8(8 - 4)(8 - 5)(8 - 7)} = \sqrt{8(4)(3)(1)} = \sqrt{96} = 4\sqrt{6} \text{ in.}^2.$$

∎

**P1.1** Two points determine one, and only one, line.

**P1.2** A line segment has only one midpoint.

**P1.3** Every angle has exactly one angle bisector. (The bisector of an angle is unique.)

**P1.4** Every line segment has one and only one perpendicular bisector.

**P1.5** The perpendicular to a line through a point not on the line is unique.

**P1.6** If $a$, $b$, and $c$ are any three quantities, then

$$a = a \qquad \text{(The reflexive law)}$$

If $a = b$, then $b = a$ (The symmetric law)

If $a = b$ and $b = c$, then $a = c$
(The transitive law)

**P1.7** If equal quantities are added to or subtracted from equal quantities, then the results are equal quantities.

**P1.8** If equal quantities are multiplied or divided by equal quantities, then the results are equal quantities. However, there can be no division by zero.

**P1.9** (**The substitution law**) A quantity can be substituted for its equal in any mathematical expression to obtain an equivalent mathematical expression.

**P1.10** If $A$, $B$, and $C$ are three points on the same line and $B$ is between $A$ and $C$, then

$$m(\overline{AC}) = m(\overline{AB}) + m(\overline{BC}),$$
$$m(\overline{BC}) = m(\overline{AC}) - m(\overline{AB}),$$
$$m(\overline{AB}) = m(\overline{AC}) - m(\overline{BC}).$$

**P1.11** If points $A$, $B$, and $C$ determine $\angle ABC$ and if point $D$ is in the interior of the angle, then

$$m(\angle ABC) = m(\angle ABD) + m(\angle DBC),$$
$$m(\angle ABD) = m(ABC) - m(\angle DBC),$$
$$m(\angle DBC) = m(\angle ABC) - m(\angle ABD).$$

**P2.1** If three sides of one triangle are congruent, respectively, to three sides of a second triangle, then the triangles are congruent (SSS ≅ SSS).

**P2.2** If two sides and the included angle of one triangle are congruent, respectively, to two sides and the included angle of a second triangle, then the triangles are congruent (SAS ≅ SAS).

**P2.3** If two angles and the included side of one triangle are congruent, respectively, to two angles and the included side of a second triangle, then the triangles are congruent (ASA ≅ ASA).

**P2.4** If $\overline{AB}$, $\overline{CD}$, and $\overline{EF}$ are three line segments, then

$$\overline{AB} \cong \overline{AB} \qquad \text{(The reflexive law)}$$

If $\overline{AB} \cong \overline{CD}$, then $\overline{CD} \cong \overline{AB}$
(The symmetric law)

If $\overline{AB} \cong \overline{CD}$ and $\overline{CD} \cong \overline{EF}$,
then $\overline{AB} \cong \overline{EF}$. (The transitive law)

**P3.1** If two lines are perpendicular to a third line, then they are parallel.

**P3.2** Given a line $l$ and a point $P$ not on line $l$, there is only one line through point $P$ that is parallel to line $l$.

**P3.3** A polygon has the same number of sides as it has angles.

**P5.1** The **area, $A$, of a rectangle** is given by the formula $A = bh$ where $b$ is the length of the base and $h$ is the height of the rectangle.

**P5.2** Any two congruent figures have the same area.

**P5.3** **The distributive law** If $a$, $b$, and $c$ are real numbers, then $a(b + c) = ab + ac$.

**P5.4** **The additive property of areas** If line segments divide a given area into smaller areas, then the given area is equal to the sum of the smaller areas.

**P5.5** The surface area, $A$, and the volume, $V$, of a rectangular solid are given by the formulas

$$A = 2lw + 2wh + 2lh$$
$$V = lwh$$

where $l$ is its length, $w$ is its width, and $h$ is its height.

**P5.6** The volume, $V$, of a right prism is given by the formula

$$V = Bh$$

where $B$ is the area of its base and $h$ is its height.

**P5.7** The surface area, $A$, and the volume, $V$, of a sphere are given by the formulas

$$A = 4\pi r^2$$
$$V = \frac{4}{3}\pi r^3$$

where $r$ is the radius of the sphere.

**P5.8** The surface area, $A$, and the volume, $V$, of a right-circular cylinder are given by the formulas

$$A = 2\pi rh + 2\pi r^2$$
$$V = \pi r^2 h$$

where $r$ is the radius of the cylinder and $h$ is its height.

**P5.9** The surface area, $A$, and the volume, $V$, of a right-circular conical solid are given by the formulas

$$A = \pi rs + \pi r^2$$
$$V = \frac{1}{3}\pi r^2 h$$

where $r$ is the length of the radius of its base, $h$ is its height, and $s$ is the length of its slant height.

**P5.10** The volume, $V$, of a pyramid is given by the formula

$$V = \frac{1}{3}Bh$$

where $B$ is the area of its base and $h$ is its height.

**P6.1** Two triangles are similar if, and only if, two angles of one triangle are equal to two angles of the other triangle.

**P7.1** A line drawn from the center of a circle to a point of tangency is perpendicular to the tangent that passes through the point of tangency.

**P7.2** If a line is perpendicular to a radius at the point where the radius intersects a circle, then the line is a tangent to the circle.

**P7.3** If $A$, $B$, and $C$ are three points on the same arc, and $B$ is between $A$ and $C$, then

$$m(\overset{\frown}{AC}) = m(\overset{\frown}{AB}) + m(\overset{\frown}{BC})$$
$$m(\overset{\frown}{BC}) = m(\overset{\frown}{AC}) - m(\overset{\frown}{AB})$$
$$m(\overset{\frown}{AB}) = m(\overset{\frown}{AC}) - m(\overset{\frown}{BC})$$

**P8.1** **The trichotomy principle** If $a$ and $b$ are two numbers, then exactly one of these relationships is true: $a < b$, $a = b$, or $a > b$.

**P8.2** **The triangle inequality** The sum of the lengths of any two sides of a triangle is greater than the third side.

**P10.1** The area of a sector of a circle divided by the area of the circle is equal to the measure of its central angle divided by $360°$.

**P12.1** All segments of a nonvertical line have the same slope.

**P12.2** Nonvertical lines are parallel if, and only if, they have the same slope. Vertical lines are parallel.

**P12.3** Nonvertical lines are perpendicular if, and only if, their slopes are negative reciprocals. A vertical line is perpendicular to a horizontal line.

**T1.1** **(The addition theorem for line segments)** If $B$ lies between $A$ and $C$ on line segment $\overline{AC}$, $E$ lies between $D$ and $F$ on line segment $\overline{DF}$, $m(\overline{AB}) = m(\overline{DE})$, and $m(\overline{BC}) = m(\overline{EF})$, then $m(\overline{AC}) = m(\overline{DF})$.

**T1.2** **(The subtraction theorem for line segments)** If point $B$ lies between points $A$ and $C$ on line segment $\overline{AC}$, point $E$ lies between points $D$ and $F$ on line segment $\overline{DF}$, $m(\overline{AC}) = m(\overline{DF})$, and $m(\overline{BC}) = m(\overline{EF})$, then $m(\overline{AB}) = m(\overline{DE})$.

**T1.3** **(The angle-addition theorem)** If point $D$ is in the interior of $\angle ABC$, point $P$ is in the interior of $\angle RST$, $m(\angle ABD) = m(\angle RSP)$, and $m(\angle DBC) = m(\angle PST)$, then $m(\angle ABC) = m(\angle RST)$.

**T1.4** **(The angle-subtraction theorem)** If point $D$ is in the interior of $\angle ABC$, point $P$ is in the interior of $\angle RST$, $m(\angle ABC) = m(\angle RST)$, and $m(\angle ABD) = m(\angle RSP)$, then $m(\angle DBC) = m(\angle PST)$.

**T1.5** If $B$ is the midpoint of line segment $\overline{AC}$, then $B$ divides $\overline{AC}$ into two segments, each half as long as $\overline{AC}$.

**T1.6** **(The line-segment-bisector theorem)** If $B$ is the midpoint of line segment $\overline{AC}$, $E$ is the midpoint of line segment $\overline{DF}$, and $m(\overline{AC}) = m(\overline{DF})$, then $m(\overline{AB}) = m(\overline{DE})$.

**T1.7** If ray $\overrightarrow{BD}$ bisects $\angle ABC$, then it divides $\angle ABC$ into two angles, each with half the measure of $\angle ABC$.

**T1.8** **(The angle-bisector theorem)** If ray $\overrightarrow{BD}$ bisects $\angle ABC$, ray $\overrightarrow{SP}$ bisects $\angle RST$, and $m(\angle ABC) = m(\angle RST)$, then $m(\angle ABD) = m(\angle RSP)$.

**T1.9** If two angles are both supplementary and congruent, then they are right angles.

**T1.10** Supplements of the same angle are congruent.

**T1.11** Complements of the same angle are congruent.

**T1.12** Supplements of congruent angles are congruent.

**T1.13** Complements of congruent angles are congruent.

**T1.14** Two adjacent angles whose noncommon sides form a line are supplementary.

**T1.15** Vertical angles are congruent.

**T2.1** If two triangles are congruent to a third triangle, then they are congruent to each other.

**T2.2** If two sides of a triangle are congruent then the angles opposite those sides are congruent.

**T2.3** The bisector of the vertex angle of an isosceles triangle is the perpendicular bisector of the base of the triangle.

**T2.4** The perpendicular bisector of the base of an isosceles triangle passes through the vertex of its vertex angle.

**T2.5** If a triangle has two congruent angles, then the sides opposite those angles are congruent.

**T2.6** If the hypotenuse and a leg of one right triangle are congruent, respectively, to the hypotenuse and a leg of a second right triangle, then the right triangles are congruent. $(hl \cong hl)$.

**T3.1** If two lines are cut by a transversal so that a pair of alternate interior angles are congruent, then the lines are parallel.

**T3.2** If two lines are cut by a transversal so that a pair of corresponding angles are congruent, then the lines are parallel.

**T3.3** If two lines are cut by a transversal so that two interior angles on the same side of the transversal are supplementary, then the lines are parallel.

**T3.4** If a line is perpendicular to one of two parallel lines, then it is also perpendicular to the other line.

**T3.5** If two parallel lines are cut by a transversal, then all pairs of alternate interior angles are congruent.

**T3.6** If two parallel lines are cut by a transversal, then all pairs of corresponding angles are congruent.

**T3.7** If two parallel lines are cut by a transversal, then all pairs of interior angles on the same side of the transversal are supplementary.

**T3.8** The sum of the measures of the interior angles of a triangle is $180°$.

**C3.9** A triangle can have at most one right angle. A triangle can have at most one obtuse angle.

**C3.10** The acute angles of a right triangle are complementary.

**C3.11** If two angles of one triangle are congruent respectively, to two angles of a second triangle, then their third angles are congruent.

**C3.12** If two angles and any side of one triangle are congruent, respectively, to two angles and the corresponding side of a second triangle, then the triangles are congruent. (AAS $\cong$ AAS).

**C3.13** The measure of an exterior angle of a triangle is equal to the sum of the measures of the nonadjacent interior angles.

**T3.14** If the hypotenuse and an acute angle of one right triangle are congruent, respectively, to the hypotenuse and an acute angle of a second right triangle, then the triangles are congruent. ($ha \cong ha$).

**T3.15** The sum of the measures of the interior angles of a polygon is given by the formula

$$S = (n - 2)180°$$

where $n$ is the number of sides of the polygon.

**T3.16** The sum of the measures of the exterior angles of a polygon is $360°$.

**T3.17** The measure of an interior angle $a$ of a regular polygon is given by the formula $a = [(n - 2)180°]/n$ where $n$ is the number of sides of the polygon.

**T3.18** If two lines are cut by a transversal so that alternate interior angles are not congruent, then the lines are not parallel.

**T3.19** If two lines are cut by a transversal so that corresponding angles are not congruent, then the lines are not parallel.

**T3.20** If two lines are cut by a transversal so that two interior angles on the same side of the transversal are not supplementary, then the lines are not parallel.

**T3.21** If two nonparallel lines are cut by a transversal, then alternate interior angles are not congruent.

**T3.22** Two lines can intersect in at most one point.

**T3.23** If two distinct lines are parallel to a third line, then they are parallel to each other.

**T4.1** A diagonal of a parallelogram divides the parallelogram into two congruent triangles.

**C4.2** Opposite sides of a parallelogram are congruent.

**C4.3** Nonconsecutive angles of a parallelogram are congruent.

**T4.4** Consecutive angles of a parallelogram are supplementary.

**T4.5** Parallel lines are always the same distance apart.

**T4.6** If both pairs of opposite sides of a quadrilateral are congruent, then the quadrilateral is a parallelogram.

**T4.7** If two opposite sides of a quadrilateral are both parallel and congruent, then the quadrilateral is a parallelogram.

**T4.8** The diagonals of a parallelogram bisect each other.

**T4.9** If the diagonals of a quadrilateral bisect each other, then the quadrilateral is a parallelogram.

**T4.10** All angles of a rectangle are right angles.

**T4.11** The diagonals of a rectangle are congruent.

**T4.12** If the diagonals of a parallelogram are congruent, then the parallelogram is a rectangle.

**T4.13** All sides of a rhombus are congruent.

**T4.14** The diagonals of a rhombus are perpendicular.

**T4.15** If the diagonals of a parallelogram are perpendicular, then the parallelogram is a rhombus.

**T4.16** The diagonals of a rhombus bisect the angles of the rhombus.

**T4.17** If a line joins the midpoints of two sides of a triangle, then the line is parallel to the third side and is half as long as the third side.

**T4.18** If line segments are drawn connecting the midpoints of the consecutive sides of a quadrilateral, then those segments form a parallelogram.

**T4.19** The base angles of an isosceles trapezoid are congruent.

**T4.20** The median of a trapezoid is parallel to the bases and its measure is one-half the sum of the measures of the bases.

**T4.21** If three or more parallel lines cut off congruent segments on one transversal, then they cut off congruent segments on all transversals.

**T5.1** The **area of a parallelogram** is given by the formula $A = bh$, where $b$ is the length of a base, and $h$ is the corresponding height of the parallelogram.

**T5.2** The **area of a triangle** is given by the formula $A = \frac{1}{2}bh$, where $b$ is the length of a base, and $h$ is the corresponding height of the triangle.

**T5.3** The **area of a trapezoid** is given by the formula $A = \frac{1}{2}h(b + b')$ where $h$ is the height and $b$ and $b'$ are the lengths of the bases of the trapezoid.

**T5.4** The **area of a rhombus** is equal to one-half the product of the measures of its diagonals.

**T5.5** The area $A$ of a circle is given by the formula

$$A = \pi r^2$$

where $r$ is the length of the radius of the circle.

**T5.6** In a right triangle, the square of the length of the hypotenuse is equal to the sum of the squares of the lengths of the other two sides.

**T5.7** If a triangle has sides of length $a$, $b$, and $c$ and $c^2 = a^2 + b^2$, then the triangle is a right triangle.

**T5.8** In a 30°–60° right triangle, the hypotenuse is twice as long as the side opposite the 30° angle. The side opposite the 60° angle is equal to the length of the side opposite the 30° angle multiplied by $\sqrt{3}$.

**T5.9** In an isosceles right triangle, the length of the hypotenuse is equal to the length of one of the legs multiplied by $\sqrt{2}$.

**T6.1** In a proportion, the product of the extremes is equal to the product of the means. (If $a/b = c/d$, then $ad = bc$.)

**T6.2** A proportion can be written by inversion to obtain another proportion. (If $a/b = c/d$, then $b/a = d/c$.)

**T6.3** In any proportion the means can be interchanged to obtain another proportion. (If $a/b = c/d$, then $a/c = b/d$.)

**T6.4** In any proportion the extremes can be interchanged to obtain another proportion. (If $a/b = c/d$, then $d/b = c/a$.)

**T6.5** A proportion can be written by addition to obtain another proportion. (If $a/b = c/d$, then $(a + b)/b = (c + d)/d$.

**T6.6** A proportion can be written by subtraction to obtain another proportion. (If $a/b = c/d$, then $(a - b)/b = (c - d)/d$.)

**T6.7** If three terms of one proportion are equal, respectively, to three terms of a second proportion, then the fourth terms are equal.

**T6.8** If $a$, $b$, $c$, $d$, $e$, and $f$ are nonzero real numbers, and $a/b = c/d = e/f$, then $(a + c + e)/(b + d + f) = a/b$.

**T6.9** A line parallel to one side of a triangle divides the other two sides proportionally.

**T6.10** The bisector of one angle of a triangle divides the opposite side in the same ratio as the lengths of the other two sides.

**T6.11** If a line segment divides two sides of a triangle proportionally, then it is parallel to the third side.

**T6.12** All congruent triangles are similar.

**T6.13** Two triangles similar to the same triangle are similar to each other.

**T6.14** If an angle in one triangle is congruent to an angle in a second triangle, and the lengths of the respective sides including those angles are in proportion, then the triangles are similar.

**T6.15** If the lengths of three sides of the one triangle are in proportion to the lengths of three corresponding sides of a second triangle, then the triangles are similar.

**T6.16** If the altitude is drawn on the hypotenuse of a right triangle, then the length of either leg is the mean proportional between the length of the hypotenuse and the length of the segment of the hypotenuse adjacent to the leg.

**T6.17** The length of the altitude on the hypotenuse of a right triangle is the mean proportional between the lengths of the segments of the hypotenuse.

**T7.1** The length of a diameter is twice the length of any of its radii.

**T7.2** Two circles are congruent if, and only if, their diameters are congruent.

**T7.3** The measure of an inscribed angle is equal to one-half the measure of its intercepted arc.

**T7.4** The measure of an angle formed by a tangent and a chord is equal to one-half the measure of its intercepted arc.

**T7.5** If two chords intersect within a circle, the measure of each angle formed is equal to one-half the sum of the measures of its intercepted arc and the intercepted arc of its vertical angle.

**T7.6** The measure of an angle formed by the intersection of two secants outside a circle is equal to one-half the difference of the measures of the intercepted arcs.

**T7.7** The measure of an angle formed by the intersection of a tangent and a secant outside a circle is equal to one-half the difference of the measures of the intercepted arcs.

**T7.8** If a line is drawn from the center of a circle perpendicular to a chord, then it bisects the chord and its arc.

**T7.9** A line drawn from the center of circle to the midpoint of a chord that is not a diameter or to the midpoint of its arc is perpendicular to the chord.

**T7.10** The perpendicular bisector of a chord passes through the center of the circle.

**T7.11** In the same circle or congruent circles, congruent chords have congruent arcs.

**T7.12** In the same circle or congruent circles, congruent arcs have congruent chords.

**T7.13** In the same circle or congruent circles, congruent chords are equidistant from the center.

**T7.14** In the same circle or congruent circles, chords equidistant from the center are congruent.

**T7.15** If two tangent segments are drawn to a circle from a point outside the circle, then the segments are congruent.

**T7.16** If two circles are tangent, either internally or externally, their point of tangency and the center of each circle are collinear.

**T7.17** If two circles intersect in two points, then their line of centers is the perpendicular bisector of their common chord.

**T7.18** In a circle, parallel lines intercept congruent arcs.

**T7.19** If two chords intersect within a circle, then the product of the lengths of the segments of one chord is equal to the product of the lengths of the segments of the other chord.

**T7.20** If a secant and a tangent are drawn to a circle, then the length of the tangent is the mean proportional between the lengths of the secant and its external segment.

**T7.21** If two secants are drawn to a circle from a point outside the circle, then the products of the lengths of the secants and their external segments are equal.

**T7.22** The opposite angles of an inscribed quadrilateral are supplementary.

**T7.23** If a parallelogram is inscribed within a circle, then it is a rectangle.

**T8.1** If the same quantity is added to both sides of an inequality, then the sums are unequal and in the same order. (If $a < b$, then $a + c < b + c$.)

**T8.2** If equal quantities are added to unequal quantities, then the sums are unequal and in the same order. (If $a < b$ and $c = d$, then $a + c < b + d$.)

**T8.3** If unequal quantities are subtracted from equal quantities, then the differences are unequal but in the *opposite* order. (If $a < b$ and $c = d$, then $c - a > d - b$.)

**T8.4** If both sides of an inequality are multiplied by a positive number, then the products are unequal and in the same order. (If $a < b$ and $c$ is a positive number, then $ac < bc$.)

**T8.5** If both sides of an inequality are multiplied by a negative number, then the products are unequal but in the *opposite* order. (If $a < b$ and $c$ is a negative number, then $ac > bc$.)

**T8.6** The relation "<" is transitive. (If $a < b$ and $b < c$, then $a < c$.)

**T8.7** If unequal quantities are added to unequal quantities in the same order, then the sums are unequal quantities in the same order. (If $a < b$ and $c < d$, then $a + c < b + d$.)

**T8.8** In a triangle an exterior angle is greater than either nonadjacent interior angle.

**T8.9** If the lengths of two sides of a triangle are unequal, then the measures of the angles opposite those sides are unequal in the same order.

**T8.10** If the measures of two angles of a triangle are unequal, then the lengths of the sides opposite those angles are unequal in the same order.

**T8.11** If two sides of one triangle are congruent to two sides of a second triangle, and the included angle of the first is greater than the included angle of the second, then the third side of the first triangle is longer than the third side of the second triangle.

**T8.12** If two sides of one triangle are congruent to two sides of a second triangle, and if the third side of the first is longer than the third side of the second, then the angle opposite the third side of the first triangle is greater than the angle opposite the third side of the second triangle.

**T8.13** In the same circle or in congruent circles, the greater of two central angles will intercept the greater arc.

**T8.14** In the same circle or in congruent circles, the greater of two arcs will be intercepted by the greater of the two central angles.

**T8.15** In the same circle or in congruent circles, the greater of two chords intercepts the greater minor arc.

**T8.16** In the same circle or in congruent circles, the greater of two minor arcs has the greater chord.

**T8.17** If two congruent chords form an inscribed angle within a circle, then the shorter chord is the farther from the center of the circle.

**T8.18** In the same circle or in congruent circles, the shorter of two chords is the greater distance from the center of the circle.

**T9.1** The locus of a point at a given distance from a given line is a pair of lines, one on each side of the given line, parallel to the given line and at the given distance from it.

**T9.2** The locus of all points equidistant from two points is the perpendicular bisector of the segment joining the two points.

**T9.3** The locus of all points equidistant from the sides of an angle is the angle bisector.

**T9.4** The locus of the vertex of the right angle of a right triangle with fixed hypotenuse is a circle with the hypotenuse as diameter.

**T9.5** The perpendicular bisectors of the sides of a triangle meet at a point that is equally distant from the vertices of the triangle.

**T9.6** The angle bisectors of a triangle meet in a point that is equidistant from the sides of the triangle.

**T9.7** The altitudes of a triangle (or their extensions) meet in a point.

**T9.8** The medians of a triangle meet in a point.

**T9.9** The point of intersection of the medians of a triangle is two-thirds of the way from the vertex to the midpoint of the opposite side.

**T.10.1** If a circle is divided into $n$ congruent arcs, then the chords of the arcs form a regular polygon ($n > 2$).

**T10.2** If a circle is divided into $n$ congruent arcs and tangents are drawn to the circle at the endpoints of these arcs, then the figure formed by the tangents will be a regular polygon ($n > 2$).

**T10.3** A circle can be circumscribed about any regular polygon.

**T10.4** A circle can be inscribed in any regular polygon.

**T10.5** The center of a circle that is circumscribed about a regular polygon is the center of the circle that is inscribed within the regular polygon.

**T10.6** The measure $a°$ of a central angle $a$ of a regular polygon is given by the formula $a = \dfrac{360°}{n}$, where $n$ is the number of sides of the polygon.

**T10.7** An apothem of a regular polygon bisects its respective side.

**T10.8** The area of a regular polygon is equal to the product of one-half the length of its apothem and its perimeter. ($A = \frac{1}{2}ap$).

**T10.9** The area of a circle is given by the formula $A = \pi r^2$, where $r$ is the radius of the circle and $\pi = 3.14159 \ldots .$

**T11.1** $p \Leftrightarrow \sim(\sim p)$

**T11.2** $\sim(p \wedge q) \Leftrightarrow (\sim p \vee \sim q)$
$\sim(p \vee q) \Leftrightarrow (\sim p \wedge \sim q)$

**T11.3** $(p \Rightarrow q) \Leftrightarrow (\sim q \Rightarrow \sim p)$

**T11.4** $[(p \Rightarrow q) \wedge p] \Rightarrow q$
$[(p \Rightarrow q) \wedge \sim q] \Rightarrow \sim p$
$[(p \vee q) \wedge \sim p] \Rightarrow q$
$[(p \Rightarrow q) \wedge (q \Rightarrow r)] \Rightarrow (p \Rightarrow r)$

**T12.1** The distance $d = m(\overline{PQ})$ between points $P(x_1, y_1)$ and $Q(x_2, y_2)$ is given by

$$d = \sqrt{(x_2 - x_1)^2 + (y_2 - y_1)^2}$$

**T12.2** The midpoint of the line segment with endpoints $A(x_1, y_1)$ and $B(x_2, y_2)$ has the coordinates

$$\left(\frac{x_1 + x_2}{2}, \frac{y_1 + y_2}{2}\right)$$

**T12.3** The equation of a line passing through the point $P(x_1, y_1)$ and having a slope of $m$ is given by

$$y - y_1 = m(x - x_1)$$

**T12.4** The equation of the line with slope of $m$ and $y$-intercept $b$ is

$$y = mx + b$$

# Answers to Selected Exercises

## EXERCISE 1.1 (page 4)

1. Yes, inductive reasoning.
2. No, 9 is not a prime number, inductive reasoning.
3. Yes, deductive reasoning.    4. Yes, deductive reasoning.
5. Undefined terms, defined terms, postulates, and theorems.
6. A "bucket" is a "pail." Pail, pan, tub, container, and so forth.
7. "Good" means "fine." Fine, superior, above average, and so forth.
8. It is impossible to define all terms.
9. The two ideologies accept different postulates.

## EXERCISE 1.2 (page 10)

1. Yes   2. No   3. No   4. Yes   5. No   6. Yes
7. It implies the two statements:
   (a) If a triangle is equilateral, then all three sides are of equal length.
   (b) If in a triangle, all three sides are of equal length, then the triangle is equilateral.
8. (a) If a triangle is scalene, then all three sides are of different lengths.
   (b) If a triangle has all three sides of different lengths, then it is scalene.
9. No   10. Yes   11. No   12. No   13. 24 in.   14. 32 m
15. 75 cm   16. 39 ft   17. 38.1 cm   18. $26\frac{1}{3}$ ft   19. 17.9 in.
20. 17 cm   21. One   22. Two   23. None   24. Infinitely many
25. Infinitely many   26. Infinitely many   27. One   28. 3   29. 6
30. 10   31. 3   32. 4   33. 3   34. 2   35. 1   36. 2   37. Yes
38. No   39. No   40. Yes

## EXERCISE 1.3 (page 18)

1. $\angle 1$   2. $\angle DBC$ or $\angle CBD$   3. $\angle CDB$ or $\angle BDC$   4. $\angle 4$   5. 50°
6. 180°   7. 130°   8. 80°   9. Yes   10. No
11. No; The sides are rays and extend forever.
12. No   13. Acute   14. Right   15. Obtuse   16. Straight
17. Right   18. Obtuse   19. Straight   20. Acute   21. 60°   22. 150°
23. 75°   24. 15°   25. Its measure is already greater than 90°.
26. Its measure is already greater than 180°.

**27.** 175° **28.** 25° **29.** Yes **30.** No **31.** 45° **32.** 90° **33.** 150°
**34.** 68° **35.** Yes **36.** Yes **37.** $(88 - 3x)°$ **38.** $(182 - 5x)°$
**39.** 16 **40.** 22 **41.** True
**42.** False; common side $\overrightarrow{GD}$ is not between ∠'s.
**43.** False; common side $\overrightarrow{GF}$ is not between ∠'s. **44.** True
**45.** Common side $\overrightarrow{DA}$ is not between them.
**46.** They do not have a common vertex.
**47.** Right **48.** Acute **49.** Obtuse **50.** Right **51.** 90° **52.** 120°
**53.** 180° **54.** 120°

# EXERCISE 1.4 (page 26)

**1.** A straight edge cannot be used for measuring.
**9.** They meet at a point inside the triangle.
**14.** They meet in a point. **15.** They do not intersect.

# EXERCISE 1.5 (page 34)

**1.** $x = 2$ **2.** $y = 9$ **3.** $y = 15$ **4.** $y = 7$ **5.** $x = 8$
**6.** $z = 7$ **7.** $x = 15$ **8.** $y = 18$ **9.** $u = 16$ **10.** $t = 15$
**11.** $r = 2$ **12.** $s = 3$

# EXERCISE 1.6 (page 41)

**1.** ∠$a$ and ∠$c$, ∠$b$ and ∠$d$
**2.** ∠$a$ and ∠$d$, ∠$d$ and ∠$c$, ∠$c$ and ∠$b$, ∠$b$ and ∠$a$
**3.** 130° **4.** 50° **5.** 30° **6.** 150° **7.** 130° **8.** 50° **9.** 230°
**10.** 260° **11.** 100° **12.** 80° **13.** 40° **14.** 140° **15.** No **16.** No
**17.** Yes **18.** Yes **19.** Yes **20.** Yes

# CHAPTER ONE REVIEW EXERCISES (page 43)

**1.** Yes, inductive reasoning. **2.** Yes, deductive reasoning.
**3.** Yes **4.** No **5.** 5 **6.** 2 **7.** 3 **8.** 6 **9.** 35 cm **10.** 15 m
**11.** True **12.** False **13.** False **14.** False **15.** True **16.** True
**17.** True **18.** True **19.** True **20.** False **21.** True **22.** False
**23.** ∠1; ∠2 **24.** ∠4 **25.** ∠3 **26.** ∠$AEB$
**27.** ∠1 and ∠4, ∠1 and ∠2, ∠2 and ∠3, ∠3 and ∠4
**28.** There are no intersecting lines **29.** 55° **30.** 145° **31.** No
**32.** $x = 20$ **39.** 90° **40.** 30° **41.** 60° **42.** 60° **43.** Yes **44.** No

## CHAPTER ONE TEST   (page 45)

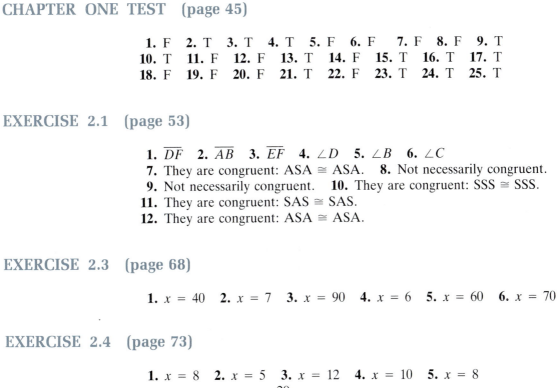

**1.** F   **2.** T   **3.** T   **4.** T   **5.** F   **6.** F   **7.** F   **8.** F   **9.** T
**10.** T   **11.** F   **12.** F   **13.** T   **14.** F   **15.** T   **16.** T   **17.** T
**18.** F   **19.** F   **20.** F   **21.** T   **22.** F   **23.** T   **24.** T   **25.** T

## EXERCISE  2.1   (page 53)

**1.** $\overline{DF}$   **2.** $\overline{AB}$   **3.** $\overline{EF}$   **4.** $\angle D$   **5.** $\angle B$   **6.** $\angle C$
**7.** They are congruent: ASA $\cong$ ASA.   **8.** Not necessarily congruent.
**9.** Not necessarily congruent.   **10.** They are congruent: SSS $\cong$ SSS.
**11.** They are congruent: SAS $\cong$ SAS.
**12.** They are congruent: ASA $\cong$ ASA.

## EXERCISE  2.3   (page 68)

**1.** $x = 40$   **2.** $x = 7$   **3.** $x = 90$   **4.** $x = 6$   **5.** $x = 60$   **6.** $x = 70$

## EXERCISE  2.4   (page 73)

**1.** $x = 8$   **2.** $x = 5$   **3.** $x = 12$   **4.** $x = 10$   **5.** $x = 8$
**6.** $x = 64.6$   **7.** $x = \dfrac{20}{3}$   **8.** $x = 6$

## EXERCISE  2.5   (page 78)

**1.** A *hypotenuse* is the side opposite the right angle in a right triangle.
A *hypothesis* is the information following the word *if* in an "If . . . ,
then . . . ." sentence.

## CHAPTER  TWO  REVIEW  EXERCISES   (page 79)

**1.** $\overline{DF}$   **2.** $\overline{AB}$   **3.** $\overline{EF}$   **4.** $\angle D$   **5.** $\angle B$
**6.** $\angle C$   **7.** $SSS \cong SSS$; $SAS \cong SAS$; $ASA \cong ASA$
**8.** Corresponding parts of congruent triangles are congruent.
**9.** They are congruent: $ASA \cong ASA$.
**10.** Not necessarily congruent.   **11.** Not necessarily congruent.
**12.** They are congruent: $SSS \cong SSS$.
**13.** They are congruent: $SAS \cong SAS$.
**14.** They are congruent: $ASA \cong ASA$.   **17.** $x = 70$   **18.** $x = 5$

**19.** $x = 90$  **20.** $x = 11$  **21.** $x = 11$  **22.** $x = 30$  **23.** $x = 120$
**24.** $x = 12$  **27.** They are congruent: $hl \cong hl$.  **28.** Not necessar
congruent.  **29.** Not not necessarily congruent.  **30.** They are
congruent: $hl \cong hl$.

## CHAPTER TWO TEST  (page 82)

**1.** T  **2.** F  **3.** T  **4.** F  **5.** F  **6.** T  **7.** T  **8.** F  **9.** F
**10.** T  **11.** F  **12.** T  **13.** T  **14.** F  **15.** T  **16.** F  **17.** F
**18.** T  **19.** T  **20.** F

## EXERCISE 3.1  (page 89)

**1.** $\angle b$ and $\angle g$, $\angle a$ and $\angle h$
**2.** $\angle a$ and $\angle e$, $\angle b$ and $\angle f$, $\angle c$ and $\angle g$, $\angle d$ and $\angle h$
**3.** $\angle d$ and $\angle e$, $\angle c$ and $\angle f$  **4.** $\angle b$, $\angle c$, $\angle g$, and $\angle f$
**5.** Yes  **6.** Not necessarily  **7.** Yes  **8.** Not necessarily  **9.** No
**10.** No  **11.** Yes  **12.** No  **13.** Yes  **14.** No  **15.** No  **16.** Yes
**17.** No  **18.** No

## EXERCISE 3.2  (page 95)

**1.** 150°  **2.** 30°  **3.** 30°  **4.** 150°  **5.** 60°  **6.** 60°  **7.** 120°
**8.** 120°  **9.** 110°  **10.** 70°  **11.** 70°  **12.** 70°  **13.** 50°  **14.** 50°
**15.** 50°  **16.** 130°  **17.** 50°  **18.** 50°  **19.** 80°  **20.** 50°  **21.** 90°
**22.** 60°  **23.** 60°  **24.** 90°

## EXERCISE 3.3  (page 103)

**1.** 105°  **2.** 45°  **3.** 105°  **4.** 30°  **5.** 40°  **6.** 50°  **7.** 40°
**8.** 140°  **9.** 20°  **10.** 50°  **11.** 40°  **12.** 100°
**13.** 40°  **14.** 80°  **15.** 60°  **16.** 100°  **17.** 60°, 60°, 60°
**18.** 45°, 45°, 90°

## EXERCISE 3.4  (page 109)

**1.** 360°  **2.** 540°  **3.** 720°  **4.** 1080°  **5.** 1440°  **6.** 1800°
**7.** 90°  **8.** 108°  **9.** 120°  **10.** 135°  **11.** 140°  **12.** 144°
**13.** 18 sides  **14.** 17 sides  **15.** No polygon exists.
**16.** No polygon exists.  **17.** 24 sides  **18.** 11 sides  **19.** 22 sides
**20.** No polygon exists.  **21.** 12 sides  **22.** 8 sides  **23.** 9 sides

**24.** No polygon exists.  **25.** No polygon exists.  **26.** 20 sides
**27.** 15 sides  **28.** 16 sides  **29.** 360°  **30.** Yes, for a 24-sided figure.
**31.** 24°  **33.** 6 sides  **34.** 6 sides  **35.** 5 sides

## CHAPTER THREE REVIEW EXERCISES  (page 114)

**1.** ∠1 and ∠4; ∠2 and ∠3  **2.** ∠1 and ∠5; ∠2 and ∠6, ∠7 and ∠3,
∠8 and ∠4  **3.** ∠7 and ∠6; ∠8 and ∠5
**4.** ∠8, ∠4, ∠5  **5.** ∠7, ∠3, ∠6  **6.** ∠1, ∠4, ∠5, ∠8
**7.** Yes, alternate interior angles are congruent.
**8.** Yes, alternate exterior angles are congruent.
**9.** No  **10.** Yes  **11.** 58°  **12.** 58°  **13.** 122°  **14.** 58°  **17.** 50°
**18.** 130°  **19.** 65°  **20.** 90°  **21.** 20°  **22.** 40°  **23.** 10°  **24.** 10°
**25.** 2340°  **26.** 10 sides  **27.** 120°  **28.** 5 sides  **29.** 4 sides

## CHAPTER THREE TEST  (page 115)

**1.** F  **2.** F  **3.** T  **4.** F  **5.** T  **6.** T  **7.** T  **8.** T  **9.** T
**10.** T  **11.** T  **12.** T  **13.** F  **14.** F  **15.** F  **16.** T  **17.** F
**18.** T  **19.** T  **20.** T  **21.** T  **22.** T  **23.** T  **24.** T  **25.** T

## EXERCISE 4.1  (page 123)

**1.** No  **2.** Yes  **3.** No  **4.** Yes  **5.** 75°  **6.** 60°  **7.** 75°  **8.** 45°
**9.** 60°  **10.** 70°  **11.** 50°  **12.** 20°  **13.** 100°  **14.** 80°  **15.** 30°
**16.** 5 in.  **22.** 60°, 120°, 60°, and 120°

## EXERCISE 4.2  (page 131)

**1.** 10 in.  **2.** 40°  **3.** 50°  **4.** 14 cm  **5.** 5 cm  **6.** Yes  **7.** 120°
**8.** Yes  **9.** 20 m  **10.** 5 in.  **11.** 10 cm  **12.** 5 yd  **25.** No
**26.** Yes  **27.** No  **28.** Yes  **29.** Yes  **30.** No

## EXERCISE 4.3  (page 138)

**1.** 120°  **2.** 70°  **3.** 14  **4.** 28  **5.** 10  **6.** 16  **7.** 70°  **8.** 100°
**9.** 12 cm  **10.** 10 cm  **11.** 105°  **12.** 95°

## CHAPTER FOUR REVIEW EXERCISES  (page 139)

**1.** Yes  **2.** Yes  **3.** 5 cm  **4.** 8 cm  **5.** 20°  **6.** 30°  **7.** 130°
**8.** 20°  **11.** T  **12.** F  **13.** F  **14.** T  **15.** T  **16.** T  **17.** F

**18.** F   **19.** 12 cm   **20.** 65°   **21.** 56 cm   **22.** 80°   **23.** 10 m
**24.** 10 ft   **25.** 60°   **26.** 68 in.   **27.** 7 cm   **28.** 6 ft   **29.** 110°
**30.** 110°

## CHAPTER FOUR TEST   (page 141)

**1.** T   **2.** F   **3.** F   **4.** F   **5.** F   **6.** F   **7.** T   **8.** T   **9.** F
**10.** F   **11.** F   **12.** T   **13.** F   **14.** F   **15.** F   **16.** T   **17.** T
**18.** T   **19.** F   **20.** F

## EXERCISE 5.1   (page 154)

**1.** 15 in.²   **2.** 14 ft²   **3.** 16 cm²   **4.** 6 in.²   **5.** 25 in.²
**6.** 24 cm²   **7.** 169 mm²   **8.** 112 ft²   **9.** 12 in.²   **10.** 23 m²
**11.** 52.5 ft²   **12.** 96 in.²   **13.** 56 ft²   **14.** 17.5 ft²   **15.** 54 in.²
**16.** 70.5 in.²   **17.** 32 m²   **18.** 85 in.²   **19.** 174 in.²   **20.** 56 m²
**21.** If $s$ is the side of the smallest square, then the area is $341s^2$.
**22.** $\frac{73}{27}s$   **23.** It is multiplied by 4.   **24.** 8, 2; 4, 1   **25.** 144 in.²
**26.** 24.5 in.²   **27.** 12 cm   **28.** 45 m²   **29.** $1008   **30.** $361.20
**31.** 11 gal   **32.** 1950 in.²   **33.** $80   **34.** 400 m²   **35.** $\frac{33}{4}$ in.
**36.** 78 in².   **40.** 10 in.   **41.** 16 ft

## EXERCISE 5.2   (page 161)

**1.** $9\pi$ in.²   **2.** $49\pi$ ft²   **3.** $243\pi$ cm²   **4.** $(120 + 25\pi)$ m²
**5.** $(18 + 4.5\pi)$ mi²   **6.** $(117 - 25\pi)$ m²   **7.** $32\pi$ ft²
**8.** $75\pi$ ft²   **9.** $(40 - 4\pi)$ in.²   **10.** $(64 - 16\pi)$ in.²
**11.** $(135 - 27\pi/4)$ cm²   **12.** $(135 + 9\pi/4)$ cm²   **13.** $4\pi$ in.²
**14.** $12.5\pi$ ft²   **15.** $\pi$ mi²   **16.** $2\pi$ mi   **17.** $\frac{40}{\pi}$ or approximately 12.7
times   **18.** $72,394.44   **19.** Approximately 1.59 ft   **20.** $4a\sqrt{\pi}$
**21.** $s\sqrt{\pi}/\pi$   **22.** Four times the area.   **23.** Nine times the area.
**24.** It is doubled.

## EXERCISE 5.3   (page 168)

**1.** 5   **2.** 13   **3.** 6   **4.** 8   **5.** 24   **6.** 9   **7.** Yes   **8.** Yes   **9.** No
**10.** No   **11.** Yes   **12.** Yes   **13.** $2\sqrt{2}$ m   **14.** 60 in.²   **15.** Yes
**16.** No   **17.** Yes   **18.** $\sqrt{3}$ ft   **19.** 6 ft by 6 ft   **20.** 12 ft
**21.** Approximately 453 ft plus waste.   **22.** 53 ft   **23.** 5.8 mi

**24.** $14\sqrt{2}$ in.   **25.** $16\sqrt{2}$ ft   **26.** 16 ft   **27.** $\dfrac{\sqrt{13}}{2}$ ft   **28.** $\frac{3}{2}$ ft

**29.** $(90\sqrt{2} - 60.5)$ ft   **30.** $30\sqrt{10}$ ft   **31.** 13   **32.** 12

## EXERCISE 5.4   (page 174)

**1.** 40   **2.** 60   **3.** $14\sqrt{3}$   **4.** $12/\sqrt{3}$ or $4\sqrt{3}$   **5.** $2\sqrt{2}/3$   **6.** 1
**7.** $5/2$   **8.** $2\sqrt{3}$   **9.** $\sqrt{3}/3$   **10.** 1   **11.** $20\sqrt{2}$   **12.** $30\sqrt{2}$
**13.** 14   **14.** 12   **15.** $\sqrt{6}$   **16.** 2   **17.** $5/\sqrt{2}$ or $5\sqrt{2}/2$
**18.** $4/\sqrt{2}$ or $2\sqrt{2}$   **19.** $5\sqrt{3}/\sqrt{2}$ or $5\sqrt{6}/2$
**20.** $5\sqrt{3}/\sqrt{2}$ or $5\sqrt{6}/2$   **21.** $155\sqrt{2}$ mi   **22.** $220\sqrt{2}$ mi
**23.** $\dfrac{230}{\sqrt{2}}$ or $\dfrac{230\sqrt{3}}{3}$ ft   **24.** 106 ft   **25.** 2 dm and $2\sqrt{3}$ dm
**26.** $2\sqrt{3}$ dm²   **27.** $2\sqrt{2}$ units and $2\sqrt{6}$ units
**28.** Each leg is $4/\sqrt{2}$ units or $2\sqrt{2}$ units   **29.** 4 units
**30.** $\sqrt{3}$ m   **31.** $\sqrt{3}$ dm²   **32.** $16\sqrt{3}/3$ in.   **33.** $64\sqrt{3}/3$ in.²

## EXERCISE 5.5   (page 183)

**1.** $A = 94$ cm², $V = 60$ cm³   **2.** $A = 340$ m², $V = 400$ m³
**3.** $A = 108$ m², $V = 48$ m³   **4.** $A = 360$ ft², $V = 300$ ft³
**5.** $A = 324\pi$ in.², $V = 972\pi$ in.³   **6.** $A = 100\pi$ ft², $V = \dfrac{500\pi}{3}$ ft³
**7.** $A = 216\pi$ m², $V = 432\pi$ m³   **8.** $A = 234\pi$ m², $V = 324\pi$ m³
**9.** $A = 90\pi$ cm², $V = 100\pi$ cm³   **10.** $A = 36\pi$ in.², $V = 16\pi$ in.³
**11.** $A = 360$ m², $V = 400$ m³   **12.** $A = 96$ in.²; $V = 48$ in.³
**13.** 576 cm³   **14.** $180\pi$ cm³   **15.** $\frac{320}{3}\pi$ in.³   **16.** 52 in.³   **17.** $\frac{1}{8}$ in.³
**18.** 10,800 ft³   **19.** $63\pi$ ft³   **20.** $3\pi$ in.³   **21.** $100\pi$ ft²   **22.** 24 yd²
**23.** $\frac{32,000}{3}\pi$ ft³   **24.** 100 in.³   **25.** $V = lwh = s \cdot s \cdot s = s^3$
**26.** 1728 in.³   **27.** 27 ft³   **28.** 46,656 in.³

## CHAPTER FIVE REVIEW EXERCISES   (page 184)

**1.** 49 cm²   **2.** 590.49 m²   **3.** 180 ft²   **4.** 156 m²   **5.** 240 cm²
**6.** 128 in.²   **7.** 80 in.²   **8.** 152.5 in.²   **9.** 72 m²   **10.** 277.22 cm²
**11.** 48 cm²   **12.** 10.24 cm   **13.** 9 ft²   **14.** 144 in.²   **15.** 113.04 ft
**16.** 18 ft   **17.** $56.25\pi$ cm²   **18.** 10 cm   **19.** 17 m   **20.** Yes   **21.** 750 ft
**22.** Yes   **23.** 6 ft   **24.** $\dfrac{10\sqrt{3}}{3}$ in.   **25.** $8\sqrt{2}$ cm   **26.** $25\sqrt{3}$ m²
**27.** 1350 cm²   **28.** 315 m³   **29.** 254.52 in.³   **30.** $100\pi$ m²
**31.** $672\pi$ in.²   **32.** $2304\pi$ in.³   **33.** 128.29 cm²   **34.** $112.5\pi$ cm³

**35.** 93,312 in.³  **36.** 5 ft³  **37.** 576 in.³  **38.** 23 pounds  **39.** 1614 ft²
**40.** 8.25π ft³  **41.** 533⅓ yd³  **42.** 9 cm²  **45.** Approximately .0013 in.
**46.** $V = d^3$  **47.** $\dfrac{3\sqrt{7}}{2}$ in.

## CHAPTER FIVE TEST  (page 186)

**1.** T  **2.** F  **3.** T  **4.** F  **5.** F  **6.** F  **7.** F  **8.** F  **9.** T
**10.** T  **11.** F  **12.** F  **13.** T  **14.** F  **15.** T  **16.** T  **17.** T
**18.** T  **19.** F  **20.** T  **21.** T  **22.** T  **23.** F  **24.** T  **25.** F

## EXERCISE 6.1  (page 198)

**1.** $\frac{5}{7}$  **2.** $\frac{3}{5}$  **3.** $\frac{1}{2}$  **4.** $\frac{1}{2}$  **5.** $\frac{2}{3}$  **6.** $\frac{2}{3}$  **7.** $\frac{1}{3}$  **8.** $\frac{1}{5}$  **9.** $\frac{1}{5}$  **10.** $\frac{1}{2}$
**11.** $\frac{3}{7}$  **12.** $\frac{1}{2}$  **13.** $\frac{1}{18}$  **14.** $\frac{1}{1}$  **15.** $\frac{3}{4}$  **16.** $\frac{5}{1}$  **17.** $\frac{6}{5}$
**18.** $\frac{1}{4}$  **19.** $\frac{25}{1}$  **20.** $\frac{4}{1}$  **21.** No  **22.** Yes  **23.** Yes  **24.** Yes  **25.** No
**26.** No  **27.** Yes  **28.** No  **29.** 4  **30.** 4  **31.** 6  **32.** 4  **33.** 3
**34.** 16  **35.** 9  **36.** 6  **37.** 0  **38.** $\frac{5}{3}$  **39.** −3  **40.** 5  **41.** 65
**42.** 102  **43.** 51  **44.** 50  **45.** 9  **46.** 5  **47.** $17  **48.** $62.50
**49.** $6.50  **50.** 20  **51.** 7.5 gal  **52.** 204 mi  **53.** $309  **54.** 65.25 ft
**55.** 49.29 ft  **56.** 32 ft  **57.** 162 teachers  **58.** 80 ft
**59.** 48 to 1 close enough  **60.** 33, 30, $\frac{11}{10}$

## EXERCISE 6.2  (page 206)

**1.** T  **2.** F  **3.** F  **4.** F  **5.** T  **6.** T  **7.** $AA \cong AA$
**8.** $AA \cong AA$  **9.** $AA \cong AA$  **10.** $AA \cong AA$  **11.** 8¾ cm
**12.** 5⅗ cm  **13.** 6$\frac{18}{25}$ cm  **14.** 18 cm  **15.** 8  **16.** 7  **17.** 144
**18.** 12  **19.** 44  **20.** 8  **21.** 50°  **22.** 30°  **27.** $\frac{9}{2}$  **28.** $\frac{16}{5}$  **29.** 4
**30.** $\frac{48}{7}$  **31.** m($\overline{AD}$) = 12, m($\overline{DB}$) = 20  **32.** 18  **33.** 60  **34.** 20
**35.** $\frac{20}{3}$  **36.** 9  **37.** $\frac{16}{3}$  **38.** 4  **39.** 37½ ft  **40.** 32$\frac{8}{11}$ ft  **41.** 60 ft
**42.** 1466⅔ ft  **43.** 48 ft  **44.** 9 m  **45.** 528 ft  **46.** 733⅓ ft  **47.** 605½ ft
**48.** 174 ft

## EXERCISE 6.3  (page 217)

**1.** Yes  **2.** No  **3.** Yes  **4.** No  **5.** 8  **6.** 16$\sqrt{2}$  **7.** 9 and −9
**8.** 20 and −20  **9.** 8  **10.** 9  **11.** m($\overline{DB}$) = 12 ft
**12.** m($\overline{CD}$) = 8 ft  **13.** m($\overline{DB}$) = 35 yd  **14.** m($\overline{AD}$) = 48 in.
**15.** m($\overline{AB}$) = 34 dm  **16.** m($\overline{CD}$) = 6 cm  **17.** m($\overline{AC}$) = 2$\sqrt{26}$ in.

**18.** $m(\overline{CD}) = 2\sqrt{6}$ ft   **19.** $m(\overline{CD}) = 2\sqrt{15}$ ft   **20.** $12\sqrt{10}$ m
**21.** 80 yd² **22.** 60 ft² **23.** 12 m² **24.** $24\sqrt{3}$ in.² **25.** $37\frac{1}{2}$ ft
**26.** $23\frac{1}{3}$ m **27.** 30 cm² **28.** Approximately 1799 ft **29.** 360 cm²
**30.** $d = \dfrac{yz}{x}$ **31.** 150 ft

## EXERCISE 6.4    (page 222)

**1.** 5.5 ft **2.** 33.9 in. **3.** 87.8 cm **4.** 17.1 m **5.** 1393 mi
**6.** 30.6 in. **7.** 26.6 ft **8.** 322 cm **9.** 7.3 ft **10.** 136,000 m
**11.** 231 ft **12.** 837 ft **13.** 57.5 ft **14.** 54.3 ft **15.** 43 ft **16.** 124 ft
**17.** 213 ft **18.** 5.27 ft **19.** 18 ft² **20.** 244 ft² **21.** $\frac{1}{3}$ **22.** $16\sqrt{3}$ cm²
**23.** $\tan 30° = \sqrt{3}/3$ **24.** $\tan 60° = \sqrt{3}$

## EXERCISE 6.5    (page 227)

**1.** 64.3 ft **2.** 107.6 in. **3.** 413.6 cm **4.** 29.3 m **5.** 25.6 mi
**6.** 320.6 in. **7.** 887 ft **8.** 552 cm **9.** 12 ft **10.** 94 m
**11.** 34.1 ft **12.** 52.5 in. **13.** 126 cm **14.** 27.5 m
**15.** 61.0 mi **16.** 45.2 in. **17.** 949 ft **18.** 4910 cm **19.** 0.175 ft
**20.** 7451 m **21.** 106 ft **22.** 300 ft **23.** 89.4 ft **24.** 241 ft
**25.** 22.6 in. **26.** 31.7 ft **27.** 0.21 mi or 1100 ft
**28.** 118 cm **29.** 15.3 in. **30.** $\sin 60° = \sqrt{3}/2$
**31.** $\sin 45° = \sqrt{2}/2$ **32.** 158 ft **33.** 165 ft **34.** 905 ft
**35.** 53.7 nautical mi **36.** 15.3 in. **37.** 24.1 ft **38.** 18.8 ft
**39.** 203 cm **40.** 6.8 in. **41.** $\cos 60° = \frac{1}{2}$ **42.** $\cos 45° = \sqrt{2}/2$
**43.** Yes **44.** Yes

## CHAPTER SIX REVIEW EXERCISES    (page 229)

**1.** $\frac{5}{12}$ **2.** $\frac{7}{32}$ **3.** No **4.** No **5.** 3 **6.** $\frac{85}{12}$ **7.** $\frac{19}{2}$ **8.** $-18$ **10.** 5
**11.** 18 **12.** 9 **13.** 5 **14.** 44 **15.** 50° **16.** 90° **17.** 40° **18.** 50°
**19.** 6 **20.** 8 **21.** 90 ft **22.** $21\frac{9}{11}$ ft **23.** 3 ft
**24.** 9 ft **25.** $12\sqrt{2}$ **26.** $-25$ **27.** 4 **28.** $\frac{3}{4}$ **29.** 20 **30.** 6 **31.** $\frac{3}{4}$
**32.** $\frac{3}{5}$ **33.** $\frac{4}{5}$ **34.** $\frac{3}{5}$ **35.** $\frac{4}{5}$ **36.** $\frac{4}{3}$ **37.** $\frac{1}{2}$ **38.** $\dfrac{\sqrt{3}}{3}$ **39.** $\dfrac{\sqrt{3}}{2}$
**40.** $\sqrt{3}$ **41.** $\dfrac{\sqrt{2}}{2}$ **42.** $\dfrac{\sqrt{2}}{2}$ **43.** 1 **44.** $\frac{1}{2}$ **45.** 0.9397
**46.** 0.9397 **47.** 5.6713 **48.** 0.7660 **49.** 0.4540 **50.** 0.6018
**51.** 2.1445 **52.** 0.3249 **53.** 4842 ft **54.** 1769 ft **55.** about 28 ft
**56.** about 52 ft **57.** 12.7 ft **58.** 377 ft

## CHAPTER SIX TEST  (page 232)

**1.** T  **2.** T  **3.** T  **4.** F  **5.** F  **6.** T  **7.** T  **8.** F  **9.** F
**10.** T  **11.** T  **12.** F  **13.** T  **14.** T  **15.** T  **16.** F  **17.** F
**18.** F  **19.** T  **20.** T  **21.** F  **22.** T  **23.** T  **24.** T  **25.** T
**26.** F  **27.** F  **28.** T  **29.** T  **30.** T

## EXERCISE 7.1  (page 246)

**1.** 45°  **2.** 22$\frac{1}{2}$°  **3.** 35°  **4.** 170°  **5.** 70°  **6.** 100°  **7.** 50°
**8.** 80°  **9.** 45°  **10.** 90°  **11.** 45°  **12.** 85°  **13.** 20°  **14.** 140°
**15.** 36°  **16.** 15°  **17.** 100°  **18.** 50°  **19.** 70°  **20.** 120°
**21.** 70°  **22.** 110°  **23.** 90°  **24.** 35°  **25.** 130°  **26.** 50°,  30°
**27.** 30°, 15°  **28.** 40°

## EXERCISE 7.2  (page 254)

**1.** 6 in.  **2.** 60°  **3.** 125°  **4.** 80°  **5.** 8 in.  **6.** 1 cm  **7.** 80°
**8.** 160°  **9.** 140°  **10.** 20°  **11.** 120°  **12.** 160°  **13.** 240°
**14.** 75°

## EXERCISE 7.3  (page 258)

**1.** 2  **2.** An unlimited number.  **3.** 0, 1, or 2  **4.** 0, 1, or 2
**5.** 12 cm  **6.** 40°  **7.** 150°  **8.** 20°  **9.** 100°  **10.** 50°  **11.** 160°
**12.** 100°  **13.** 10 cm  **14.** 60°  **15.** 130°  **16.** 10°

## EXERCISE 7.4  (page 268)

**1.** 3 cm  **2.** 13 cm  **3.** 13 in.  **4.** 2 in.  **5.** 12 cm  **6.** 6 m
**7.** 32 yd  **8.** 125 ft  **9.** $4\sqrt{10}$ cm  **10.** 60 cm  **11.** 78$\frac{3}{4}$ in.
**12.** $10\sqrt{5}$ yd  **13.** 25 cm  **14.** 10 ft  **15.** 16 cm
**16.** 11 mi  **17.** 11 m  **18.** 30 cm  **19.** 100°  **20.** 160°
**21.** 80°  **22.** 60 cm²  **23.** 130°

## CHAPTER SEVEN REVIEW EXERCISES  (page 270)

**1.** 40°  **2.** 40°  **3.** 40°  **4.** 40°  **5.** 100°  **6.** 50°  **7.** 70°
**8.** 90°  **9.** 85°  **10.** 95°  **11.** 70°  **12.** 20°  **13.** 120°  **14.** 80°
**15.** 35°  **16.** 60°  **17.** 27.5°  **18.** 70°  **19.** 16 cm  **20.** 60°
**21.** 20°  **22.** 280°  **23.** 8 m  **24.** 50°  **25.** 140°  **26.** 20°  **27.** 2$\frac{2}{5}$
**28.** 13$\frac{2}{3}$  **29.** 10$\frac{3}{4}$  **30.** 4  **31.** 3$\frac{3}{5}$  **32.** 6  **33.** 5  **34.** 1

## CHAPTER SEVEN TEST    (page 272)

**1.** T   **2.** F   **3.** T   **4.** F   **5.** F   **6.** T   **7.** F   **8.** F   **9.** T
**10.** F   **11.** F   **12.** F   **13.** F   **14.** T   **15.** T   **16.** T   **17.** T
**18.** F   **19.** T   **20.** F

## EXERCISE 8.1    (page 281)

**1.** $x < 4$   **2.** $x > -2$   **3.** $x < -10/3$   **4.** $x > -10$   **5.** $x < 1$
**6.** $x > 2$   **7.** No solutions.   **8.** $x < 10$   **9.** $x < \frac{2}{5}$
**10.** All values of $x$.   **11.** $x > \frac{30}{19}$   **12.** $x > \frac{10}{3}$   **13.** $x > 0$   **14.** $x < 2$

## EXERCISE 8.2    (page 288)

**1.** $\angle A$   **2.** $\angle B$   **3.** $\overline{AB}$   **4.** $\overline{AB}$   **5.** No   **6.** Yes   **7.** $\angle C$
**8.** $\overline{DE}$   **9.** A positive number.   **10.** $\overline{AC}$   **11.** $\angle 1$   **12.** $\angle DOB$
**13.** $\widehat{DAB}$   **14.** $\angle 2$

## EXERCISE 8.3    (page 292)

**1.** $\widehat{AB}$   **2.** $\overline{EC}$   **3.** $\widehat{DC}$   **4.** $\overline{CD}$   **5.** $\overline{AB}$   **6.** $\widehat{CAB}$

## CHAPTER EIGHT REVIEW EXERCISES    (page 293)

**1.** $x < 4$   **2.** $x > 15$   **3.** $y > 5$   **4.** $y < 4$   **5.** $t < 3$   **6.** $t > 3$
**7.** $a > 56$   **8.** $b > -15$   **9.** $y < 20$   **10.** $c > -\frac{24}{15}$
**11.** $x < 1$   **12.** $z > 4$   **13.** $\angle C$   **14.** $\angle 1$   **15.** $\angle 3$   **16.** $\angle A$   **17.** No
**18.** Yes   **19.** $\angle 3$   **20.** $\angle 4$   **21.** $\angle 2$   **22.** $\widehat{DC}$   **23.** $\widehat{BC}$   **24.** $\widehat{BC}$
**25.** $\overline{OA}$   **26.** $\angle 1$   **27.** $\widehat{AB}$   **28.** $\angle 2$   **29.** $\overline{CD}$   **30.** $\overline{AD}$   **31.** $\overline{AB}$
**32.** $m(\widehat{AB})$   **33.** $m(\angle 2)$   **34.** $\overline{CD}$   **35.** $m(\overline{CD}), m(\overline{BC}), m(\overline{AB})$
**36.** $m(\overline{OR}), m(\overline{OQ}), m(\overline{OP})$   **37.** $m(\overline{AB}), m(\overline{BC}), m(\overline{CD})$
**38.** $m(\overline{OP}), m(\overline{OQ}), m(\overline{OR})$

## CHAPTER EIGHT TEST    (page 295)

**1.** F   **2.** T   **3.** F   **4.** T   **5.** F   **6.** T   **7.** T   **8.** T   **9.** T
**10.** T   **11.** F   **12.** F   **13.** T   **14.** T   **15.** T   **16.** T   **17.** F
**18.** T   **19.** T   **20.** F

## EXERCISE 9.1 (page 304)

**1.** A parallel line midway between them.
**2.** Along the arc of a circle.     **3.** Along the arc of a circle.
**4.** A point.     **11.** A right circular cylinder with the given line as axis.
**12.** Two planes, parallel to and on opposite sides of the given plane.
**13.** A plane perpendicular to and bisecting the line segment joining the two points.
**14.** A cone.

## EXERCISE 9.2 (page 311)

**8.** No     **11.** Yes     **12.** $\frac{4}{3}$ in.     **13.** $\frac{8}{3}$ in.     **14.** 9 in.

## CHAPTER NINE REVIEW EXERCISES (page 312)

**5.** A line perpendicular to the plane of the circle and passing through the center of the circle.
**6.** A line perpendicular to the plane of the three points, and passing through the center of the circle that contains the points.

## CHAPTER NINE TEST (page 313)

**1.** F   **2.** T   **3.** T   **4.** T   **5.** T   **6.** F   **7.** T   **8.** T   **9.** F
**10.** F   **11.** T   **12.** T   **13.** T   **14.** T   **15.** T   **16.** F   **17.** T
**18.** F   **19.** F   **20.** T

## EXERCISE 10.1 (page 322)

**8.** Yes   **9.** Yes   **10.** No   **11.** 45°   **12.** 36°   **13.** 10 cm
**14.** $6\sqrt{2}$ in.

## EXERCISE 10.2 (page 329)

**1.** 1600 cm²   **2.** $48\sqrt{3}$ in.²   **3.** $150\sqrt{3}$ m²   **4.** 12 yd
**5.** 225π cm²   **6.** 30π cm   **7.** 25π/12 ft²   **8.** 12 units
**9.** $48\sqrt{3}$ in².   **10.** 750 in².   **11.** 245π/36 in².
**12.** $r = 2$ units   **14.** 2500π yd²
**15.** $(18\pi - 36)$ cm² or approximately 38.5 cm²
**16.** $\left(\dfrac{400\pi}{3} - 100\sqrt{3}\right)$ cm² or approximately 245.67 cm²

# CHAPTER TEN REVIEW EXERCISES   (page 330)

**5.** 60°  **6.** 18°  **7.** 12 m  **8.** 24.5 cm  **9.** 309 ft²  **10.** $8\sqrt{3}$ cm²

**11.** 7.95 m  **12.** 144π in².  **13.** $\dfrac{50\pi}{3}$ ft²  **14.** 6 cm  **15.** 540 in².

**16.** 4π m²  **17.** 3 in.  **18.** 6 ft  **19.** $4\sqrt{\pi}$ m

# CHAPTER TEN TEST   (page 331)

**1.** F  **2.** T  **3.** T  **4.** T  **5.** F  **6.** F  **7.** T  **8.** T  **9.** F
**10.** T  **11.** F  **12.** F  **13.** T  **14.** F  **15.** T  **16.** F  **17.** T
**18.** F  **19.** F  **20.** T

# EXERCISE 11.1   (page 339)

**1.** Yes  **2.** No  **3.** Yes  **4.** No  **5.** No  **6.** No  **7.** Yes  **8.** Yes
**9.** Today is not Tuesday.  **10.** The crops will survive.  **11.** The earth
is not round.  **12.** The moon is not made of green cheese.
**13.** $4 \neq 7$  **14.** $7 = 7$  **15.** Spaghetti is not slimy.  **16.** Crime pays.
**17.** Bill and Sue are not friends.  **18.** Bill is not older than Sue.
**19.** F  **20.** T  **21.** F  **22.** T  **23.** F  **24.** T  **25.** F  **26.** F
**27.** T  **28.** T  **29.** F  **30.** T  **31.** T  **32.** F  **33.** F  **34.** T
**35.** T  **36.** F  **37.** T  **38.** T  **39.** F  **40.** F  **41.** F  **42.** T
**43.** Bob did not catch a fish.
**44.** Jim did not catch a cold.
**45.** Bob caught a fish or Jim caught a cold.
**46.** Bob caught a fish and Jim caught a cold.
**47.** If Bob caught a fish, then Jim caught a cold.
**48.** If Jim caught a cold, then Bob caught a fish.
**49.** Bob caught a fish and Jim did not catch a cold.
**50.** Jim didn't catch a cold or Bob caught a fish.
**51.** If Bob didn't catch a fish, then Jim caught a cold.
**52.** If Bob caught a fish, then Jim didn't catch a cold.
**53.** Bob caught a fish if, and only if, Jim caught a cold.
**54.** Bob didn't catch a fish if, and only if, Jim didn't catch a cold.
**55.** A triangle does not have three sides and a pentagon does not have
five sides.
**56.** A quadrilateral does not have four sides or a pentagon does not have
five sides.
**57.** A triangle has three sides and a quadrilateral has four sides and a
pentagon has five sides.
**58.** A triangle has three sides or a quadrilateral has four sides or a pentagon
has five sides.

**59.** If a triangle has three sides, then both a pentagon has five sides and a quadrilateral has four sides.

**60.** A pentagon has five sides and either a triangle has three sides or a quadrilateral has four sides.

**61.** If a triangle has three sides, then a pentagon has five sides, and a quadrilateral has four sides.

**62.** Either a pentagon has five sides and a triangle has three sides, or a quadrilateral has four sides.

**63.** If a triangle has three sides and a quadrilateral has four sides, then a pentagon has five sides.

**64.** A triangle has three sides and, if a quadrilateral has four sides, then a pentagon has five sides.

**65.** $p \wedge q$   **66.** $p \vee r$   **67.** $\sim p \vee r$   **68.** $p \Rightarrow q$   **69.** $\sim q \Rightarrow \sim r$

**70.** $p \Rightarrow (q \vee r)$   **71.** $(p \vee q) \Rightarrow r$   **72.** $(\sim p \wedge \sim q) \vee r$

**73.** $r \Rightarrow (p \wedge q)$   **74.** $(\sim p \vee \sim q) \Rightarrow \sim r$   **75.** $\sim p \vee (\sim q \Rightarrow \sim r)$

**76.** $(\sim p \wedge \sim q) \Rightarrow \sim r$   **77.** $p \Leftrightarrow q$   **78.** $p \Leftrightarrow q$

**79.** *Converse*: If you pass, then you will study.
*Inverse*: If you don't study, then you won't pass.
*Contrapositive*: If you don't pass, then you don't study.

**80.** *Converse*: If Sue is disappointed, then she was not invited.
*Inverse*: If Sue is invited, then she will not be disappointed.
*Contrapositive*: If Sue is not disappointed, then she is invited.

**81.** *Converse*: If a triangle is equiangular, then it is equilateral.
*Inverse*: If a triangle is not equilateral, then it is not equiangular.
*Contrapositive*: If a triangle is not equiangular, then it is not equilateral.

**82.** *Converse*: If you take five classes, then you are a full-time student.
*Inverse*: If you are not a full-time student, they you will not take five classes.
*Contrapositive*: If you do not take five classes, then you are not a full-time student.

**83.** *Converse*: If you just wait a minute, then you do not like the weather.
*Inverse*: If you like the weather, then don't wait a minute.
*Contrapositive*: If you don't wait a minute, then you like the weather.

**84.** *Converse*: If I missed Friday's exam, then today is Saturday.
*Inverse*: If today isn't Saturday, then I didn't miss Friday's exam.
*Contrapositive*: If I didn't miss Friday's exam, then today is not Saturday.

**85.** *Converse*: If they grow, then you watered the plants.
*Inverse*: If you didn't water the plants, then they won't grow.
*Contrapositive*: If they won't grow, then you didn't water the plants.

**86.** *Converse*: If you can, then you think you can.
*Inverse*: If you don't think you can, then you can't.
*Contrapositive*: If you can't, then you don't think you can.

**87.** *Converse*: If we have a hot summer, then it's a mild spring.
*Inverse*: If it's not a mild spring, then we won't have a hot summer.

*Contrapositive*: If we don't have a hot summer, then it's not a mild spring.

**88.** *Converse*: If you can't have dessert, then you don't finish your vegetables.

*Inverse*: If you finish your vegetables, then you can have dessert.

*Contrapositive*: If you can have dessert, then you finish your vegetables.

**89.** *Converse*: If you would do better, then you finish your homework.

*Inverse*: If you do not finish your homework, then you won't do better.

*Contrapositive*: If you don't do better, then you don't do your homework.

**90.** *Converse*: If you don't succeed, then you don't try.

*Inverse*: If you try, then you will succeed.

*Contrapositive*: If you succeed, then you try.

**91.** *Converse*: If you don't understand the situation, then you can keep your head while others around you are losing theirs.

*Inverse*: If you can't keep your head while others around you are losing theirs, then you do understand the situation.

*Contrapositive*: If you do understand the situation, then you don't keep your head while others around you are losing theirs.

**92.** *Converse*: If people don't learn to trust you, then you don't tell the truth

*Inverse*: If you tell the truth, then people will learn to trust you.

*Contrapositive*: If people learn to trust you, then you tell the truth.

**97.** If you don't like mosquitoes, then you won't love camping.

**98.** If Sue does not stay, then Sam will not stay.

## EXERCISE 11.2  (page 345)

**1.**

| $p$ | $\sim p$ | $\sim p \vee p$ |
|---|---|---|
| T | F | T |
| F | T | T |

**2.**

| $p$ | $\sim p$ | $p \wedge \sim p$ |
|---|---|---|
| T | F | F |
| F | T | F |

**3.**

| $p$ | $q$ | $\sim p$ | $\sim p \vee q$ |
|---|---|---|---|
| T | T | F | T |
| T | F | F | F |
| F | T | T | T |
| F | F | T | T |

**4.**

| $p$ | $q$ | $\sim p$ | $\sim q$ | $p \wedge \sim q$ |
|---|---|---|---|---|
| T | T | F | F | F |
| T | F | F | T | T |
| F | T | T | F | F |
| F | F | T | T | F |

**5.**

| $p$ | $q$ | $\sim p$ | $\sim p \Leftrightarrow q$ |
|---|---|---|---|
| T | T | F | F |
| T | F | F | T |
| F | T | T | T |
| F | F | T | F |

**6.**

| $p$ | $q$ | $\sim q$ | $p \Leftrightarrow \sim q$ |
|---|---|---|---|
| T | T | F | F |
| T | F | T | T |
| F | T | F | T |
| F | F | T | F |

**7.**

| $p$ | $q$ | $p \Rightarrow q$ | $\sim(p \Rightarrow q)$ |
|---|---|---|---|
| T | T | T | F |
| T | F | F | T |
| F | T | T | F |
| F | F | T | F |

**8.**

| $p$ | $q$ | $p \Leftrightarrow q$ | $\sim(p \Leftrightarrow q)$ |
|---|---|---|---|
| T | T | T | F |
| T | F | F | T |
| F | T | F | T |
| F | F | T | F |

**9.**

| $p$ | $q$ | $p \vee q$ | $p \wedge (p \vee q)$ |
|---|---|---|---|
| T | T | T | T |
| T | F | T | T |
| F | T | T | F |
| F | F | F | F |

**10.**

| $p$ | $q$ | $p \wedge q$ | $p \vee (p \wedge q)$ |
|---|---|---|---|
| T | T | T | T |
| T | F | F | T |
| F | T | F | F |
| F | F | F | F |

**11.**

| $p$ | $q$ | $p \vee q$ | $p \Rightarrow (p \vee q)$ |
|---|---|---|---|
| T | T | T | T |
| T | F | T | T |
| F | T | T | T |
| F | F | F | T |

**12.**

| $p$ | $q$ | $p \wedge q$ | $q$ | $(p \wedge q) \Rightarrow q$ |
|---|---|---|---|---|
| T | T | T | T | T |
| T | F | F | F | T |
| F | T | F | T | T |
| F | F | F | F | T |

**13.**

| $p$ | $q$ | $\sim p$ | $p \Leftrightarrow q$ | $\sim p \vee (p \Leftrightarrow q)$ |
|---|---|---|---|---|
| T | T | F | T | T |
| T | F | F | F | F |
| F | T | T | F | T |
| F | F | T | T | T |

**14.**

| $p$ | $q$ | $\sim p$ | $(p \Leftrightarrow \sim p)$ | $(p \Leftrightarrow \sim p) \Rightarrow q$ |
|---|---|---|---|---|
| T | T | F | F | T |
| T | F | F | F | T |
| F | T | T | F | T |
| F | F | T | F | T |

**15.**

| $p$ | $\sim p$ | $\sim p \vee p$ |
|---|---|---|
| T | F | T |
| F | T | T |

**16.**

| $p$ | $\sim p$ | $p \wedge \sim p$ | $\sim (p \wedge \sim p)$ |
|---|---|---|---|
| T | F | F | T |
| F | T | F | T |

**17.**

| $p$ | $p \vee p$ | $(p \vee p) \Leftrightarrow p$ |
|---|---|---|
| T | T | T |
| F | F | T |

**18.**

| $p$ | $p \wedge p$ | $p \Leftrightarrow (p \wedge p)$ |
|---|---|---|
| T | T | T |
| F | F | T |

**19.**

| $p$ | $q$ | $p \vee q$ | $p \Rightarrow (p \vee q)$ |
|---|---|---|---|
| T | T | T | T |
| T | F | T | T |
| F | T | T | T |
| F | F | F | T |

**20.**

| $p$ | $q$ | $p \wedge q$ | $(p \wedge q) \Rightarrow p$ |
|---|---|---|---|
| T | T | T | T |
| T | F | F | T |
| F | T | F | T |
| F | F | F | T |

**21.**

| $p$ | $q$ | $p \vee q$ | $q \vee p$ | $(p \vee q) \Leftrightarrow (q \vee p)$ |
|---|---|---|---|---|
| T | T | T | T | T |
| T | F | T | T | T |
| F | T | T | T | T |
| F | F | F | F | T |

**22.**

| $p$ | $q$ | $(p \wedge q)$ | $(q \wedge p)$ | $(p \wedge q) \Leftrightarrow (q \wedge p)$ |
|---|---|---|---|---|
| T | T | T | T | T |
| T | F | F | F | T |
| F | T | F | F | T |
| F | F | F | F | T |

**23.**

| $p$ | $q$ | $p \vee q$ | $p \wedge (p \vee q)$ | $p \wedge (p \vee q) \Leftrightarrow p$ |
|---|---|---|---|---|
| T | T | T | T | T |
| T | F | T | T | T |
| F | T | T | F | T |
| F | F | F | F | T |

**24.**

| p | q | p ∧ q | p ∨ (p ∧ q) | p ∨ (p ∧ q) ⇔ p |
|---|---|-------|-------------|-----------------|
| T | T | T | T | T |
| T | F | T | T | T |
| F | T | F | F | T |
| F | F | F | F | T |

**25.**

| p | q | p ⇒ q | ~p | ~p ∨ q | (p ⇒ q) ⇔ (~p ∨ q) |
|---|---|-------|----|--------|--------------------|
| T | T | T | F | T | T |
| T | F | F | F | F | T |
| F | T | T | T | T | T |
| F | F | T | T | T | T |

**26.**

| p | q | (p ⇒ q) | ~(p ⇒ q) | ~q | p ∧ ~q | ~(p ⇒ q) ⇔ (p ∧ ~q) |
|---|---|---------|----------|----|--------|----------------------|
| T | T | T | F | F | F | T |
| T | F | F | T | T | T | T |
| F | T | T | F | F | F | T |
| F | F | T | F | T | F | T |

**27.**

| p | q | p ⇔ q | ~(p ⇔ q) | ~q | p ⇔ ~q | ~(p ⇔ q) ⇔ (p ⇔ ~q) |
|---|---|-------|----------|----|--------|----------------------|
| T | T | T | F | F | F | T |
| T | F | F | T | T | T | T |
| F | T | F | T | F | T | T |
| F | F | T | F | T | F | T |

**28.**

| p | q | p ⇔ q | p ⇒ q | q ⇒ p | [(p ⇒ q) ∧ (q ⇒ p)] | (p ⇔ q) ⇔ [(p ⇒ q) ∧ (q ⇒ p)] |
|---|---|-------|-------|-------|----------------------|--------------------------------|
| T | T | T | T | T | T | T |
| T | F | F | F | T | F | T |
| F | T | F | T | F | F | T |
| F | F | T | T | T | T | T |

**33.** It's not hot or it's not humid.

**34.** The moon is not full or the kids are not wild.

**35.** We are not late or my watch is not broken.

**36.** I will not watch the late movie and I will not go to bed.

**37.** I didn't hear a voice and I am not imagining things.

**38.** Tim is not fat or he is not forty.

**39.** You don't buy me a ring and I won't break the engagement.

**40.** Wishes are horses and beggars don't ride.

**41.** It tastes good and it's not fattening.

**42.** You made a mess and you neither cleaned it up nor said you were sorry.

# EXERCISE 11.3  (page 351)

**1.** I shouldn't drive.   **2.** I am not tired.   **3.** No valid conclusion.
**4.** No valid conclusion.   **5.** I'll take a nap.

**6.** No valid conclusion.   **7.** No valid conclusion.
**8.** I will not come home early.   **9.** She has fleas.
**10.** No valid conclusion.   **11.** No valid conclusion.
**12.** She should be bathed.   **13.** Tomorrow is a holiday.
**14.** No valid conclusion.   **15.** No valid conclusion.
**16.** Today is not Friday.   **17.** It is not raining.
**18.** No valid conclusion.   **19.** It is cloudy.
**20.** No valid conclusion.   **21.** It is raining.
**22.** No valid conclusion.   **23.** No valid conclusion.
**24.** No valid conclusion.   **25.** It will rain.
**26.** Wishes are not horses.   **27.** No valid conclusion.
**28.** It is not hot and humid.   **29.** I will pass geometry.
**30.** No valid conclusion.   **31.** Purple is not yellow.
**32.** Red is not green.   **33.** I will pass geometry.
**34.** No valid conclusion.   **35.** The mail is late.
**36.** I have the measles.   **37.** I am sick.   **38.** I am sick.
**39.** It's Friday.   **40.** Geometry is fun.   **41.** It's not cloudy.
**42.** Purple is not yellow.   **43.** Red is purple.
**44.** If my parakeet gets lonesome, then he will sing.
**45.** If I don't get a diamond, then my mother will be annoyed.
**46.** If red is green, then I'll adjust the TV.
**47.** If Frank quits the game, then the game is over.
**48.** No valid conclusion.
**49.** If the door is unlocked, then the security is not good.
**50.** If the temperatures have been below zero, then I invite my friends.

**51.**

| $p$ | $q$ | $p \vee q$ | $\sim p$ | $(p \vee q) \wedge \sim p$ | $[(p \vee q) \wedge \sim p] \Rightarrow q$ |
|---|---|---|---|---|---|
| T | T | T | F | F | T |
| T | F | T | F | F | T |
| F | T | T | T | T | T |
| F | F | F | T | F | T |

**52.**

| $p$ | $q$ | $r$ | $p \Rightarrow q$ | $q \Rightarrow r$ | $(p \Rightarrow q) \wedge (q \Rightarrow r)$ | $p \Rightarrow r$ | $[(p \Rightarrow q) \wedge (q \Rightarrow r)] \Rightarrow (p \Rightarrow r)$ |
|---|---|---|---|---|---|---|---|
| T | T | T | T | T | T | T | T |
| T | T | F | T | F | F | F | T |
| T | F | T | F | T | F | T | T |
| T | F | F | F | T | F | F | T |
| F | T | T | T | T | T | T | T |
| F | T | F | T | F | F | T | T |
| F | F | T | T | T | T | T | T |
| F | F | F | T | T | T | T | T |

## EXERCISE 11.4   (page 355)

**1.** Jean will pass geometry.   **2.** No conclusion.   **3.** Bob will flunk.
**4.** No conclusion.   **5.** Jim will not succeed.   **6.** George is not lazy.

7. All my friends are not successful students.
8. None of my friends are lazy.    **9.** Some good cars are expensive.
10. All foreign cars are expensive.    **11.** No conclusion.
12. Math teachers are rich.    **13.** No conclusion.
14. Ostriches are illiterate.    **15.** Alligators are not good teachers.
16. All alligators can be helped.    **17.** No aardvark is a good companion.
18. No kittens are reptiles.    **19.** No conclusion.
20. All gunners' mates are enlisted sailors.    **21.** No cats are reptiles.
22. No babies are biologists.    **23.** None of my sisters can sing.
24. No lizard needs a hairbrush.    **25.** Babies cannot manage crocodiles.
26. These apples were not grown in the shade.
27. All pawnbrokers are honest.

## EXERCISE 11.5    (page 359)

1. No. All great circles intersect.    **2.** No. The segment can be either part of the great circle.    **3.** Yes    **4.** Yes    **5.** Yes
6. less than 540°    **7.** No    **8.** No

## CHAPTER ELEVEN REVIEW EXERCISES    (page 360)

1. Today is not Wednesday.    **2.** $8 = 9$    **3.** I am not going to the store and I am not going to a movie.    **4.** Jim is not shopping or Jill is not bored.    **5.** I take a nap and I will not go out tonight.
6. You have the time and you don't call me.    **7.** T    **8.** F
9. T    **10.** T    **11.** T    **12.** T    **13.** T    **14.** F
15. It is not true that both Fred is fat and Sam is sloppy.    **16.** Fred is not fat or Ted is not tired.    **17.** Fred is fat or Ted is not tired, and Sam is sloppy.    **18.** Either Fred is fat or, Ted is not tired and Sam is sloppy.
19. If Fred is fat and Ted is tired, then Sam is sloppy.
20. Fred is fat and if Ted is tired, then Sam is sloppy.
21. $q \lor p$    **22.** $r \Rightarrow q$    **23.** $q \Rightarrow {\sim}p$    **24.** $({\sim}r \land {\sim}q) \Rightarrow {\sim}p$
25. *Converse*: If the car stops, then you applied the brakes.
    *Inverse*: If you don't apply the brakes, then the car will not stop.
    *Contrapositive*: If the car doesn't stop, then you didn't apply the brakes.
26. *Converse*: If you are shoplifting, then you did not pay for that.
    *Inverse*: If you pay for that, then you are not shoplifting.
    *Contrapositive*: If you are not shoplifting, then you will pay for that.
27. *Converse*: If a triangle has two congruent angles, then it is isosceles.
    *Inverse*: If a triangle isn't isosceles, then it doesn't have two congruent angles.
    *Contrapositive*: If a triangle doesn't have two congruent angles, then it isn't isosceles.

**28.** *Converse*: If you would just listen, then you would understand me better.

*Inverse*: If you wouldn't understand me better, then you wouldn't listen.

*Contrapositive*: If you wouldn't listen, then you wouldn't understand me better.

**29.**

| $p$ | $q$ | $p \wedge q$ |
|---|---|---|
| T | T | T |
| T | F | F |
| F | T | F |
| F | F | F |

**30.**

| $p$ | $q$ | $p \vee q$ |
|---|---|---|
| T | T | T |
| T | F | T |
| F | T | T |
| F | F | F |

**31.**

| $p$ | $q$ | $p \Rightarrow q$ |
|---|---|---|
| T | T | T |
| T | F | F |
| F | T | T |
| F | F | T |

**32.**

| $p$ | $q$ | $p \Leftrightarrow q$ |
|---|---|---|
| T | T | T |
| T | F | F |
| F | T | F |
| F | F | T |

**33.**

| $p$ | $q$ | $\sim p$ | $p \vee q$ | $\sim p \wedge (p \vee q)$ |
|---|---|---|---|---|
| T | T | F | T | F |
| T | F | F | T | F |
| F | T | T | T | T |
| F | F | T | F | F |

**34.**

| $p$ | $q$ | $\sim p$ | $\sim q$ | $\sim p \wedge \sim q$ | $p \vee (\sim p \wedge \sim q)$ |
|---|---|---|---|---|---|
| T | T | F | F | F | T |
| T | F | F | T | F | T |
| F | T | T | F | F | F |
| F | F | T | T | T | T |

**35.**

| $p$ | $q$ | $p \Leftrightarrow q$ | $p \vee (p \Leftrightarrow q)$ |
|---|---|---|---|
| T | T | T | T |
| T | F | F | T |
| F | T | F | F |
| F | F | T | T |

**36.**

| $p$ | $q$ | $p \Leftrightarrow q$ | $(p \Leftrightarrow p) \Rightarrow q$ |
|---|---|---|---|
| T | T | T | T |
| T | F | T | F |
| F | T | T | T |
| F | F | T | F |

**37.**

| $p$ | $q$ | $p \wedge q$ | $q \wedge p$ | $(p \wedge q) \Leftrightarrow (q \wedge p)$ |
|---|---|---|---|---|
| T | T | T | T | T |
| T | F | F | F | T |
| F | T | F | F | T |
| F | F | F | F | T |

**38.**

| $p$ | $q$ | $p \wedge q$ | $(p \wedge q) \Rightarrow p$ |
|---|---|---|---|
| T | T | T | T |
| T | F | F | T |
| F | T | F | T |
| F | F | F | T |

**39.**

| $p$ | $q$ | $p \wedge q$ | $p \vee q$ | $(p \wedge q) \Rightarrow (p \vee q)$ |
|---|---|---|---|---|
| T | T | T | T | T |
| T | F | F | T | T |
| F | T | F | T | T |
| F | F | F | F | T |

**40.**

| $p$ | $q$ | $p \Leftrightarrow q$ | $p \Rightarrow q$ | $q \Rightarrow p$ | $(p \Rightarrow q) \wedge (q \Rightarrow p)$ | $(p \Leftrightarrow q) \Leftrightarrow [(p \Rightarrow q) \wedge (q \Rightarrow p)]$ |
|---|---|---|---|---|---|---|
| T | T | T | T | T | T | T |
| T | F | F | F | T | F | T |
| F | T | F | T | F | F | T |
| F | F | T | T | T | T | T |

**41.** I'll watch a movie. **42.** No valid conclusion. **43.** It isn't snowing.
**44.** No valid conclusion. **45.** No valid conclusion.
**46.** I will not work hard. **47.** We will ski. **48.** It is not snowing.
**49.** The corresponding parts are congruent. **50.** No valid conclusion.

**51.** If two angles of a triangle are congruent, then the measure of two sides are equal.

**52.** If two angles are complements of the same angle, then their measures are equal.

**53.** If I'm not suspended from school, then I will not go to jail.

**54.** If the fire trucks don't come, then the cow is not a careless smoker.

**55.** Reading books is not a worthy activity.   **56.** No valid conclusion.

**57.** All beagles are trusted friends.   **58.** No valid conclusion.

**59.** No valid conclusion.   **60.** No freshman are upperclassmen.

**61.** All mathematics professors do not drive too fast.

**62.** No blueberries are purple mushrooms.


## CHAPTER ELEVEN TEST   (page 362)

**1.** F   **2.** F   **3.** T   **4.** F   **5.** T   **6.** T   **7.** T   **8.** F   **9.** F

**10.** T   **11.** T   **12.** F   **13.** F   **14.** T   **15.** T   **16.** F   **17.** V

**18.** I   **19.** V   **20.** V   **22.** V   **23.** I   **24.** I   **25.** V


## EXERCISE 12.1   (page 370)

**1.** QI   **2.** QII   **3.** QIV   **4.** QII   **5.** QIII   **6.** QIII   **21.** Vertical; 2

**22.** Horizontal; 3   **23.** Horizontal; 6   **24.** Vertical; 14

**25.** Horizontal; 3   **26.** Horizontal; 4   **27.** 5   **28.** 5

**29.** 4   **30.** 13   **31.** 10   **32.** 10   **33.** $3\sqrt{2}$   **34.** 4   **35.** 20

**36.** 26   **37.** 5   **38.** 13   **39.** $\sqrt{26}$   **40.** $3\sqrt{2}$   **41.** $(\frac{11}{2}, 6)$

**42.** $(\frac{3}{2}, \frac{3}{2})$   **43.** $(0, 9)$   **44.** $(-8, -\frac{25}{2})$   **45.** $(7, -6)$   **46.** $(-10, -4)$

**47.** $(1, \frac{7}{6})$   **48.** $(-\frac{1}{2}, \frac{1}{2})$   **49.** $(8, 4)$   **50.** $(3, 0)$   **51.** $(-1, 4)$

**52.** $(-5, -13)$   **53.** $(1, 4)$   **54.** $(-10, 8)$   **55.** $(2, 3)$   **56.** $(0, 3)$

**57.** $(1, 2)$   **58.** $(\sqrt{3}, 1)$ or $(-\sqrt{3}, 1)$   **62.** $5\sqrt{5}$   **63.** $\frac{169}{2}$ square units

**64.** $(0, 4); (0, -10)$


## EXERCISE 12.2   (page 380)

**1.**

**2.**

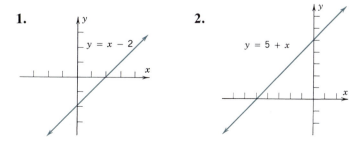

**3.**

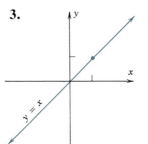

$y = x$

**4.**

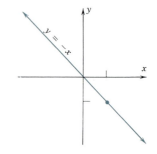

$y = -x$

**5.**

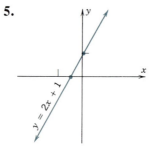

$y = 2x + 1$

**6.**

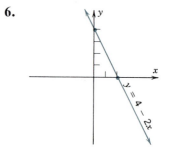

$y = 4 - 2x$

**7.**

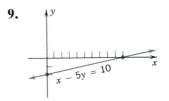

$2x + 4y = 8$

**8.**

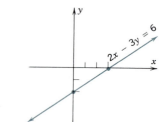

$2x - 3y = 6$

**9.**

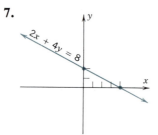

$x - 5y = 10$

**10.**

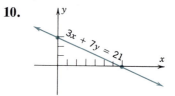

$3x + 7y = 21$

**11.**

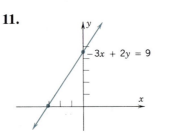

$-3x + 2y = 9$

**12.**

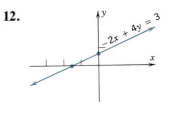

$-2x + 4y = 3$

**13.**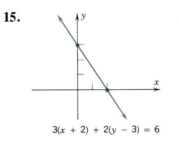

**14.**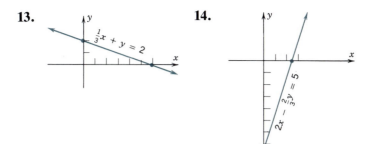

**15.**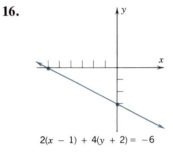

$3(x + 2) + 2(y - 3) = 6$

**16.**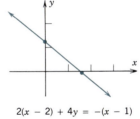

$2(x - 1) + 4(y + 2) = -6$

**17.**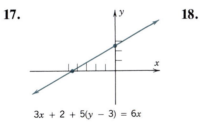

$3x + 2 + 5(y - 3) = 6x$

**18.**

$2(x - 2) + 4y = -(x - 1)$

**19.**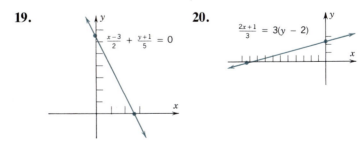

$\frac{x - 3}{2} + \frac{y + 1}{5} = 0$

**20.**

$\frac{2x + 1}{3} = 3(y - 2)$

**21.** $-1$  **22.** $-3$  **23.** $0$  **24.** $-2$  **25.** Undefined  **26.** $1$  **27.** $-\frac{3}{5}$
**28.** $\frac{1}{2}$  **29.** $-\frac{5}{4}$  **30.** $-2$  **31.** $y = x + 1$  **32.** $y = 2x - 1$
**33.** $y = x + 4$  **34.** $y = x - 3$  **35.** $y = x$  **36.** $y = \frac{5}{2}x - \frac{19}{2}$
**37.** $y = -x + \frac{1}{6}$  **38.** $y = -2x + \frac{43}{15}$  **39.** $y = 3x - 3$

**40.** $y = -3x + 11$   **41.** $y = 4x - 1$   **42.** $y = -4x - 11$
**43.** $y = 0$   **44.** $y = -x$   **45.** $y = -\frac{1}{2}x - \frac{1}{4}$
**46.** $y = -7x + \frac{2}{7}$   **47.** $y = 3x + 2$   **48.** $y = -2x + 3$
**49.** $y = -5x + 7$   **50.** $y = 7x - 5$   **51.** $y = \frac{3}{5}x + 5$
**52.** $y = -\frac{5}{3}x - 9$   **53.** $y = -\frac{1}{2}$   **54.** $y = -\frac{1}{2}x$   **55.** Neither
**56.** Perpendicular   **57.** Perpendicular   **58.** Neither
**59.** Neither   **60.** Parallel   **61.** $y = 3x - 1$   **62.** $y = -2x + 5$
**63.** $y = 5x + 8$   **64.** $y = -2x - 6$   **65.** $y = \frac{5}{7}x$
**66.** $y = -\frac{7}{9}x - \frac{2}{9}$   **67.** $y = -\frac{1}{3}x + \frac{7}{3}$   **68.** $y = \frac{1}{2}x$
**69.** $y = -\frac{1}{5}x + \frac{14}{5}$   **70.** $y = \frac{1}{2}x - 1$   **71.** $y = -\frac{7}{5}x$
**72.** $y = \frac{9}{7}x - \frac{16}{7}$   **73.** Parallel   **74.** Neither   **75.** Neither
**76.** Perpendicular   **77.** Perpendicular   **78.** Neither   **79.** Parallel
**80.** Perpendicular   **81.** $y = 2x + 10$   **82.** $y = -\frac{3}{5}x + 3$
**83.** $y = 3x - 10$   **84.** $y = -\frac{1}{3}x$

## EXERCISE 12.3   (page 387)

**1.** Is a solution.   **2.** Is a solution.   **3.** Is not a solution.
**4.** Is a solution.   **5.** Is a solution.   **6.** Is not a solution.

**7.**                                  **8.**

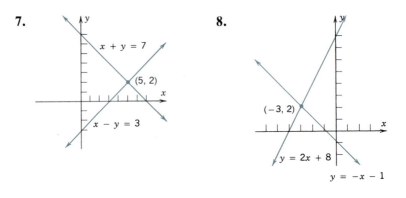

**9.** $(3, -2)$   **10.** $(\frac{8}{7}, -\frac{5}{7})$   **11.** $(9, 0)$   **12.** $(\frac{17}{5}, \frac{16}{5})$
**13.** $(3, 18)$   **14.** $(0, 0)$   **15.** $(1, \frac{1}{3})$   **16.** $(2, 1)$   **17.** $(-6, 1)$
**18.** $(1, -2)$   **19.** $(0, 0)$   **20.** $(1, 1)$

## EXERCISE 12.4   (page 394)

**15.** $\frac{45}{2}$ square units   **16.** 18 square units.   **17.** 16 square units
**18.** 8 square units

## CHAPTER TWELVE REVIEW EXERCISES   (page 395)

**1.–4.**

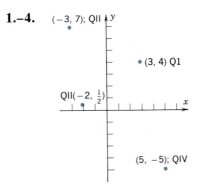

**5.** 7; horizontal   **6.** 10; vertical   **7.** 10   **8.** 13   **9.** $\sqrt{194}$
**10.** 15   **11.** (5, 7)   **12.** $(\frac{5}{2}, 9)$   **13.** $(-4, -2)$   **14.** $(-10, 2)$
**15.** $Q(11, 6)$   **16.** $Q(13, -3)$   **17.** (2, 2); $(-3, -3)$
**18.** $(2, 2\sqrt{3})$; $(2, -2\sqrt{3})$

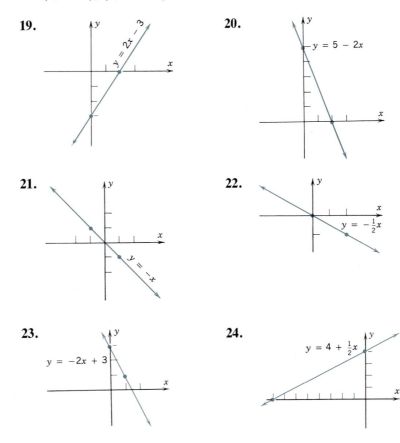

**19.**   **20.**

**21.**   **22.**

**23.**   **24.**

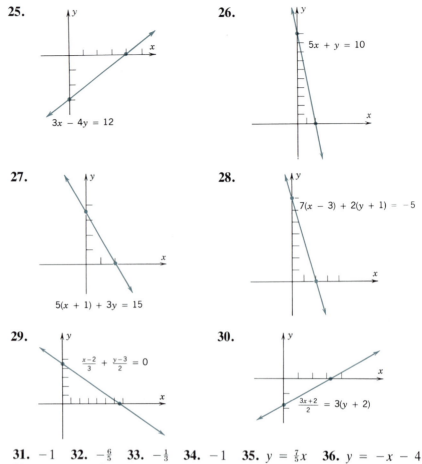

**25.**   $3x - 4y = 12$

**26.**   $5x + y = 10$

**27.**   $5(x + 1) + 3y = 15$

**28.**   $7(x - 3) + 2(y + 1) = -5$

**29.**   $\frac{x-2}{3} + \frac{y-3}{2} = 0$

**30.**   $\frac{3x+2}{2} = 3(y + 2)$

**31.** $-1$   **32.** $-\frac{6}{5}$   **33.** $-\frac{1}{3}$   **34.** $-1$   **35.** $y = \frac{7}{5}x$   **36.** $y = -x - 4$
**37.** $y = 6x - \frac{4}{3}$   **38.** $y = -4x + \frac{22}{21}$   **39.** $y = 2x - 5$
**40.** $y = -2x - 7$   **41.** $y = 5x + 1$   **42.** $y = -\frac{1}{2}x - 2$
**43.** $y = 2x + 1$   **44.** $y = -3x + 2$   **45.** $y = -2x - 7$
**46.** $y = -7x + 2$   **47.** Neither   **48.** Neither   **49.** Parallel
**50.** Perpendicular   **51.** $y = 5x - 5$   **52.** $y = -5x - 7$
**53.** $y = \frac{2}{3}x - \frac{11}{3}$   **54.** $y = -\frac{3}{2}x - \frac{13}{2}$   **55.** $y = -\frac{1}{5}x + \frac{27}{5}$
**56.** $y = \frac{1}{5}x + \frac{17}{5}$   **57.** $y = -\frac{3}{2}x - \frac{3}{2}$   **58.** $y = \frac{2}{3}x$   **59.** Parallel
**60.** Parallel   **61.** Parallel   **62.** Perpendicular
**63.** $y = \frac{2}{7}x + \frac{45}{14}$   **64.** $y = -\frac{1}{7}x$   **65.** $(3, -7)$   **66.** $(3, 1)$
**67.** $(3, -5)$   **68.** $(17, 8)$   **69.** $\left(\frac{31}{3}, \frac{10}{9}\right)$   **70.** $(-9, 14)$

# CHAPTER TWELVE TEST   (page 397)

**1.** F   **2.** T   **3.** T   **4.** F   **5.** T   **6.** F   **7.** T   **8.** T   **9.** T
**10.** F   **11.** T   **12.** T   **13.** T   **14.** F   **15.** T   **16.** F   **17.** F
**18.** F   **19.** F   **20.** T   **21.** T   **22.** F   **23.** T   **24.** F   **25.** F
**26.** T   **27.** F   **28.** T

## Table A Powers and Roots

| $n$ | $n^2$ | $\sqrt{n}$ | $n^3$ | $\sqrt[3]{n}$ | $n$ | $n^2$ | $\sqrt{n}$ | $n^3$ | $\sqrt[3]{n}$ |
|---|---|---|---|---|---|---|---|---|---|
| 1 | 1 | 1.000 | 1 | 1.000 | 51 | 2,601 | 7.141 | 132,651 | 3.708 |
| 2 | 4 | 1.414 | 8 | 1.260 | 52 | 2,704 | 7.211 | 140,608 | 3.733 |
| 3 | 9 | 1.732 | 27 | 1.442 | 53 | 2,809 | 7.280 | 148,877 | 3.756 |
| 4 | 16 | 2.000 | 64 | 1.587 | 54 | 2,916 | 7.348 | 157,464 | 3.780 |
| 5 | 25 | 2.236 | 125 | 1.710 | 55 | 3,025 | 7.416 | 166,375 | 3.803 |
| 6 | 36 | 2.449 | 216 | 1.817 | 56 | 3,136 | 7.483 | 175,616 | 3.826 |
| 7 | 49 | 2.646 | 343 | 1.913 | 57 | 3,249 | 7.550 | 185,193 | 3.849 |
| 8 | 64 | 2.828 | 512 | 2.000 | 58 | 3,364 | 7.616 | 195,112 | 3.871 |
| 9 | 81 | 3.000 | 729 | 2.080 | 59 | 3,481 | 7.681 | 205,379 | 3.893 |
| 10 | 100 | 3.162 | 1,000 | 2.154 | 60 | 3,600 | 7.746 | 216,000 | 3.915 |
| 11 | 121 | 3.317 | 1,331 | 2.224 | 61 | 3,721 | 7.810 | 226,981 | 3.936 |
| 12 | 144 | 3.464 | 1,728 | 2.289 | 62 | 3,844 | 7.874 | 238,328 | 3.958 |
| 13 | 169 | 3.606 | 2,197 | 2.351 | 63 | 3,969 | 7.937 | 250,047 | 3.979 |
| 14 | 196 | 3.742 | 2,744 | 2.410 | 64 | 4,096 | 8.000 | 262,144 | 4.000 |
| 15 | 225 | 3.873 | 3,375 | 2.466 | 65 | 4,225 | 8.062 | 274,625 | 4.021 |
| 16 | 256 | 4.000 | 4,096 | 2.520 | 66 | 4,356 | 8.124 | 287,496 | 4.041 |
| 17 | 289 | 4.123 | 4,913 | 2.571 | 67 | 4,489 | 8.185 | 300,763 | 4.062 |
| 18 | 324 | 4.243 | 5,832 | 2.621 | 68 | 4,624 | 8.246 | 314,432 | 4.082 |
| 19 | 361 | 4.359 | 6,859 | 2.668 | 69 | 4,761 | 8.307 | 328,509 | 4.102 |
| 20 | 400 | 4.472 | 8,000 | 2.714 | 70 | 4,900 | 8.367 | 343,000 | 4.121 |
| 21 | 441 | 4.583 | 9,261 | 2.759 | 71 | 5,041 | 8.426 | 357,911 | 4.141 |
| 22 | 484 | 4.690 | 10,648 | 2.802 | 72 | 5,184 | 8.485 | 373,248 | 4.160 |
| 23 | 529 | 4.796 | 12,167 | 2.844 | 73 | 5,329 | 8.544 | 389,017 | 4.179 |
| 24 | 576 | 4.899 | 13,824 | 2.884 | 74 | 5,476 | 8.602 | 405,224 | 4.198 |
| 25 | 625 | 5.000 | 15,625 | 2.924 | 75 | 5,625 | 8.660 | 421,875 | 4.217 |
| 26 | 676 | 5.099 | 17,576 | 2.962 | 76 | 5,776 | 8.718 | 438,976 | 4.236 |
| 27 | 729 | 5.196 | 19,683 | 3.000 | 77 | 5,929 | 8.775 | 456,533 | 4.254 |
| 28 | 784 | 5.292 | 21,952 | 3.037 | 78 | 6,084 | 8.832 | 474,552 | 4.273 |
| 29 | 841 | 5.385 | 24,389 | 3.072 | 79 | 6,241 | 8.888 | 493,039 | 4.291 |
| 30 | 900 | 5.477 | 27,000 | 3.107 | 80 | 6,400 | 8.944 | 512,000 | 4.309 |
| 31 | 961 | 5.568 | 29,791 | 3.141 | 81 | 6,561 | 9.000 | 531,441 | 4.327 |
| 32 | 1,024 | 5.657 | 32,768 | 3.175 | 82 | 6,724 | 9.055 | 551,368 | 4.344 |
| 33 | 1,089 | 5.745 | 35,937 | 3.208 | 83 | 6,889 | 9.110 | 571,787 | 4.362 |
| 34 | 1,156 | 5.831 | 39,304 | 3.240 | 84 | 7,056 | 9.165 | 592,704 | 4.380 |
| 35 | 1,225 | 5.916 | 42,875 | 3.271 | 85 | 7,225 | 9.220 | 614,125 | 4.397 |
| 36 | 1,296 | 6.000 | 46,656 | 3.302 | 86 | 7,396 | 9.274 | 636,056 | 4.414 |
| 37 | 1,369 | 6.083 | 50,653 | 3.332 | 87 | 7,569 | 9.327 | 658,503 | 4.431 |
| 38 | 1,444 | 6.164 | 54,872 | 3.362 | 88 | 7,744 | 9.381 | 681,472 | 4.448 |
| 39 | 1,521 | 6.245 | 59,319 | 3.391 | 89 | 7,921 | 9.434 | 704,969 | 4.465 |
| 40 | 1,600 | 6.325 | 64,000 | 3.420 | 90 | 8,100 | 9.487 | 729,000 | 4.481 |
| 41 | 1,681 | 6.403 | 68,921 | 3.448 | 91 | 8,281 | 9.539 | 753,571 | 4.498 |
| 42 | 1,764 | 6.481 | 74,088 | 3.476 | 92 | 8,464 | 9.592 | 778,688 | 4.514 |
| 43 | 1,849 | 6.557 | 79,507 | 3.503 | 93 | 8,649 | 9.644 | 804,357 | 4.531 |
| 44 | 1,936 | 6.633 | 85,184 | 3.530 | 94 | 8,836 | 9.695 | 830,584 | 4.547 |
| 45 | 2,025 | 6.708 | 91,125 | 3.557 | 95 | 9,025 | 9.747 | 857,375 | 4.563 |
| 46 | 2,116 | 6.782 | 97,336 | 5.583 | 96 | 9,216 | 9.798 | 884,736 | 4.579 |
| 47 | 2,209 | 6.856 | 103,823 | 3.609 | 97 | 9,409 | 9.849 | 912,673 | 4.595 |
| 48 | 2,304 | 6.928 | 110,592 | 3.634 | 98 | 9,604 | 9.899 | 941,192 | 4.610 |
| 49 | 2,401 | 7.000 | 117,649 | 3.659 | 99 | 9,801 | 9.950 | 970,299 | 4.626 |
| 50 | 2,500 | 7.071 | 125,000 | 3.684 | 100 | 10,000 | 10.000 | 1,000,000 | 4.642 |

## Table B  Values of the Trigonometric Ratios

| Angle | Sine | Cosine | Tangent |
|---|---|---|---|
| 0 | 0.0000 | 1.0000 | 0.0000 |
| 1 | 0.0175 | 0.9998 | 0.0175 |
| 2 | 0.0349 | 0.9994 | 0.0349 |
| 3 | 0.0523 | 0.9986 | 0.0524 |
| 4 | 0.0698 | 0.9976 | 0.0699 |
| 5 | 0.0872 | 0.9962 | 0.0875 |
| 6 | 0.1045 | 0.9945 | 0.1051 |
| 7 | 0.1219 | 0.9925 | 0.1228 |
| 8 | 0.1392 | 0.9903 | 0.1405 |
| 9 | 0.1564 | 0.9877 | 0.1584 |
| 10 | 0.1736 | 0.9848 | 0.1763 |
| 11 | 0.1908 | 0.9816 | 0.1944 |
| 12 | 0.2079 | 0.9781 | 0.2126 |
| 13 | 0.2250 | 0.9744 | 0.2309 |
| 14 | 0.2419 | 0.9703 | 0.2493 |
| 15 | 0.2588 | 0.9659 | 0.2679 |
| 16 | 0.2756 | 0.9613 | 0.2867 |
| 17 | 0.2924 | 0.9563 | 0.3057 |
| 18 | 0.3090 | 0.9511 | 0.3249 |
| 19 | 0.3256 | 0.9455 | 0.3443 |
| 20 | 0.3420 | 0.9397 | 0.3640 |
| 21 | 0.3584 | 0.9336 | 0.3839 |
| 22 | 0.3746 | 0.9272 | 0.4040 |
| 23 | 0.3907 | 0.9205 | 0.4245 |
| 24 | 0.4067 | 0.9135 | 0.4452 |
| 25 | 0.4226 | 0.9063 | 0.4663 |
| 26 | 0.4384 | 0.8988 | 0.4877 |
| 27 | 0.4540 | 0.8910 | 0.5095 |
| 28 | 0.4695 | 0.8829 | 0.5317 |
| 29 | 0.4848 | 0.8746 | 0.5543 |
| 30 | 0.5000 | 0.8660 | 0.5774 |
| 31 | 0.5150 | 0.8572 | 0.6009 |
| 32 | 0.5299 | 0.8480 | 0.6249 |
| 33 | 0.5446 | 0.8387 | 0.6494 |
| 34 | 0.5592 | 0.8290 | 0.6745 |
| 35 | 0.5736 | 0.8192 | 0.7002 |
| 36 | 0.5878 | 0.8090 | 0.7265 |
| 37 | 0.6018 | 0.7986 | 0.7536 |
| 38 | 0.6157 | 0.7880 | 0.7813 |
| 39 | 0.6293 | 0.7771 | 0.8098 |
| 40 | 0.6428 | 0.7660 | 0.8391 |
| 41 | 0.6561 | 0.7547 | 0.8693 |
| 42 | 0.6691 | 0.7431 | 0.9004 |
| 43 | 0.6820 | 0.7314 | 0.9325 |
| 44 | 0.6947 | 0.7193 | 0.9657 |
| 45 | 0.7071 | 0.7071 | 1.000 |

## Table B (*Continued*)

| Angle | Sine | Cosine | Tangent |
|-------|------|--------|---------|
| 46 | 0.7193 | 0.6947 | 1.036 |
| 47 | 0.7314 | 0.6820 | 1.072 |
| 48 | 0.7431 | 0.6691 | 1.111 |
| 49 | 0.7547 | 0.6561 | 1.150 |
| 50 | 0.7660 | 0.6428 | 1.192 |
| 51 | 0.7771 | 0.6293 | 1.235 |
| 52 | 0.7880 | 0.6157 | 1.280 |
| 53 | 0.7986 | 0.6018 | 1.327 |
| 54 | 0.8090 | 0.5878 | 1.376 |
| 55 | 0.8192 | 0.5736 | 1.428 |
| 56 | 0.8290 | 0.5592 | 1.483 |
| 57 | 0.8387 | 0.5446 | 1.540 |
| 58 | 0.8480 | 0.5299 | 1.600 |
| 59 | 0.8572 | 0.5150 | 1.664 |
| 60 | 0.8660 | 0.5000 | 1.732 |
| 61 | 0.8746 | 0.4848 | 1.804 |
| 62 | 0.8829 | 0.4695 | 1.881 |
| 63 | 0.8910 | 0.4540 | 1.963 |
| 64 | 0.8988 | 0.4384 | 2.050 |
| 65 | 0.9063 | 0.4226 | 2.145 |
| 66 | 0.9135 | 0.4067 | 2.246 |
| 67 | 0.9205 | 0.3907 | 2.356 |
| 68 | 0.9272 | 0.3746 | 2.475 |
| 69 | 0.9336 | 0.3584 | 2.605 |
| 70 | 0.9397 | 0.3420 | 2.747 |
| 71 | 0.9455 | 0.3256 | 2.904 |
| 72 | 0.9511 | 0.3090 | 3.078 |
| 73 | 0.9563 | 0.2924 | 3.271 |
| 74 | 0.9613 | 0.2756 | 3.487 |
| 75 | 0.9659 | 0.2588 | 3.732 |
| 76 | 0.9703 | 0.2419 | 4.011 |
| 77 | 0.9744 | 0.2250 | 4.331 |
| 78 | 0.9781 | 0.2079 | 4.705 |
| 79 | 0.9816 | 0.1908 | 5.145 |
| 80 | 0.9848 | 0.1736 | 5.671 |
| 81 | 0.9877 | 0.1564 | 6.314 |
| 82 | 0.9903 | 0.1392 | 7.115 |
| 83 | 0.9925 | 0.1219 | 8.144 |
| 84 | 0.9945 | 0.1045 | 9.514 |
| 85 | 0.9962 | 0.0872 | 11.43 |
| 86 | 0.9976 | 0.0698 | 14.30 |
| 87 | 0.9986 | 0.0523 | 19.08 |
| 88 | 0.9994 | 0.0349 | 28.64 |
| 89 | 0.9998 | 0.0175 | 57.29 |
| 90 | 1.0000 | 0.0000 | Undef |

# Photo Credits

# Index